DESIGNING ELECTRONIC CIRCUITS

Also by the Author

New Ways to Use Test Meters: A Modern Guide to Electronic Servicing

New Handbook of Troubleshooting Techniques for Microprocessors & Microcomputers

Troubleshooting Electronic Equipment Without Service Data

New Digital Troubleshooting Techniques: A Complete Illustrated Guide

DESIGNING ELECTRONIC CIRCUITS

A Manual of Procedures and Essential Reference Data

Robert G. Middleton

Prentice-Hall, Inc.
Business & Professional Division
Englewood Cliffs, New Jersey

Prentice-Hall International (UK) Limited, *London*
Prentice-Hall of Australia, Pty. Ltd., *Sydney*
Prentice-Hall Canada, Inc., *Toronto*
Prentice-Hall of India Private Ltd., *New Delhi*
Prentice-Hall of Japan, Inc., *Tokyo*
Prentice-Hall of Southeast Asia Pte. Ltd., *Singapore*
Whitehall Books, Ltd., Wellington, *New Zealand*
Editora Prentice-Hall do Brasil Ltda., *Rio de Janeiro*
Prentice-Hall Hispanoamericana, S.A., *Mexico*

First Printing...October, 1985

Editor: George E. Parker

Library of Congress Cataloging Publication Data

Middleton, Robert Gordon
Designing electronic circuits.

Includes index.
1. Electronic circuit design. 2. Electronic circuit design—Data processing. I. Title.
TK7867.M438 1985 621.3815'3 85-9531

ISBN 0-13-200650-2

Printed in the United States of America.

A Word from the Author on the Unique, Practical Value This Book Offers

This forward-looking manual of electronic circuit design techniques, tables, and formulas includes computer programs (with illustrative RUNS) to facilitate preliminary design procedures.

The text starts with an explanation of basic amplifier design procedure with bipolar transistors using hybrid parameters for preliminary device and component evaluation. I give practical examples for the CE, CC, and CB modes of operation. The introductory discussion concludes with a helpful computer program that enables the designer to quickly "punch out" basic amplifier performance figures relative to tentative device and component values.

Procedural topics continue with detailed examination of bias stabilization requirements in basic amplifier design with bipolar transistors. Stabilizing circuitry includes resistor, diode, thermistor, and transistor arrangements. Voltage and current stability factors are derived, and various practical examples are cited. Bias stabilization considerations conclude with effective computer programs for facilitating design analyses.

Amplifier discussion continues with considerations of noise factor and its optimization. I treat audio preamplifier design from an applications viewpoint and optional trade-offs. There are practical examples of preamplifier classes and performance characteristics. I review coupling circuitry for both utility applications and for high-fidelity operation. I examine negative-feedback circuitry and present helpful computer programs to speed up circuit design procedures.

The next examination is of driver and power-amplifier design considerations for bipolar transistor audio circuitry. There is a review of various trade-off options from the standpoint of production economy. I discuss both utility and high-fidelity applications, with various practical examples. I present an introduction to reliability evaluation, pointing out common design pitfalls. These topics conclude with appropriate computer programs to facilitate the design procedure.

Tolerance is a "dirty word" in some areas of design. Admittedly, design tolerances are often difficult (and sometimes impossible) to calculate with a high level of confidence. The text coverage illustrates how a heuristic programming approach can sometimes simplify tolerance calculations for optimal design. There is also a new chained graphical technique that facilitates evaluation of tolerance requirements.

Next, I review the basics of tuned-amplifier design with an examination of coupling circuitry characteristics. There is a discussion of various options from the viewpoint of production economy. And, I discuss additional tolerance factors and control of "drift" factors, together with practical examples. There is an outline of gain-control requirements. You will find appropriate computer programs to facilitate coupling-circuit design procedures.

The next description of wide-band bipolar transistor amplifier arrangements includes a review of frequency compensation, direct coupling, RC coupling, negative feedback, and tolerance considerations, all from the standpoint of practical design procedures. You will find typical trade-offs and design options, again with practical examples. A helpful computer program to assist in calculation of design tolerances (worst-case performance characteristics) is part of this section.

The next subsection recaps the foregoing chapters with respect to MOSFET design procedures. There is a detailed description of similarities and differences between bipolar and MOS amplifier circuitry with their relevant advantages and disadvantages. You will read about design procedures, which are exemplified for common-source, common-drain, and common-gate configurations.

I point out applications that are facilitated by very high input resistance, wide dynamic range, and very high forward transconductance.

There are various computer programs with the design procedures to speed up development of prototype models. You will find heuristics to compute optimal parameters in preliminary design procedures. I give practical consideration to worst-case analyses.

Next, I discuss principles of oscillator design with respect to both low-frequency and high-frequency operation. I analyze the RC oscillator in some detail again and provide various computer programs to facilitate preliminary design procedures. I review operating stability, practical production-engineering requirements, and factors relating to harmonic generation.

Up to this point, the focus of attention is directed to basic electronic circuit building blocks and their characteristics. Since a building block such as a parallel-resonant circuit may be energized by a voltage source with significant internal resistance, its parameters are modified accordingly. Again, the resonant circuit may also supply its current demand to a load that has significant conductance. Accordingly, its parameters are further modified. In turn, the focus of attention is redirected to contextual circuit analysis of both RC circuitry and LCR circuitry.

You will find computer programs for calculation of the input and output impedances of unloaded and loaded RC differentiating and integrating circuits. I consider single-section and two-section arrangements; the two-section arrangements include both symmetrical and unsymmetrical configurations. The book contains practical graphs for rapid determination of output/input voltage and phase relations. Survey programs that print out extensive frequency response data supplement the design graphs.

I then discuss basic filter design procedures, insofar as they relate to the needs of the practical design engineer. I review simple RC filters and illustrate the effect of component tolerances. Passive and active tone controls, equalizers, bandpass and loaded bandpass, and low-pass filters with low-frequency cut provisions are described and illustrated. Computer programs are provided to speed up preliminary design procedures.

The next chapter is concerned with the principles of transient circuit design. Nonsinusoidal oscillator circuitry is included. Emphasis is placed on mathematical treatment of transient circuit operation and the effects of component tolerances. I provide various computer programs to ease the task of mathematical analysis.

You will find an example of creative design procedure in the final chapter, entitled Elements of Linear RC Voltage Bootstrap Circuitry. It shows step by step, with practical examples, how to design an RC-integrating or differentiating network that develops an output voltage greater than its input voltage. Then, I show cascaded arrangements that produce an output voltage that approaches a limiting value, followed by cascaded configurations that develop an output voltage that approaches infinity as the number of RC sections is indefinitely increased. This is more of a tutorial chapter focusing on the creative process than on practical applications of known network theory.

The book includes useful appendices for explanation of program conversions among the more popular types of personal computers, for ready reference to most-often-used mathematical relations, and for speedy retrieval of electronic circuit data. The cross-referenced index facilitates localization of any topic that the designer may need to look up.

Professional electronic designers and troubleshooters know that time is money and that knowledge is power. Your success in practice of your profession is limited only by the horizons of your technical know-how. The unique and practical design approaches, application techniques, procedural data, examples, computer-aided design procedures, and technical reference data in this manual provide key stepping stones to your goal.

ROBERT G. MIDDLETON

Contents

DESIGNING ELECTRONIC CIRCUITS

1

BASIC AMPLIFIERS

COMMON-EMITTER AMPLIFIER

The simplest practical common-emitter amplifier arrangement is shown in Figure 1–1. To calculate the voltage gain, power gain, current gain, input resistance, and output resistance for this configuration, it may be skeletonized, as depicted in Figure 1–2. Since the transistor is a nonlinear device, small-signal class-A operation is discussed at this time. Calculations are made with respect to the equivalent circuit shown in Figure 1–3.

Voltage gain is calculated from the formula:*

$$A_v = \frac{-\alpha_{fe}R_L}{(h_{ie}h_{oe} - \alpha_{fe}\mu_{re})R_L + h_{ie}}$$

EXAMPLE:

$R_L = 15$ kilohms $h_{ie} = 1500$ ohms $h_{oe} = 20 \times 10^{-6}$ Siemens

$\alpha_{fe} = 50$ $\mu_{re} = 5 \times 10^{-4}$

$A_v = -476$

*A short computer proram for rapid calculation of amplifier parameters is provided at the end of this chapter.

Current gain is calculated from the formula:

$$A_i = \frac{-\alpha_{fe}}{h_{oe}R_L + 1}$$

EXAMPLE:

$R_L = 15$ kilohms $\quad h_{oe} = 20 \times 10^{-6}$ Siemens

$\alpha_{fe} = 50$

$$A_i = -38.4$$

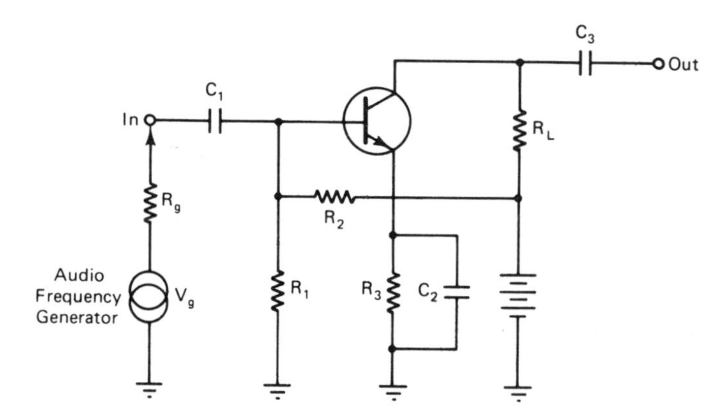

Figure 1–1. Simplest practical common-emitter amplifier arrangement.

NOTE: The hybrid parameters are frequency-dependent in this example, because R3 is bypassed by C2. At zero frequency (dc), C2 is effectively open; at high frequencies, C2 is effectively a short circuit for ac. The hybrid parameters are unaffected by the values of C1 and C3.

In preliminary design procedures, it is usually helpful to keep the requirements in mind for a low noise figure. Thus, a collector-supply voltage less than two volts, an emitter current less than one milliampere, and a generator (source) internal resistance in the 300- to 3,000-ohms range are practical guidelines for an audio-input stage.

Transistor data sheets ordinarily specify hybrid-parameter values for average values of collector voltage and current. Accordingly, adjusted values of hybrid parameters should be used when calculating amplifier performance at low values of collector voltage and current.

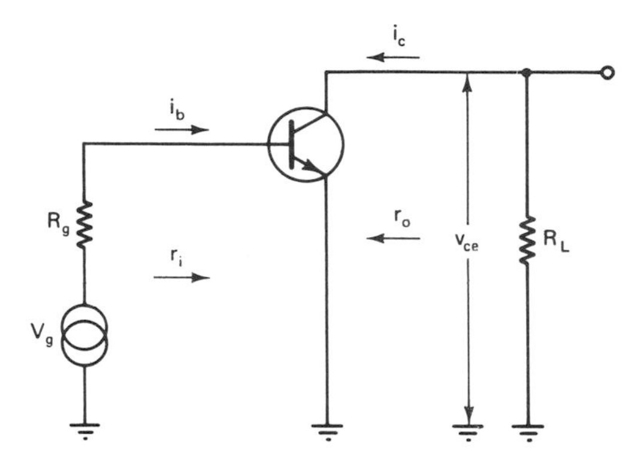

Figure 1–2. Skeletonized version of the configuration in Figure 1–1.

NOTE: In some circumstances, the designer has no choice of generator resistance. If R_g is very high, the proper procedure is to reduce the emitter current to the lowest practical level and to select a transistor type that has a maximum forward current-transfer ratio (h_{fe}). These considerations will minimize the noise figure in an audio-input stage, for example.

The input resistance r_i is the e/i ratio that the generator "sees," looking into the base of the transistor. The output resistance r_o is the e/i ratio that the load "sees," looking back into the collector of the transistor.

The predominant design goal for small-signal audio amplifiers is generally maximized power gain. However, the designer seldom has a "free hand" and must make judicious trade-offs with respect to production costs and product "price tag."

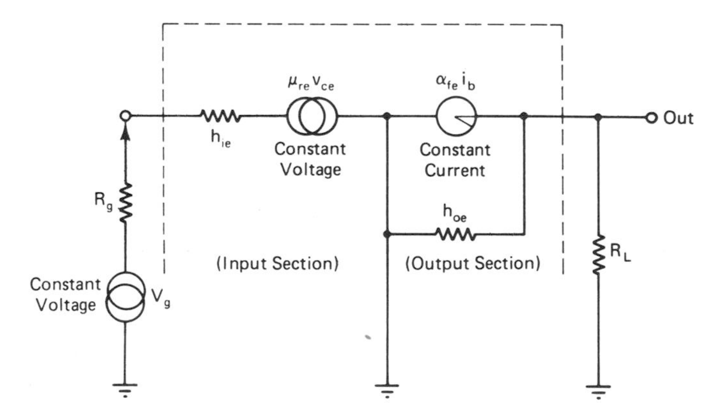

Figure 1–3. Equivalent circuit for the arrangement in Figure 1–2.

NOTE: The hybrid parameter h_{ie} is equal to the input impedance of the transistor with output short-circuited. The hybrid parameter μ_{re}, or h_{re}, is equal to the reverse open-circuit voltage amplification factor. The hybrid parameter α_{fe}, or h_{fe}, is equal to the forward short-circuit current amplification factor. The hybrid parameter h_{oe} is equal to the output admittance of the transistor with open-circuit input. V_{ce} is the collector-to-ground voltage, and i_b is the base current of the transistor.

Power gain is calculated from the formula:

$$G_p = A_v A_i$$

or,

$$G_p = \frac{(\alpha_{fe})^2 R_L}{(h_{oe}R_L + 1)\left[(h_{ie}h_{oe} - \alpha_{fe}\mu_{re})R_L + h_{ie}\right]}$$

EXAMPLE:

$A_v = -476$ $A_i = -38.4$

$G_p = 18{,}278$

$R_L = 15$ kilohms $\alpha_{fe} = 50$ $h_{oe} = 20 \times 10^{-6}$ Siemens

$h_{ie} = 1500$ ohms $\mu_{re} = 5 \times 10^{-4}$

$G_p = 18{,}315$ $G_p(\text{dB}) = 10 \log 18{,}315 = 10 \times 4.26 = 42.6$ dB

Input resistance is calculated from the formula:

$$r_i = \frac{h_{ie} + (h_{oe}h_{ie} - \alpha_{fe}\mu_{re})R_L}{1 + h_{oe}R_L}$$

EXAMPLE:

$R_L = 15$ kilohms $\alpha_{fe} = 50$ $h_{oe} = 20 \times 10^{-6}$ Siemens

$h_{ie} = 1500$ ohms $\mu_{re} = 5 \times 10^{-4}$

$r_i = 1212$ ohms

Output resistance is calculated from the formula:

$$r_o = \frac{h_{ie} + R_g}{h_{oe}h_{ie} - \mu_{re}\alpha_{fe} + h_{oe}R_g}$$

EXAMPLE:

R_g 1500 $h_{oe} = 20 \times 10^{-6}$ Siemens

$h_{ie} = 1500$ ohms $\mu_{re} = 5 \times 10^{-4}$ $\alpha_{fe} = 50$

$r_o = 85{,}714$ ohms

Observe in the foregoing examples that A_v is negative because the output voltage has reversed phase compared to the input voltage. A_i is negative because the output current has reversed phase compared to the input current.

HYBRID PARAMETERS

Hybrid parameters are employed in the foregoing examples. The hybrid parameters for the common-emitter configuration are:

h_{ie} or h_{11e} = Input impedance with output short-circuited
h_{oe} or h_{22e} = Output admittance with open-circuit input
h_{re} or h_{12e} = Reverse open-circuit voltage amplification factor
h_{fe} or h_{21e} = Forward short-circuit current amplification factor

Note that h_{re} is also called μ_{re} and h_{fe} is also called α_{fe}.

As shown in Figures 1–2 and 1–3, R_g denotes the generator resistance or the internal resistance of the signal source.

HYBRID PARAMETER DEFINITIONS

The input impedence of a transistor with its output short-circuited (h_{ie}) is defined as shown in Figure 1–4. This is a plot of base voltage versus base current, with collector voltage as the running parameter, for a typical bipolar transistor. Since the V_{BE}/I_B characteristics are nonlinear, the value of h_{ie} will depend on choice of the operating point. The operating point in this example is indicated by X. The corresponding value of h_{ie} is equal to the limiting ratio of $\Delta V_{BE}/\Delta I_B$ for a chosen value of collector voltage (in this example, V_{CE} = 7.5 volts). This ratio has a value in the limit of 608 ohms for the particular transistor and for the specified operating (quiescent) point.

The reverse open-circuit voltage amplification factor of a transistor is defined as shown in Figure 1–5. This is a plot of base voltage versus collector voltage, with base current as the running parameter, for a typical bipolar transistor. Inasmuch as the V_{BE}/V_{CE} characteristics are nonlinear, the value of h_{re} depends on choice of the operating point. In this example, the operating point is indicated by X.

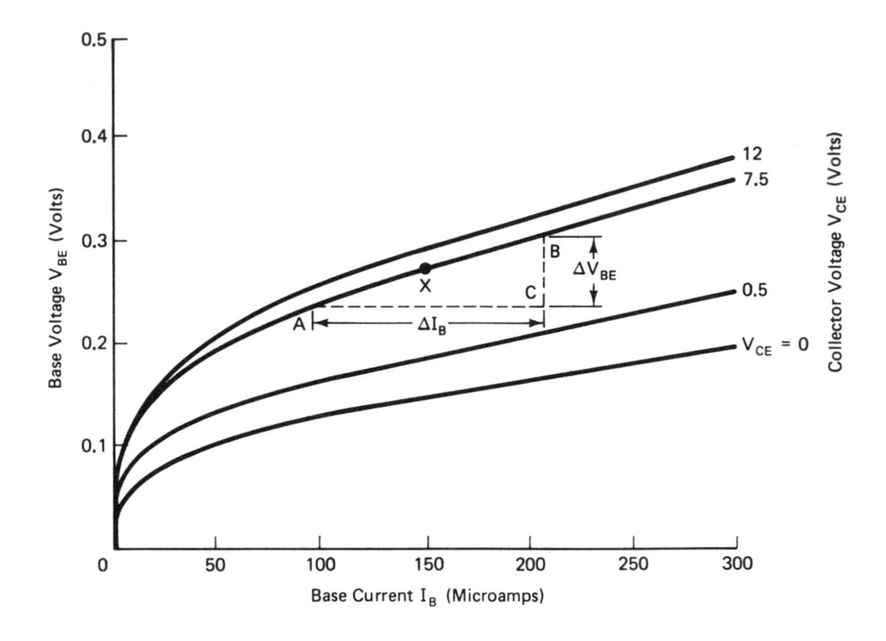

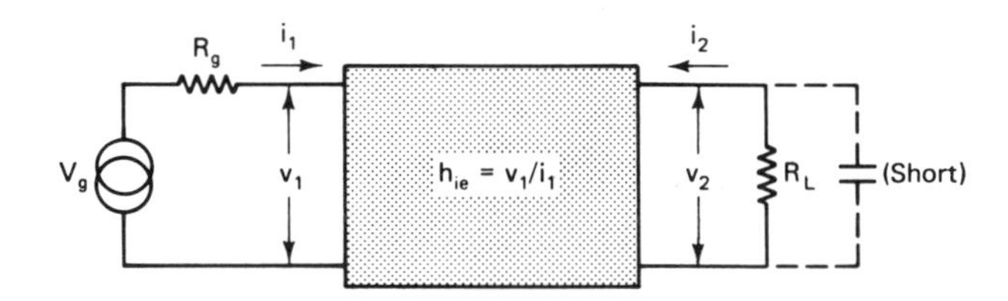

Figure 1–4. Definition of the input impedance of a transistor with its output short-circuited (h_{ie}).

NOTE: H parameters are slopes of the static characteristic curves for the particular transistor.

In turn, the corresponding value of h_{re} is equal to the slope of the tangent at point *X* on the 150-μA curve. This slope has a value of 14×10^{-4} for the particular transistor and for the specified operating point. Note that the diagram in Figure 1–5 depicts reverse transfer (feedback) characteristic curves for the common-emitter configuration. Stated otherwise, a collector-voltage variation ΔV_{CE} will appear at the input as a base-voltage variation ΔV_{BE}.

The forward short-circuit current amplification factor (h_{fe}) for a transistor is defined as shown in Figure 1–6. This is a plot of

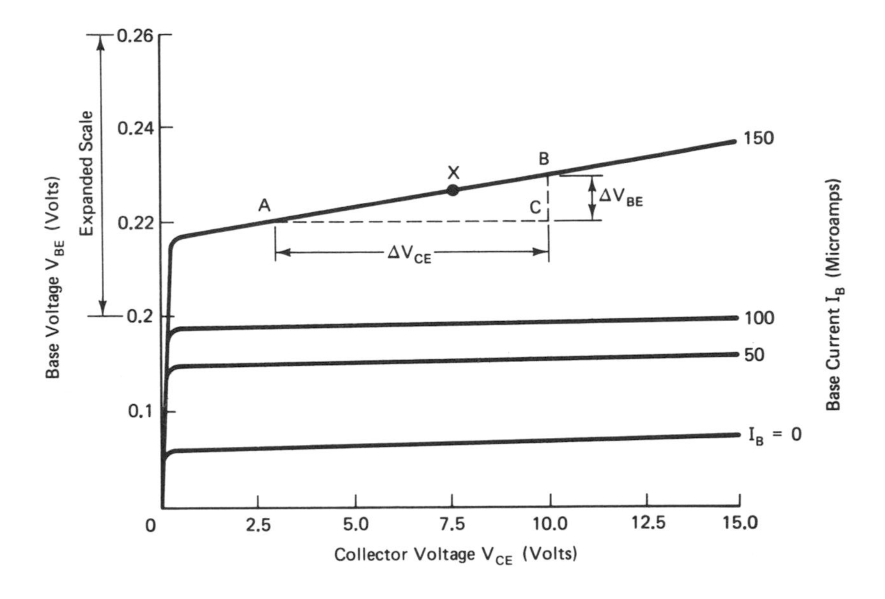

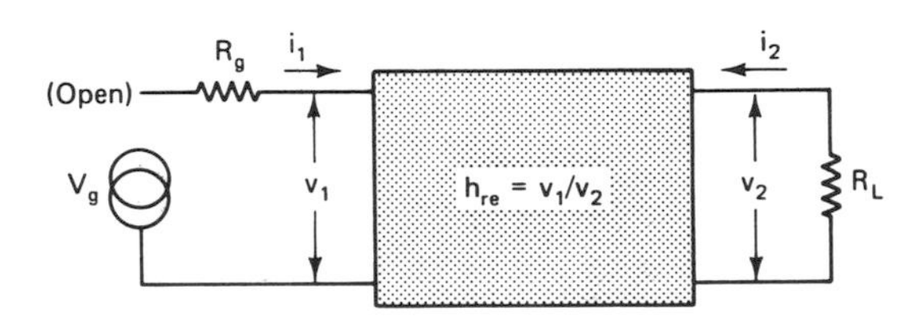

Figure 1–5. Definition of reverse open-circuit voltage amplification factor.

NOTE: The h parameters apply directly to calculation of audio amplifier characteristics. Observe that when the input frequency exceeds 20 or 30 kHz, additional parameters must be taken into account.

collector current versus base current, with collector voltage as the running parameter. The slope of the curves is not constant, and the value of h_{fe} depends on choice of the operating point. In this example, the operating point is indicated by X. The corresponding ratio of $\Delta I_C/\Delta I_B$ is 27 for the particular transistor and the specified operating point. Note that the diagram in Figure 1–6 exemplifies typical forward transfer characteristics for the common-emitter configuration.

The output admittance with open-circuit input (h_{oe}) for a transistor is defined as shown in Figure 1–7. This is a plot of

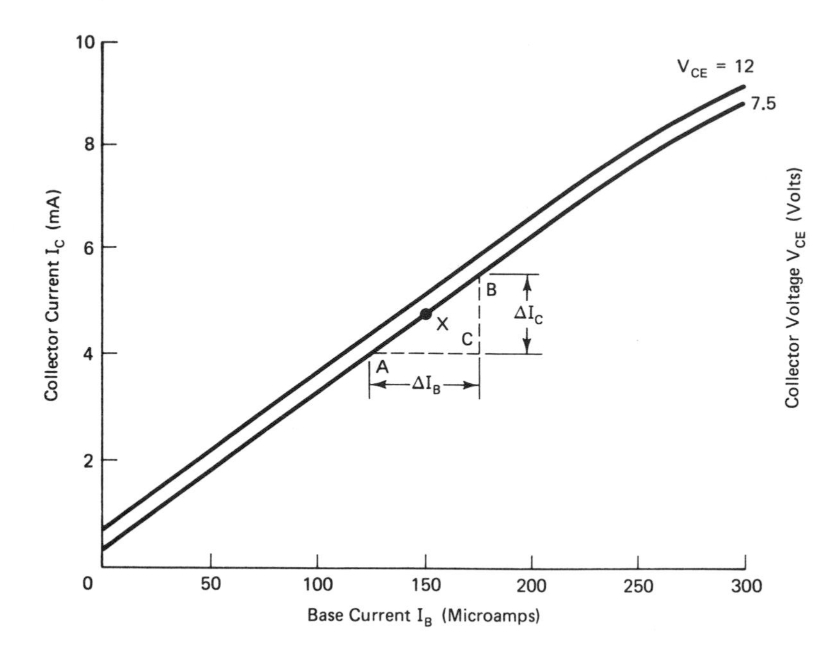

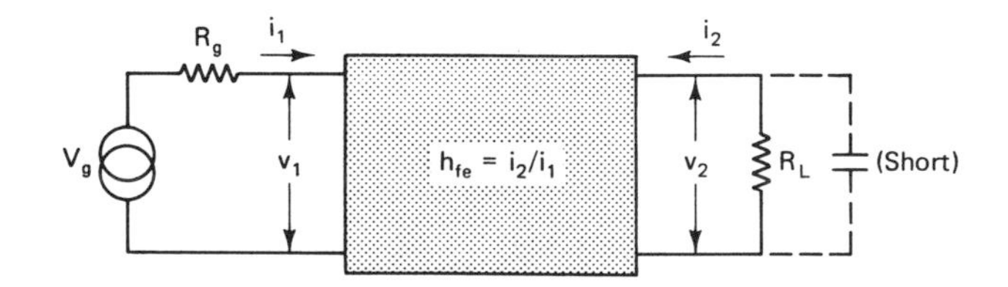

Figure 1–6. Definition of forward short-circuit current amplification factor.

NOTE: If hybrid parameters are not available for a particular type of transistor, the h parameters may be easily measured. Observe that a large capacitor will function as an ac short-circuit. To "open" a signal path and still maintain bias-current flow, an appropriate inductor may be connected in series with the path.

collector current versus collector voltage, with base current as the running parameter. The slope of the curves is not constant, and the value of h_{oe} depends on choice of operating point. This example specifies the operating point at X. In turn, the corresponding ratio of $\Delta I_C/\Delta V_{CE}$ is 40×10^{-6} Siemens for the particular transistor and the specified operating point (or an output resistance with input open-circuited of 25 kilohms). The diagram in Figure 1–7 exemplifies typical output (collector) characteristics for the common-emitter configuration.

EXAMPLE:*

Silicon epitaxial biopolar transistor for high-current audio and video amplification in CE configuration.

R_L = 1 kilohm	R_g = 200 ohms	h_{fe} = 200
h_{ie} = 600	h_{oe} = 75 × 10^{-6} Siemens	h_{re} = 125 × 10^{-6}

A_v = −322 (approx.)
A_i = − 186 (approx.)
G_p = 59,890 (approx.)
r_{ie} = 577 ohms (approx.)
r_{oe} = 227 kilohms (approx.)

* Reproduced by special permission of Reston Publishing Company from *Solid State Devices* by William D. Cooper.

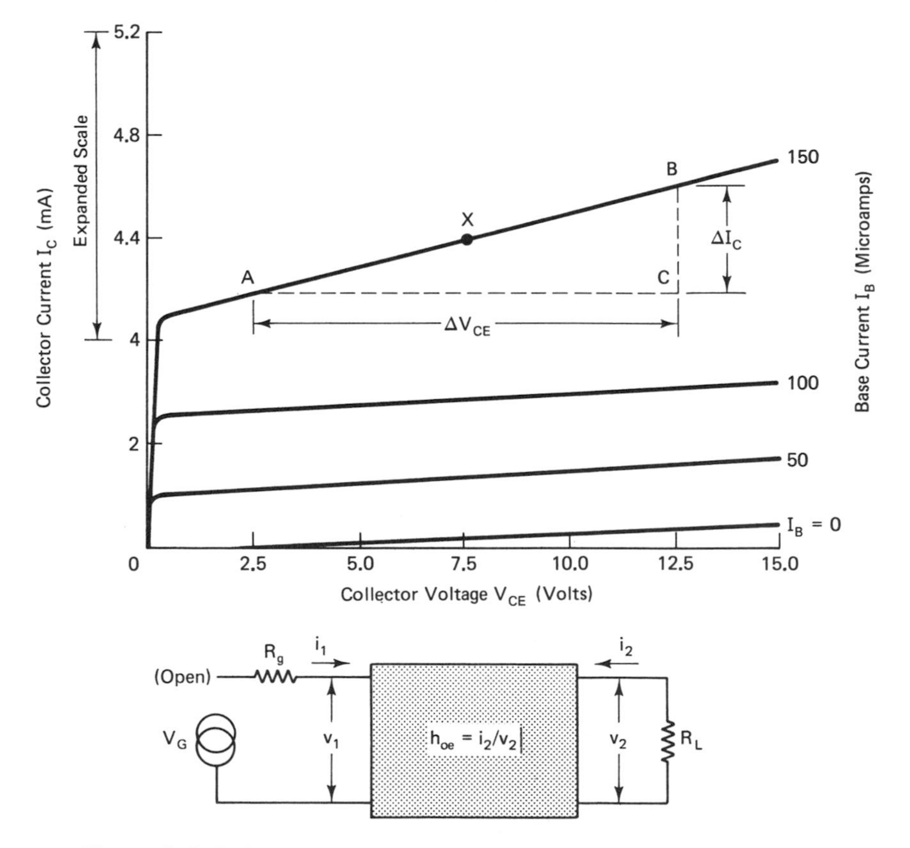

Figure 1–7. Definition of output admittance with open-circuit input.

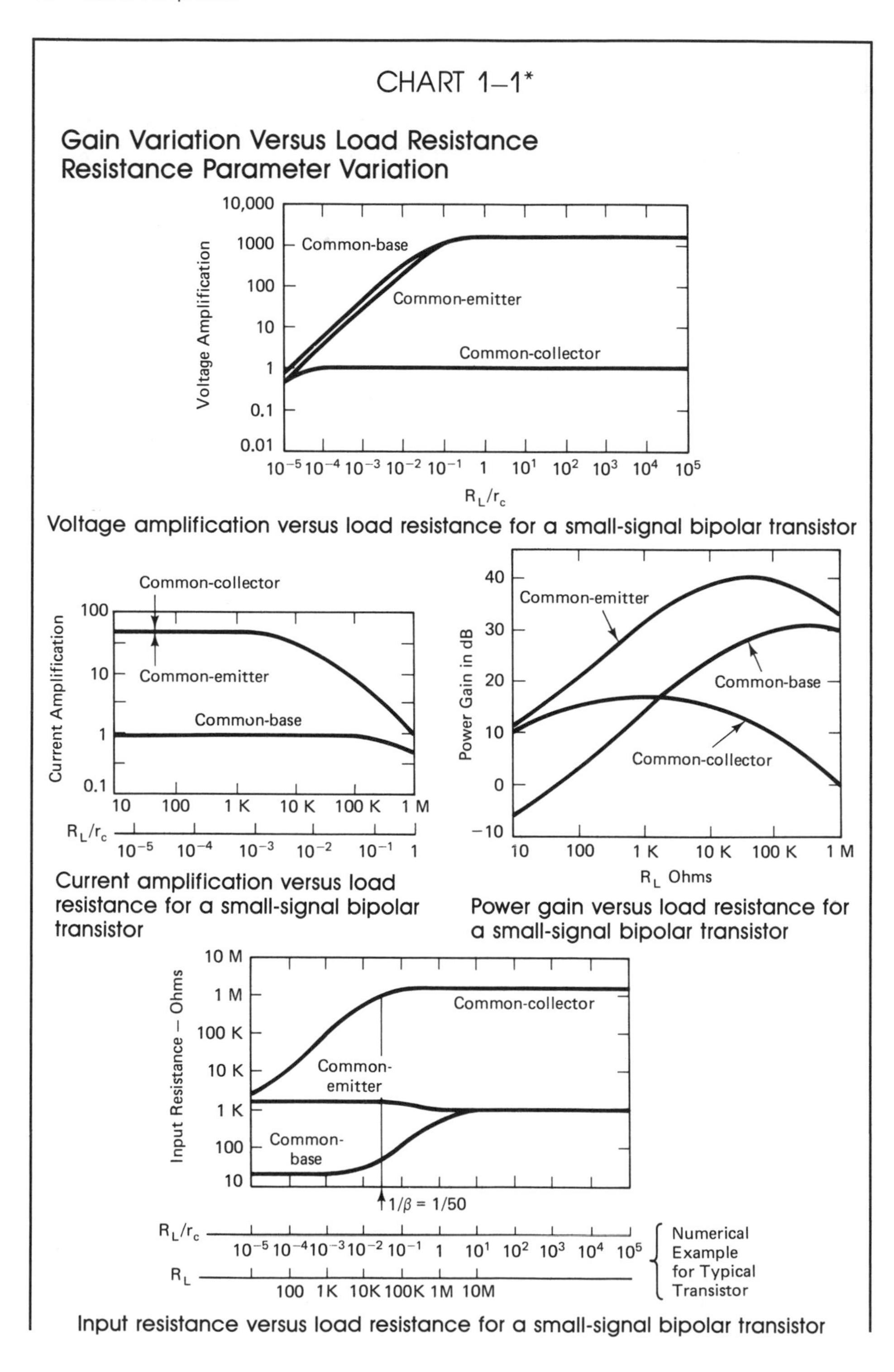

Voltage amplification versus load resistance for a small-signal bipolar transistor

Current amplification versus load resistance for a small-signal bipolar transistor

Power gain versus load resistance for a small-signal bipolar transistor

Input resistance versus load resistance for a small-signal bipolar transistor

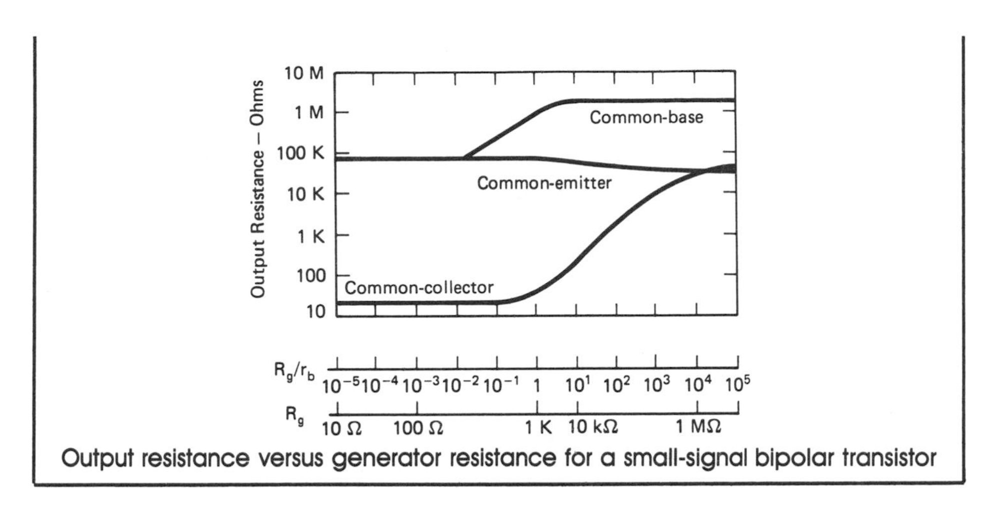

Output resistance versus generator resistance for a small-signal bipolar transistor

*Diagrams reproduced by special permission of Reston Publishing Company and Campbell Loudoun from *Handbook for Electronic Circuit Design.*

In preliminary design procedures, it is often helpful to observe the diagrams in Chart 1–1, which show typical ranges of gain variation versus load resistance and resistance parameter variation. These curves are based on the T equivalent circuit for a bipolar transistor, as depicted in Figure 1–8. Typical ranges for r_b, r_e, and r_c are also noted in Figure 1–8.

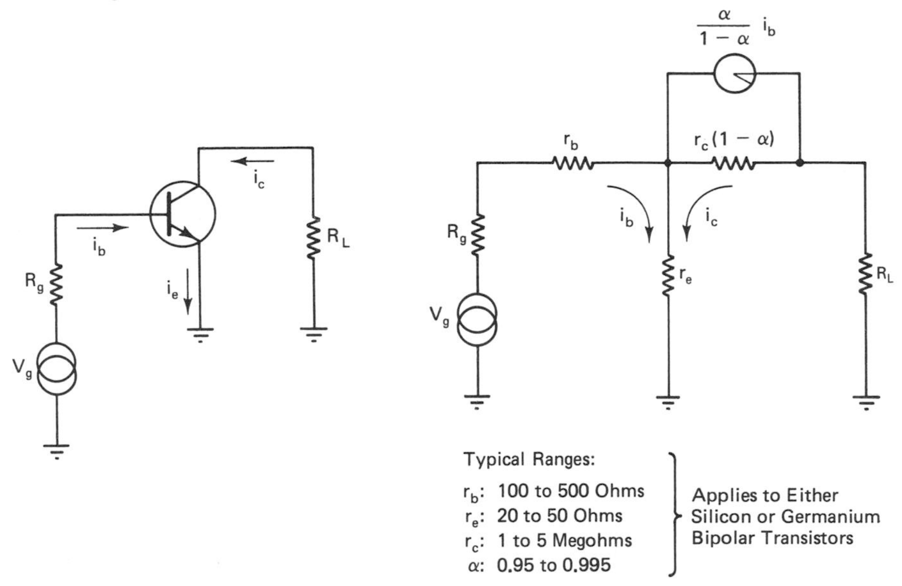

Figure 1–8. T-equivalent circuit for bipolar transistor in CE configuration with T-equivalent parameters.

THREE-TERMINAL GENERIC REPRESENTATION

A knowledgeable perspective of transistor circuit action is provided by the three-terminal generic representation of a bipolar transistor in the common-base mode, as depicted in Figure 1-9. The ratio v_{eb}/i_e represents an input resistance $r_{e'}$; the ratio v_{cb}/i_c represents an output resistance $r_{c'}$. This generic representation can be elaborated to derive a common-emitter generic version. It corresponds to a T equivalent circuit that includes two constant-current generators.

Because generic equivalent circuits are difficult to manipulate, practical design procedures employ comparatively approximate equivalent circuits, such as shown in Figure 1–8. The "r parameters" are comparatively difficult to measure, and calculations are generally made in terms of the "h parameters." H parameters regard a transistor as a generalized four-terminal network, such as exemplified in Figure 1–7.

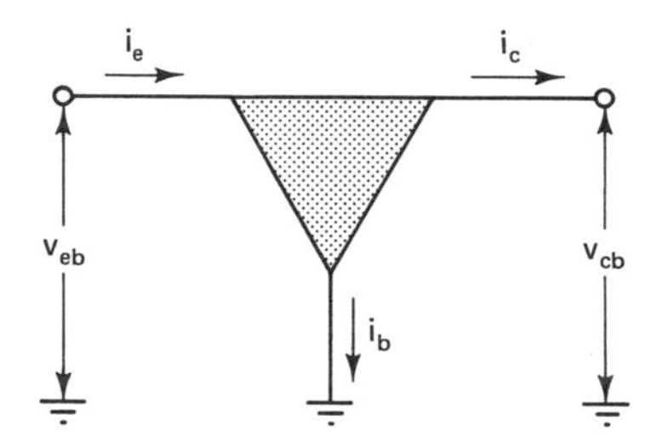

Figure 1–9. Three-terminal generic representation of a bipolar transistor in the common-base mode.

NOTE: This fundamental representation of transistor voltage-current relations is referenced to the common-base (CB) mode, with the transistor regarded as a generalized three-terminal network. Any voltage and/or current ratio can be precisely expressed only with respect to specified bias conditions and ambient temperature and is limited to small-signal operation.

COMMON-COLLECTOR AMPLIFIER

A skeletonized circuit diagram for a common-collector (emitter-follower) arrangement is shown in Figure 1–10. As illustrated in Chart 1–1, the CC (common collector) configuration has unity (or less) voltage gain. However, it has high current gain and substantial power gain. Designers usually employ the CC configuration when high input impedance is required. As indicated in the chart, it is feasible to obtain an input impedance on the order of a megohm.

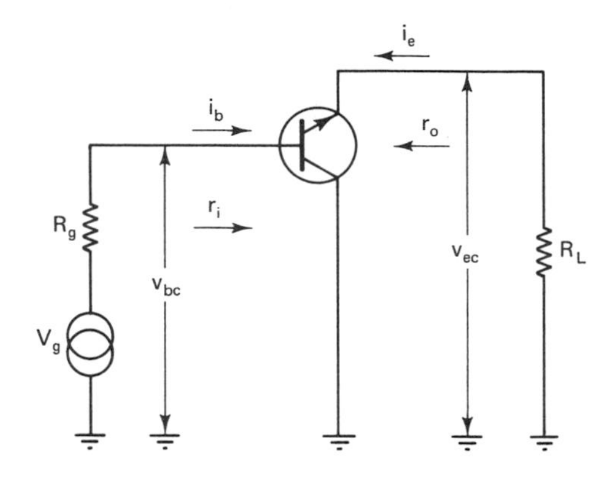

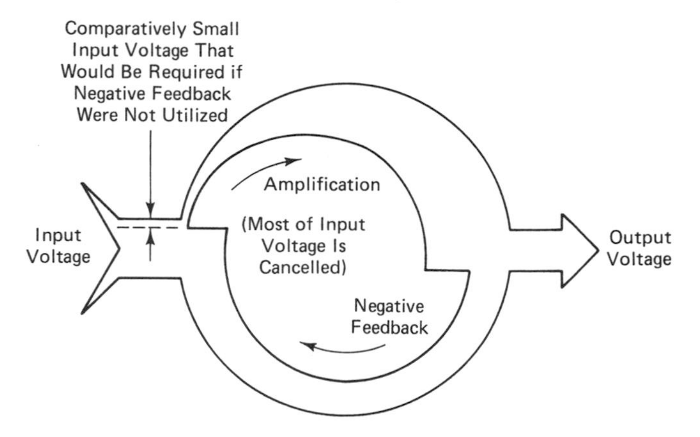

Figure 1–10. Skeletonized circuit diagram for the common-collector arrangement.

A large amount of negative feedback is inherent in the CC configuration. With reference to Figure 1–10, R_L is effectively connected in series with V_g and R_g from emitter to base of the transistor. This provides a large amount of negative feedback, which cancels most of the input signal and makes the output and input voltages approximately equal. Note that this negative feedback effectively linearizes the transfer characteristic and assists in reduction of distortion—a basic consideration in audio amplifier design.

By way of comparison, it is evident that positive feedback is present in the CE configuration (Figure 1–2) and that the effective input voltage is increased by the output-voltage drop across the base-emitter junction. It is this positive feedback that provides the high voltage gain in the CE (common emitter) configuration. By the same token, the CE transfer characteristic is not nearly as linear as the CC transfer characteristic.

Typical characteristic curves that relate to the hybrid parameters in the CC configuration are shown in Figure 1–11. Thus, h_{ic} is a V_{BC}/I_B ratio; μ_{rc} is a V_{BC}/V_{EC} ratio; α_{fc} is an I_E/I_B ratio; h_{oc} is an I_E/V_{EC} ratio. An equivalent circuit for the CC configuration is shown in Figure 1–12. Observe that the formulas that were noted for the CE configuration apply to the CC configuration when subscript e is changed into subscript c. For example, h_{ie} denotes the input impedance with output short-circuited in the CE configuration, and h_{ic} denotes the input impedance with output short-circuited in the CC configuration.

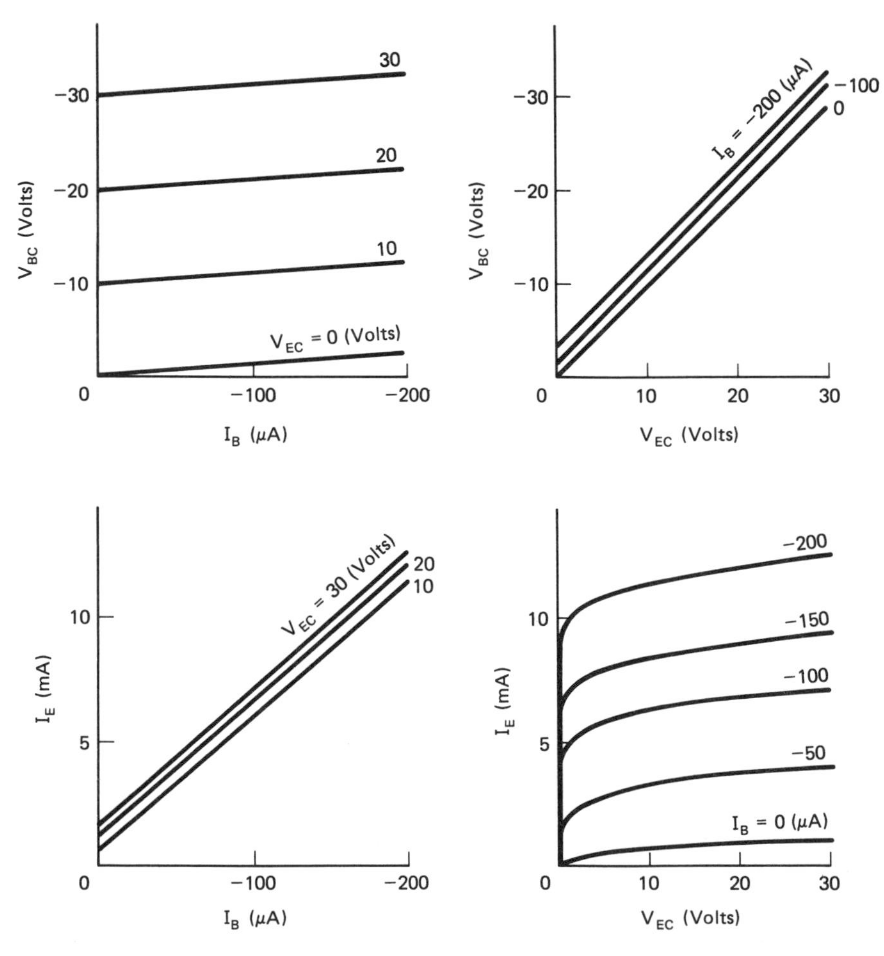

Figure 1–11. Typical characteristics curves relating to the hybrid parameters in the CC configuration.

EXAMPLE:

$\mu_{rc} = h_{rc} = 1$ $h_{ic} = 1325$ ohms $\alpha_{fc} = h_{fc} = -45$
$h_{oc} = 28.2 \times 10^{-6}$ Siemens $R_L = 10$ kilohms $R_g = 1500$ ohms

$A_v = 0.997$ $G_p = 43.6$
$r_i = 350$ kilohms $r_o = 62.9$ ohms
$A_i = 43.8$

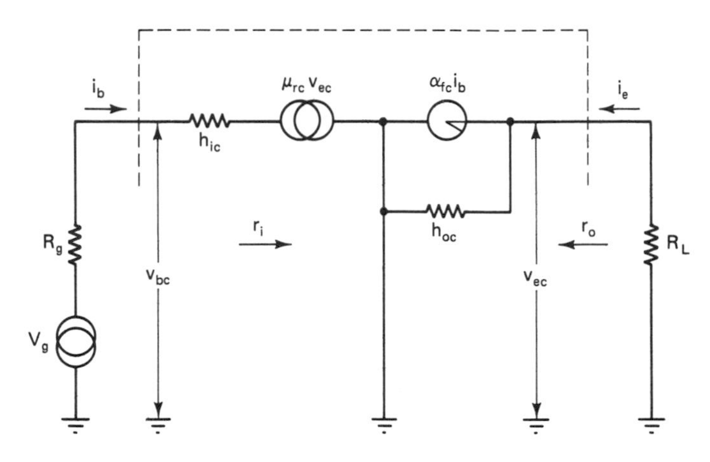

Figure 1–12. Equivalent circuit for the CC configuration.

COMMON-BASE AMPLIFIER

A skeletonized circuit diagram for a common-base amplifier is shown in Figure 1–13. As illustrated in Chart 1–1, the CB (common base) configuration has high voltage amplification and low current amplification. It may have substantial power gain. Designers usually employ the CB configuration when low input impedance and high output impedance are required.

Typical characteristic curves that relate to the hybrid parameters in the CB configuration are shown in Figure 1–14. Input and output equivalent circuits are depicted in Figure 1–13, and a complete equivalent circuit is shown in Figure 1–15. Observe that the formulas that were noted for the CE configuration apply to the CB configuration when subscript e is changed into subscript b.

PROGRAMMED FORMULAS

H parameters are easy to use in design procedures if a short program is employed, such as depicted in Figure 1–16. Note that

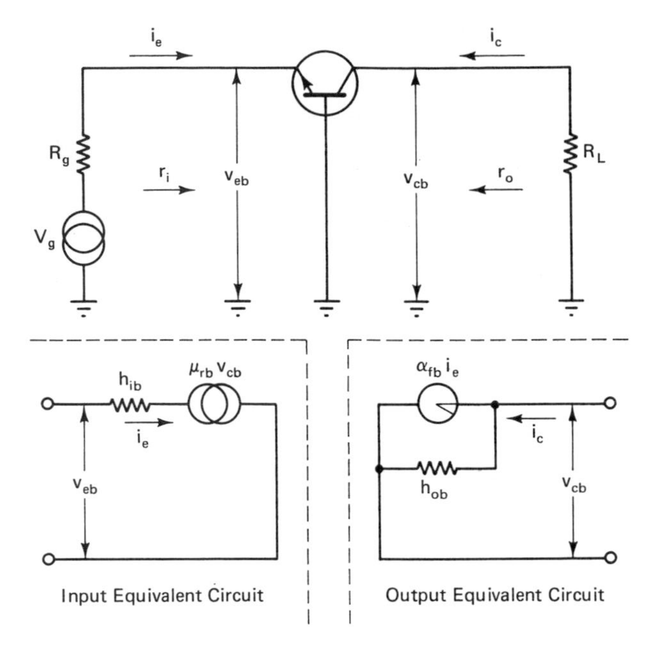

Figure 1–13. Skeletonized circuit diagram for the CB arrangement.

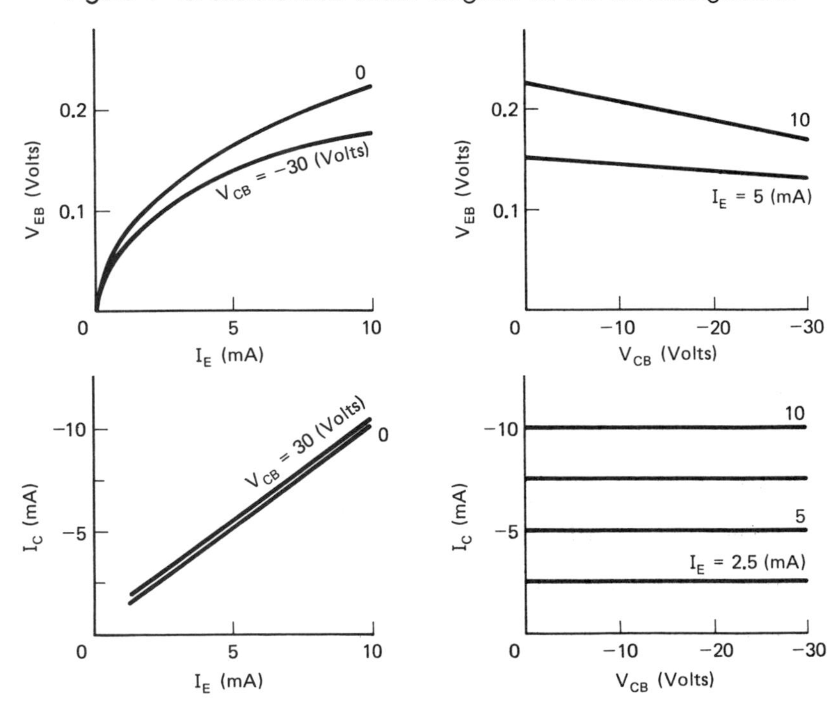

Figure 1–14. Typical characteristic curves relating to the hybrid parameters in the CB configuration.

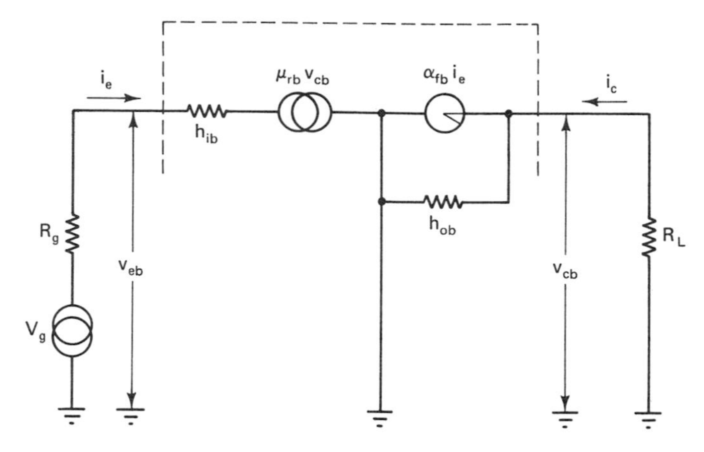

Figure 1–15. Equivalent circuit for the CB configuration.

EXAMPLE:

$h_{fb} = 0.95$ $h_{ib} = 29$ ohms $h_{rb} = 5.2 \times 10^{-4}$

$h_{ob} = 0.62 \times 10^{-6}$ Siemens $R_g = 35$ ohms $R_L =$ kilohms

$A_v = 393$ $r_i = 36$ ohms

$A_i = 0.94$ $r_o = 320$ kilohms

$G_p = 369$

this program is written for the IBM PC Computer, but can be readily converted to run on other types of computers. The program is written for common-emitter h parameters; however, it is equally useful for common-collector h parameters or for common-base h parameters. For example, h_{fc} or h_{fb} values can be substituted for h_{fe} values, and so on.

In preliminary design procedures, it is helpful to consult an overview of typical relative magnitudes in h parameter values among the CE, CC, and CB configurations, as shown in Figure 1–17. Then, if more precise conclusions are desired, the conversion formulas in Figure 1–17 may be used to convert from CE h parameters to CC or CB h parameters. (See also Chart 1–2.)

DESIGN OPTIMIZATION CONSIDERATIONS

Design optimization from the viewpoint of hybrid parameter calculations requires that variations in the values of hybrid

EXAMPLE: *

If the h parameters for the CB configuration are given, and the h parameters for the CE configuration are to be determined, look under "From CB to CE" in Figure 1–17. In turn:

$$h_{ie} = 39/(1 + (-0.98)) = 39/0.02$$
$$h_{ie} = 1950 \text{ ohms}$$

similarly:
$$h_{re} = ((39)(0.49 \times 10^{-6}))/(1 + (-0.98)) \times 380 \times 10^{-6}$$
$$h_{re} = 575 \times 10^{-6}$$

again:
$$h_{fe} = (-0.98)/(1 + (0.98))$$
$$h_{fe} = 49$$

and:
$$h_{oe} = (0.49 \times 10^{-6})/(1 + (-0.98))$$
$$h_{oe} = 24.5 \times 10^{-6} \text{ Siemens}$$

*Reproduced by special permission of Reston Publishing Company from *Solid State Devices* by William D. Cooper.

```
5 LPRINT "HYBRID PARAMETER PROGRAM":PRINT"HYBRID PARAMETER PROGRAM"
10 LPRINT"Voltage Gain, Current Gain, Power Gain, Input Resistance, and Output R
esistance":PRINT"Voltage Gain, Current Gain, Power Gain, Input Resistance, and O
utput Resistance"
15 LPRINT"(Common-Emitter Configuration)":PRINT"Common-Emitter Configuration"
20 LPRINT"":PRINT""
25 INPUT"Hfe=";A
30 LPRINT"Hfe=";A
35 INPUT"Hoe=";B
40 LPRINT"Hoe=";B
45 INPUT"Hie=";C
50 LPRINT"Hie=";C
55 INPUT"Hre=";D
60 LPRINT"Hre=";D
65 INPUT"RL (Ohms)=";E
70 LPRINT"RL (Ohms)=";E
75 INPUT"Rg (Ohms)=";F
80 LPRINT"Rg (Ohms)=";F
85 A$="######.##"
90 G=A*E/((C*B-A*D)*E+C):H=A/(B*E+1)
95 I=A*A*E/((B*E+1)*((C*B-A*D)*E+C))
100 J=(C+(B*C-A*D)*E)/(B*E+1)
105 K=(C+F)/(B*C-D*A+B*F):LPRINT"":PRINT""
110 LPRINT"Voltage Gain=";USING A$;G:PRINT"Voltage Gain=";USING A$;G
115 LPRINT"Current Gain=";USING A$;H:PRINT"Current Gain=";USING A$;H
120 LPRINT"Power Gain=";USING A$;I:PRINT"Power Gain=";USING A$;I
125 LPRINT"Input Resistance=";USING A$;J:PRINT"Input Resistance=";USING A$;J
130 LPRINT"Output Resistance=";USING A$;K:PRINT"Output Resistance=";USING A$;K
135 LPRINT"(Resistance Values in Ohms)":PRINT"(Resistance Values in Ohms)"
```

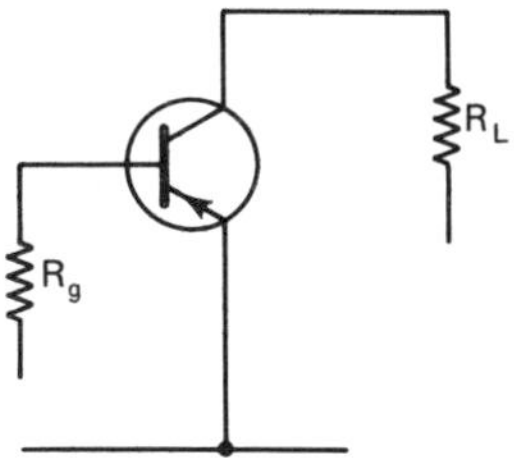

(A)

NOTE: This program is written for the IBM PC computer and will run on various other computers. Conversion programs are exemplified in the appendix.

```
HYBRID PARAMETER PROGRAM
Voltage Gain, Current Gain, Power Gain, Input Resistance, and Output Resistance
(Common-Emitter Configuration)

Hfe= 50                      Voltage Gain=   476.19
Hoe= .00002                  Current Gain=    38.46
Hie= 1500                    Power Gain= 18315.02
Hre= .0005                   Input Resistance=  1211.54
RL (Ohms)= 15000             Output Resistance= 85714.30
Rg (Ohms)= 1500              (Resistance Values in Ohms)
```

Figure 1–16A. Hybrid parameter program.

NOTE: The foregoing program is an example of a routine that includes both PRINT and LPRINT statements. It provides both video display and printout. Observe, however, that the printer must be turned on, or an error message will be displayed, and the program will not RUN. If printout is not desired, omit the LPRINT commands from the program. If video display is not desired, omit the PRINT commands from the program.

```
5 REM * POWER GAIN/INPUT RESISTANCE SURVEY PROGRAM *
10 LPRINT ""
15 INPUT "Hfe=";A
20 INPUT "Hoe=";B
25 INPUT "Hie=";C
30 INPUT "Hre=";D
35 N=0:M=0
40 M=M+1
45 N=N+5000
50 I=A*A*N/((B*N+1)*((C*B-A*D)*N+C))
55 J=(C+(B*C-A*D)*N)/(1+B*N)
60 K=10*LOG(I)/LOG(10)
65 IF M>1 THEN 75
70 LPRINT"Hfe=";A:LPRINT"Hoe=";B:LPRINT"Hie=";C:LPRINT"Hre=";D:LPRINT""
75 A$="######"
80 LPRINT"RL=";USING A$;N
85 LPRINT"Gp=";USING A$;I
90 LPRINT"Ri=";USING A$;J
95 LPRINT"Gp (dB)=";USING A$;K:LPRINT""
100 IF N>30000 THEN 110
105 GOTO 40
110 LPRINT "Note: Other Increments and Limits for RL May Be Programmed"
115 END
```

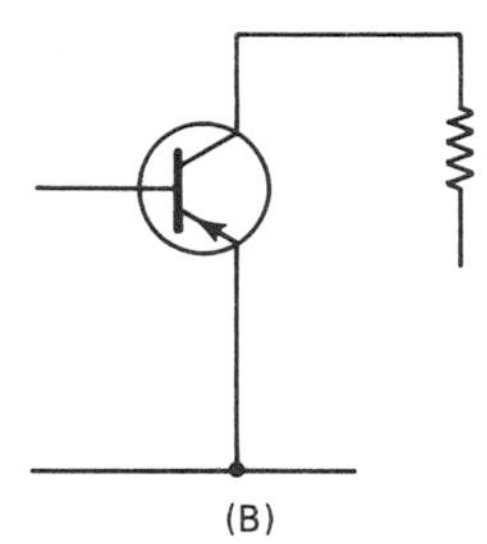

(B)

NOTE: This analysis is based on a fixed value of collector current. Stated otherwise, as the load resistance is increased, V_{cc} is to be increased as required to hold the collector current at its reference value.

Figure 1–16B. (Continues on next page.)

NOTE: Preamplifiers and driver amplifiers are usually designed for maximum feasible power gain. It is often desired to obtain high input resistance also, because the stage is then easier to drive. Both power gain and input resistance are functions of the collector load resistance RL. This program prints out power-gain and input-resistance values corresponding to a wide range of load-resistance values. Observe that as the power gain increases, the input resistance decreases. The maximum practical value of load resistance is often limited by the specified high-frequency response. This limit is a function of the load-resistance value and the total capacitance from collector to ground. The high-frequency response must usually be checked empirically.

```
Hfe= 50
Hoe= .00002
Hie= 1500
Hre= .0005

RL=  5000
Gp=  7452
Ri=  1386
Gp (dB)=     39

RL= 10000
Gp= 13441
Ri=  1292
Gp (dB)=     41

RL= 15000
Gp= 18315
Ri=  1212
Gp (dB)=     43

RL= 20000
Gp= 22321
Ri=  1143
Gp (dB)=     43

RL= 25000
Gp= 25641
Ri=  1083
Gp (dB)=     44

RL= 30000
Gp= 28409
Ri=  1031
Gp (dB)=     45

RL= 35000
Gp= 30729
Ri=   985
Gp (dB)=     45

Note: Other Increments and Limits for RL May Be Programmed
```

Figure 1–16B. Power-gain/input-resistance survey program.

parameters with respect to varying temperature and to emitter current levels be taken into account. To anticipate subsequent discussion, Figure 2–2 (page 31) shows the substantial changes that typically occur in h-parameter values over a temperature range from −50°C to +70°C. Similarly, Figure 2–4 (page 36) shows the

```
5 REM * POWER GAIN/INPUT RESISTANCE HEURISTIC ROUTINE *
10 REM * Load Resistance for Specified Power Gain and Input Resistance *
15 INPUT "Hfe=";A
17 LPRINT"Hfe=";A
20 INPUT "Hoe=";B
22 LPRINT"Hoe=";B
25 INPUT "Hie=";C
27 LPRINT"Hie=";C
30 INPUT "Hre=";D
32 LPRINT"Hre=";D
35 INPUT "Gp Minimum Limit=";E
37 LPRINT"Gp Minimum Limit=";E
40 INPUT "Ri Minimum Limit=";F
41 LPRINT"Ri Minimum Limit=";F
42 N=0:M=0:LPRINT""
45 M=M+1:N=N+1000
50 I=A*A*N/((B*N+1)*((C*B-A*D)*N+C))
55 J=(C+(B*C-A*D)*N)/(1+B*N)
60 K=10*LOG(I)/LOG(10)
65 IF (I>E)*(J*F) THEN 75
70 IF I>99999! THEN 100
71 IF J>99999! THEN 100
72 IF N>99999! THEN 100
73 GOTO 45
75 A$="######"
85 LPRINT"Gp=";USING A$;I
90 LPRINT"Ri=";USING A$;J
95 LPRINT"RL=";USING A$;N: END
100 LPRINT"COMPUTED VALUE GREATER THAN 99999; ENTER OTHER LIMITS"
```



```
Hfe= 50
Hoe= .00002
Hie= 1500
Hre= .0005
Gp Minimum Limit= 8500
Ri Minimum Limit= 1300

Gp=  8754
Ri=  1366
RL=  6000
```

Figure 1–16C. A power-gain/input-resistance heuristic routine.

NOTE: A heuristic procedure pertains to exploratory methods of solving a problem wherein a solution is discovered by evaluation of the progress made toward the final result. In this example, a particular transistor has specified h parameters, and the designer seeks the value of load resistance that will provide a power gain of no less than 8,500 times and also provide an input resistance of at least 1,300 ohms. (Amplifier operates in the CE mode.) In this example, the desired final result exists and is printed out as a power gain of 8,754 times and an input resistance of 1,366 ohms when RL has a value of 6,000 ohms.

substantial changes that typically occur in h-parameter values over an emitter-current range from 0.5 to 1.5 mA.

Worst-case conditions can often be evaluated, but only with respect to rated tolerances. In other words, there is no certainty that a rated h_{fe} value will not deteriorate with the passing of time.

Common Emitter	Common Base	Common Collector
h_{ie} = 1,950 Ohms	h_{ib} = 39 Ohms	h_{ic} = 1,950 Ohms
$\mu_{re} = 575 \times 10^{-6}$	$\mu_{rb} = 380 \times 10^{-6}$	μ_{rc} = 1
α_{fe} = 49	α_{fb} = −0.98	α_{fc} = −50
h_{oe} = 24.5 μ Siemens	h_{ob} = 0.49 μ Siemens	h_{oc} = 24.5 μ Siemens

From CE to CB	From CE to CC	From CB to CE	From CB to CC
$h_{ib} = \dfrac{h_{ie}}{1+\alpha_{fe}}$	$h_{ic} = h_{ie}$	$h_{ie} = \dfrac{h_{ib}}{1+\alpha_{fb}}$	$h_{ic} = \dfrac{h_{ib}}{1+\alpha_{fb}}$
$\mu_{rb} = \dfrac{h_{ir}h_{oe}}{1+\alpha_{fe}} - \mu_{re}$	$\mu_{rc} = 1 - \mu_{re} \sim 1$	$\mu_{re} = \dfrac{h_{ib}h_{ob}}{1+\alpha_{fe}} - \mu_{rb}$	$\mu_{rc} \cong 1$
$\alpha_{fb} = \dfrac{-\alpha_{fe}}{1+\alpha_{fe}}$	$\alpha_{fc} = -(1+\alpha_{fe})$	$\alpha_{fe} = \dfrac{-\alpha_{fb}}{1+\alpha_{fb}}$	$\alpha_{fc} = \dfrac{-1}{1+\alpha_{fb}}$
$h_{ob} = \dfrac{h_{oe}}{1+\alpha_{fe}}$	$h_{oc} = h_{oe}$	$h_{oe} = \dfrac{h_{ob}}{1+\alpha_{fb}}$	$h_{oc} = \dfrac{h_{ob}}{1+\alpha_{fb}}$

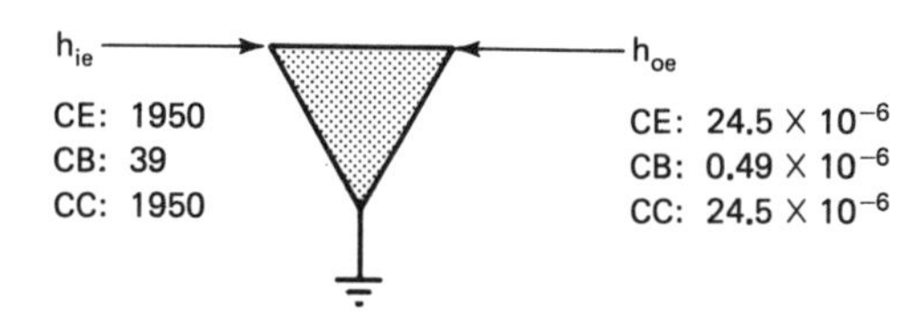

Figure 1–17. Relative magnitudes of h parameter values for typical transistor and h parameter conversion formulas.

NOTE: Alpha and beta are current-transfer ratios in the CB and CE configurations, respectively. Alpha is also denoted h_{fb}, and beta is also denoted h_{fe}.

$$\beta = \frac{\alpha}{1-\alpha}$$

$$\alpha = \frac{\beta}{1+\beta}$$

CHART 1–2

Basic Comparisons of Integrated Circuits and Transistors

Transistors are evaluated to best advantage in terms of h parameters. On the other hand, integrated circuits are evaluated to best advantage in terms of voltage and/or power gain, input resistance, and output resistance.

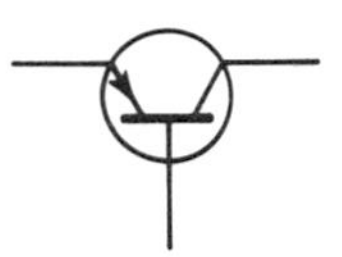

The basic distinction between integrated-circuit characteristics and transistor characteristics is in the degree of internal coupling from output to input for the two devices. In other words, a transistor has significant internal coupling from its output terminals back to its input terminals, whereas an integrated circuit has negligible internal coupling from its output terminals back to its input terminals.

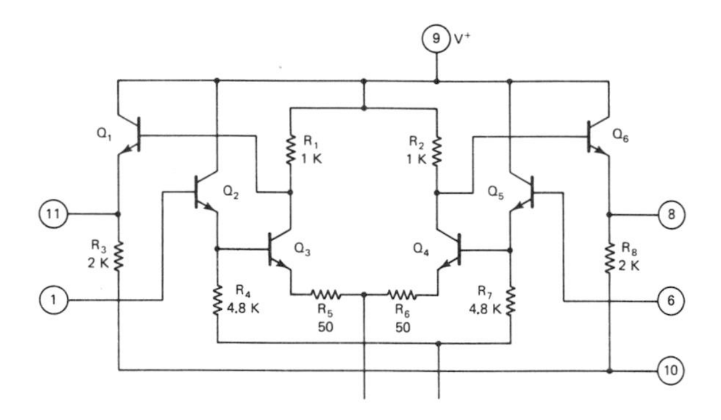

As exemplified in the diagram above, an integrated circuit comprises several transistors connected in series, whereby any reverse transfer is reduced to a negligible value. Hybrid parameters are primarily concerned with the effect of output circuit conditions upon input circuit parameters and vice versa. In the case of integrated circuits, output circuit conditions have virtually no effect upon input circuit parameters. In isolated cases, an RF integrated circuit will exhibit a small reverse transfer admittance which will be noted accordingly in the device data sheet.

Accordingly, conservative design procedure requires that appreciable reserve amplification be provided by the prototype. Similarly, there is no certainty that rated resistive tolerances will not deteriorate with the passing of time. Consequently, the conservative designer may choose to evaluate the worst-case condition on the basis of doubled resistive-tolerance ratings.

Design optimization usually involves conflicting factors. Production costs, product excellence, reliability, sales appeal, institutional image, serviceability, and product compatibility are among the basic factors that the designer will usually take into account. Since conflicting factors are often encountered, the prototype model represents a balance of judicious trade-offs.

Trade-offs associated with design optimization involve nonlinear systems from the technical viewpoint. This is just another way of saying that although components are essentially linear-circuit elements, virtually all electronic devices are nonlinear-circuit elements. Nonlinear elements are comparatively difficult to model mathematically, and various expedients are routinely employed by designers. As an illustration, a nonlinear device such as a bipolar transistor is assumed to operate linearly over a limited (small-signal) range.

When nonlinear devices are mathematically modeled over a large signal range, the designer often utilizes a graphical analysis. Although this approach is suggestive of an analytic geometry treatment, it is not—the characteristic curves used in graphical analysis are empirical, and they are not expressed as algebraic equations. In turn, graphical analyses are not precise, although they are usually adequate for preliminary design purposes. The chief value of a graphical analysis is in the perspective that it provides for the nonlinear system. It exhibits the intervals within which it is feasible to assume that a nonlinear device is operating linearly.

Note in passing that the empirical characteristic curves used in graphical analyses seldom relate to a particular device; they relate to average (bogie) device parameters and are subject to production tolerances. In some cases, these production tolerances may be very wide. It may be observed that the designer often copes with tolerances and device nonlinearities by the same means. For example, significant negative feedback is commonly employed, both to reduce the disturbance of tolerances and to reduce system nonlinearities.

Empirical characteristic curves can be mathematically modeled to any desired degree of precision by a process of curve fitting, wherein the curve is expressed as a power series. However, this is seldom a feasible procedure for the electronic design engineer. Although curve fitting is greatly facilitated by large computers, the resulting power-series equations are formidable from the manipulative standpoint. Nevertheless, special cases of curve fitting occur that are of high utility to the design engineer. As an example, the low-current interval of a MOSFET transfer characteristic is essentially a square-law curve. Again, the forward-current characteristic of a semiconductor diode is essentially a logarithmic curve.

Design procedures are greatly facilitated in many situations by recourse to computer programs. A computer is an impressive "number cruncher" and lends itself to surveys of circuit action over a wide range of component values. It also lends itself to heuristic routines wherein component values are optimized with respect to a stipulated circuit-action goal. Both survey programs and heuristic routines proceed in small increments or decrements of component values to compute a series of answers for a circuit-action equation. In turn, the optimum component values become evident.

Although the designer does not always recognize the procedure that he is following in solving a complex algebraic equation, it will become evident in retrospect that he is progressively reducing the original electrical network into its simplest possible equivalent circuit. Thus, the final equivalent circuit may consist of an impedance with an associated phase angle at a stipulated frequency of operation. Or, it may consist of an equivalent generator connected in series or shunt with an impedance or an admittance with an associated phase angle.

However, it should not be supposed that each mathematical step that is followed in the solution of a complex algebraic equation necessarily corresponds to a physically realizable equivalent circuit. It is required only that the final equivalent circuit that is derived will correspond to a physically realizable circuit and that it does not represent a spurious solution. Basically, the reason for this possible lack of step-by-step physical correspondence is that complex algebraic equations are merely mathematical models of electrical networks. Every mathematical model has its field of competence and its boundary limits.

There are two chief classes of mathematical models employed by electronic circuit designers. These are algebraic models and geometrical models. Geometrical models include principally trigonometric figures and Cartesian loci. An advantage of a geometrical model is in its felicitous visual symbolism of circuit action. This advantage is limited, however, to comparatively simple networks. Although it is readily possible to derive a geometrical model for an involved electrical network, the resulting "architecture" is likely to be prohibitively confusing. In turn, designers "instinctively" employ complex algebraic equations to model involved networks.

As detailed in Chapter 3, graphical analysis may be utilized for step-by-step network reduction in terms of geometrical models. Although this procedure lacks the precision of algebraic analysis, it is often sufficiently precise for engineering applications. Note in passing that when graphical procedures are followed, it is customary to retain the end result, such as a vector magnitude and angle, and to reject the components from which the vector was derived. On each step of the network reduction, a new vector magnitude and angle will be determined, and its preceding components will be rejected. This is a very practical method insofar as the end result is concerned, inasmuch as the designer is seeking the simplest possible equivalent circuit for the network.

It follows that since each new vector magnitude and angle is preceded by components, these components may be retained, if desired, and not rejected. When the successive components are retained in chained form, the end result is a geometrical model for the network that has a step-by-step correspondence with the associated algebraic model for the same network. This is just another way of saying that the algebraic equation models the "architecture" of the geometrical model (and vice versa) and that both are models of the physical network that is under consideration.

2

STABILIZING CIRCUIT DESIGN

TEMPERATURE EFFECTS IN AMPLIFIER OPERATION

Temperature effects in amplifier operation become of increasing concern to the designer as the signal power level is increased and as the rated ambient temperature range is increased. At collector-current levels above a few mA, the alpha or beta value of a bipolar transistor decreases at a greater rate as the operating temperature increases. As a practical note, circuit-design provisions may be required in higher power arrangements to avoid excessive power dissipation by the transistor at higher temperatures. Thus, a fuse or a current-limiting resistor may be employed.

As a useful rule of thumb, collector leakage current will double for each 8°-to-10°C increase in operating temperature. In the CE configuration, I_{CE} can vary from I_{CBO} to $h_{fe} \times I_{CBO}$. As detailed subsequently, it is helpful to minimize the circuit resistance between base and emitter, insofar as is practical. Excessive leakage current can be avoided chiefly by limiting the maximum junction temperature. Bias current or voltage variation can be reduced by limiting the minimum junction temperature.

Note that both germanium and silicon bipolar transistors have a negative temperature coefficient of approximately 2mV/°C. Large values of collector current (I_C) will assist in reduction of collector-current variation resulting from temperature changes. Note also that low values of source resistance when driving a base in the CE mode will degrade the stability factor. When diodes are used for bias stabilization, as in the following explanation, the designer should mount the diodes with the transistor on the heat sink (if used). Current and voltage stability factors are detailed.

Conservative design procedure employs a minimum h_{fe} value over the rated operating temperature range. Observe that a low stability factor improves the performance of an ac amplifier, but not of a dc amplifier. The designer should apply a suitable thermal derating factor at operating temperatures above 25°C. This derating factor is approximately 1 to 10 mW/°C for small-signal transistors, and 0.25 to 1.5 W/°C for power-type transistors. If possible, an emitter swamping resistance should be utilized to avoid the possibility of thermal runaway.

The thermal resistance rating (R_T,θ_R) for a transistor does not include the thermal resistance of an associated heat sink. Designers should keep in mind that second-source manufacturers may publish somewhat different ratings for the same type of transistor. This is a matter of little or no concern in low-level amplifier design; on the other hand, it can be an important consideration in high-level amplifier design.

BIAS VOLTAGE AND CURRENT STABILITY

Operating-point bias is established by specification of the quiescent (dc) values of collector voltage and emitter current with no signal present. Reliable transistor operation over a wide temperature range requires that bias voltage and current remain stable. Unless external temperature-compensating circuits are employed, a stable bias condition will not be realized, due to variation of reverse-bias collector current and of emitter-base junction resistance as the temperature changes.

Reverse-Bias Collector Current (I_{CBO})

Variation of saturation current versus temperature of the base-emitter junction is shown in Figure 2–1. Observe that the saturation current varies from almost zero at 10°C to over 1 mA at 125°C. On the other hand, saturation current poses no design problem at tem-

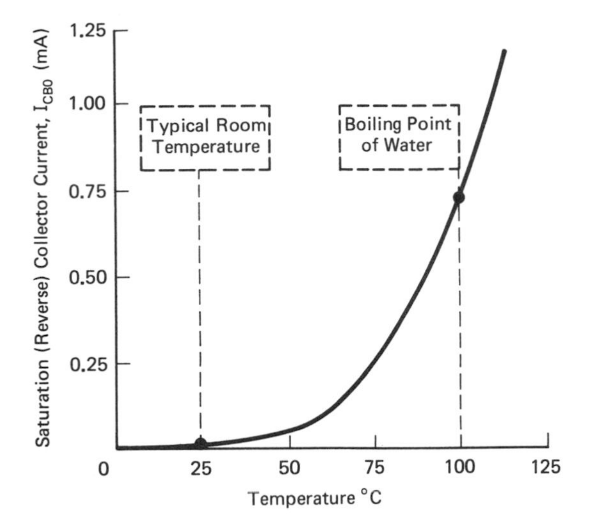

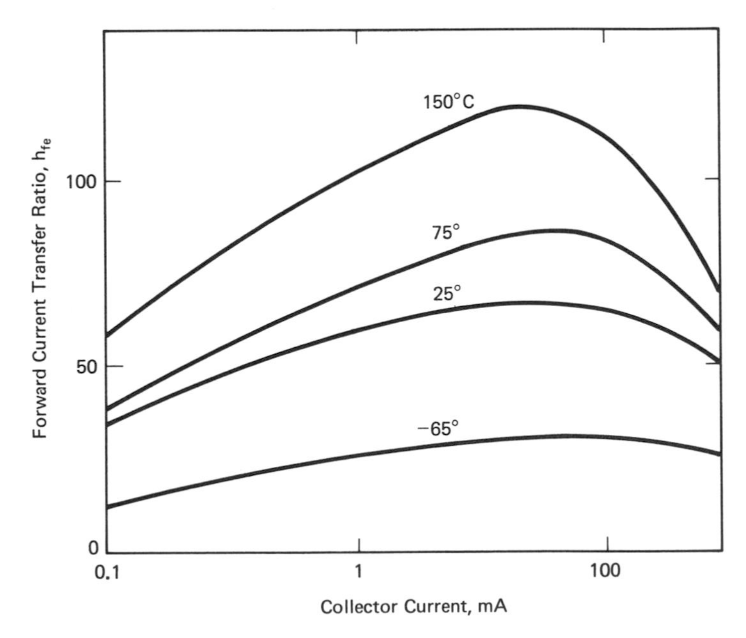

Figure 2–1. Transistor forward and reverse collector currents and h_{fe} variation versus temperature.

peratures below 10°C. High saturation current flow can "kill" amplifier action, with the possibility of thermal runaway and destruction of the transistor. This hazard can be minimized by avoiding the use of high-value resistors in the base lead.

Emitter-Base Junction Resistance

Variation of collector current with temperature is exemplified in Figure 2–2. Each curve in this family corresponds to a fixed collector-base voltage (V_{CB}) and to a fixed emitter-base voltage (V_{EB}). Note that if the variation of collector current versus temperature resulted only from saturation current, then collector-current variation would not occur below 10°C. The appreciable variation that does occur is caused by a decrease in emitter-base junction resistance as the temperature increases.

Emitter-base junction resistance has a negative temperature coefficient of resistance, and one method of reducing its effect is to include a large value of resistance in the emitter lead. This emitter resistor causes the variation in emitter-base junction resistance to be a small percentage of the total emitter-circuit resistance. An emitter resistor is often called a swamping resistor. If it is bypassed, current feedback (negative feedback) will occur with respect to dc, but not to ac.

Another design method for reduction of the effects of the negative temperature coefficient of resistance is to provide means for reduction of the emitter-base forward bias as the temperature increases. As an illustration, to maintain the collector-current value at 2 mA in Figure 2–2 while the transistor temperature varies from 10°C (X) to 30°C (Y), it is evident that the forward-bias voltage must be reduced from 200 mV (A) to 150 mV (B). The temperature difference in this example is 20°C, and the voltage difference is 50 mV. This corresponds to a variation in forward bias per °C of 2.5 mV/°C.

In other words, the collector current will not vary (in this example) with emitter-junction-resistance temperature change if the forward bias is reduced 2.5 mV/°C for increasing temperature, or increased 2.5 mV/°C for decreasing temperature.

RESISTOR STABILIZING CIRCUITS

Bias stabilizing circuits may employ resistors, thermistors, junction diodes, transistors, or breakdown diodes. With respect to the skeleton CE configuration shown in Figure 2–3, three basic configurations can be derived, as illustrated.

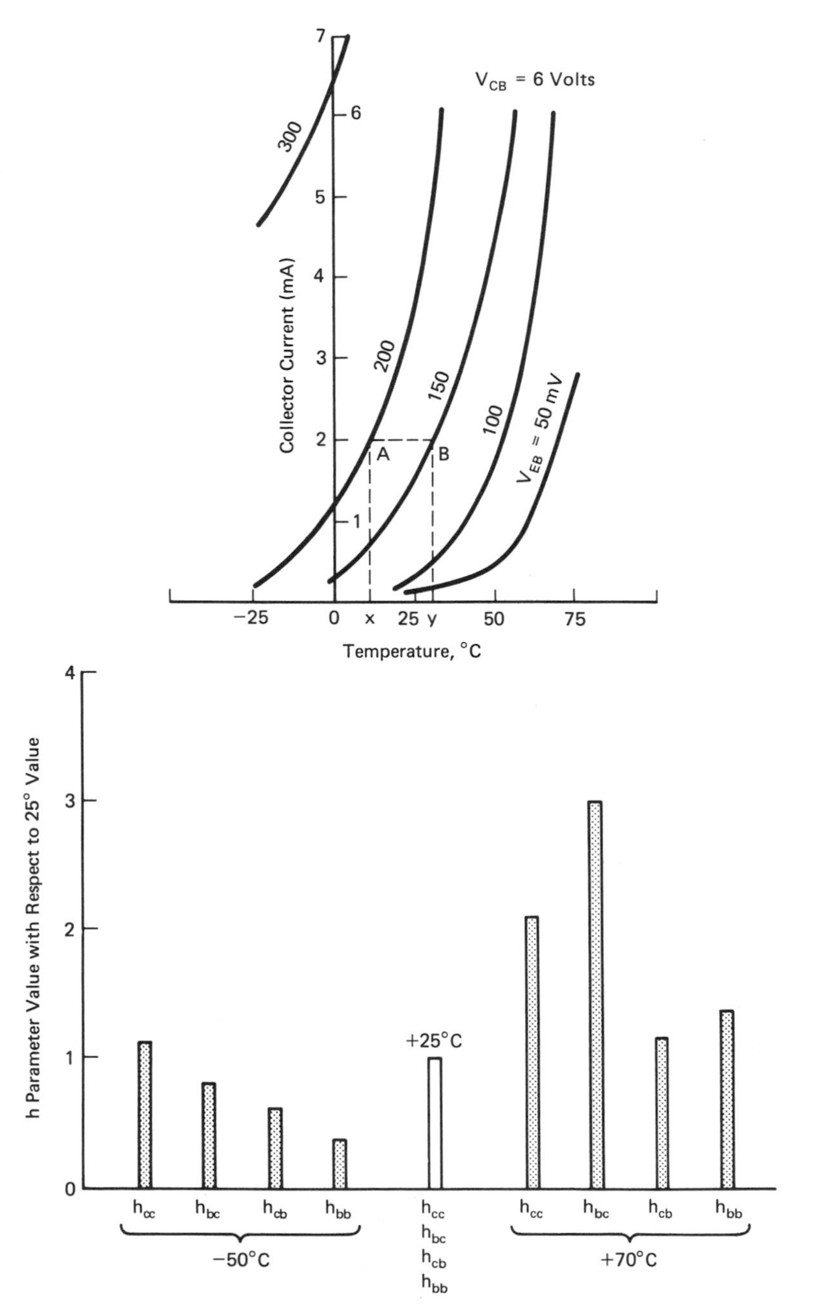

Figure 2–2. Variation of collector current versus temperature at different base-emitter biases and variation of common-emitter h parameters versus temperature.

NOTE: h_{ie} is also called h_{bb}; h_{re} is also called h_{bc}; h_{oe} is also called h_{cc}; h_{fe} is also called h_{cb}.

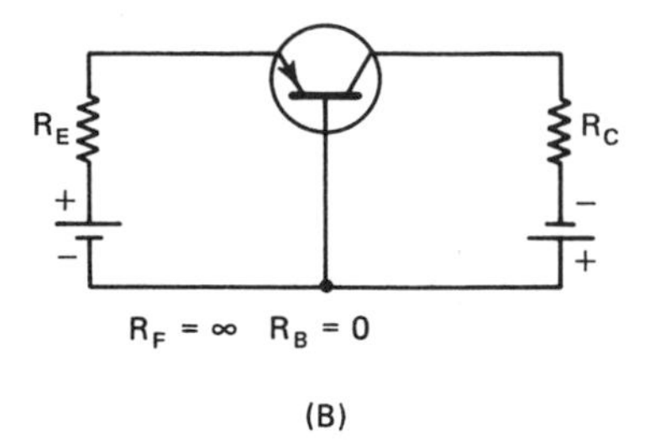

(B)

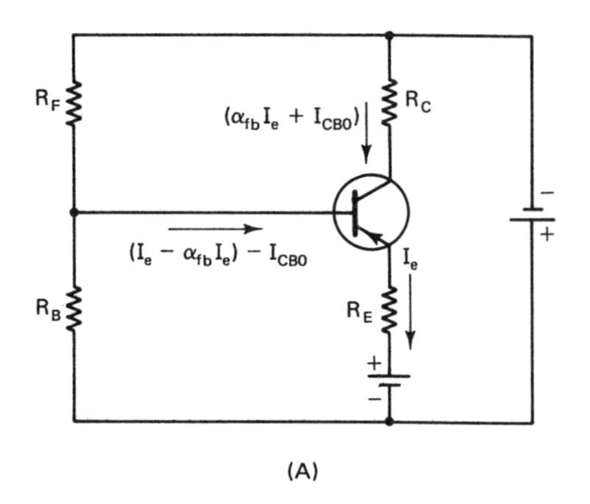

(A)

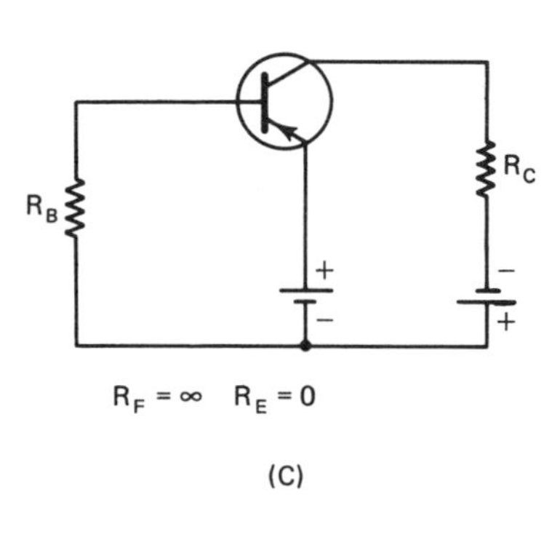

(C)

$$S_I = \frac{1}{R_E} \Big/ \frac{1}{R_B} + \frac{1}{R_F} + \frac{1-\alpha_{fb}}{R_E}$$

$$S_V = -[S_I R_E + R_C(1 + \alpha_{fb} S_I)]$$

(E)

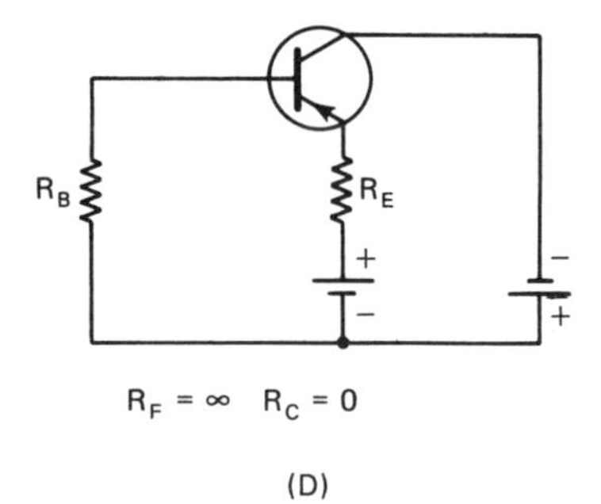

(D)

(A Short Computer Program for Calculation of Current and Voltage Stability Factors Is Provided at the End of the Chapter)

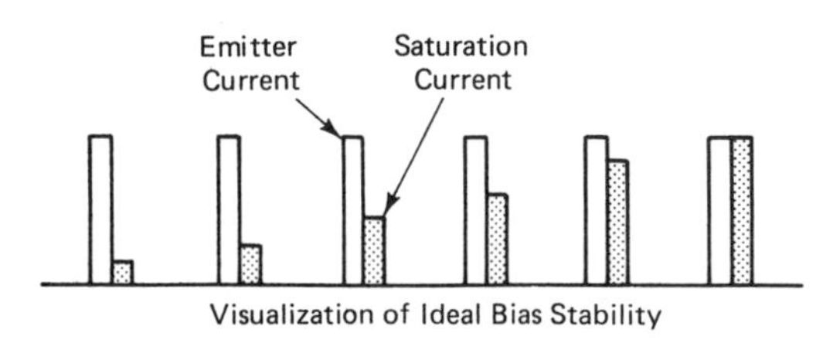

Figure 2–3. Skeleton CE configuration with derived configurations relating to the basic voltage-stability-factor formulas.

Current Stability Factor

A measure of the bias current stability in a transistor circuit is given by the ratio of a change in emitter current to a corresponding change in saturation current. It indicates the effect on the emitter current of a change in saturation current, and is called the current stability factor (S_I):

$$S_I = \frac{\Delta I_E}{\Delta I_{CBO}}$$

This stability factor is a pure number. In an ideally stabilized circuit, the current stability factor would be zero; the emitter current would be unchanged by an increase or a decrease in saturation current. With reference to Figure 2–3, the current value I_E can be formulated in terms of saturation current I_{CBO} and the resistive values. In turn, the current stability factor S_I can be formulated:

$$S_I = \frac{1}{R_B} \Bigg/ \left(\frac{1}{R_B} + \frac{1}{R_F} + \frac{1-\alpha_{fb}}{R_E} \right)$$

This formula defines the current stability factor with respect to the effectiveness of the external circuit to minimize variation in emitter current due to temperature change, employing the configuration depicted in Figure 2–3(A).

EXAMPLE:

An audio-amplifier stage with the configuration shown in Figure 2–3(A) employs the following values:

$R_E = 58k$ $\quad R_B = 7.3k$
$R_F = 120k$ $\quad \alpha_{fb} = 0.99$
$R_C = 85k$

$S_I = 0.118$

As a practical note, if R_E has a very low value, the value of S_I will be determined chiefly by the third term in the denominator. Accordingly, the value of S_I will then be very large, such as 10, 20, 30, or even greater.

Example:

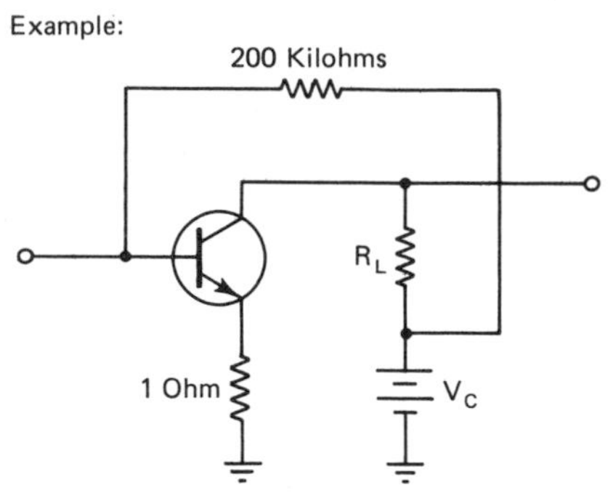

An audio-amplifier stage with this configuration provides an instructive design example. It has a current stability factor of approximately 100, and represents very poor design practice.

The only temperature compensation in this circuit is provided by the emitter resistor. However, since this resistor has a value of only 1 ohm, it contributes very little to collector-current stability. Since the base resistance is infinite, it contributes nothing to collector-current stability. This arrangement is called a "fixed-bias" CE circuit. If the 200-kilohm resistor is returned to the collector instead of V_C, the arrangement is then termed a "self-bias" CE circuit. Voltage feedback is provided which contributes appreciably to collector-current stability (at the expense of stage gain). Further improvement of collector-current stability is realized if a base resistor is employed between base and ground (at the expense of further reduction in stage gain).

Designers often accept the trade-off in gain which results from voltage feedback, finite base resistance, and a comparatively large emitter resistance, in order to obtain good collector-current stability.

Note in passing that both voltage feedback and current feedback contribute to reduction of stage distortion.

With respect to the configuration in Figure 2–3(B), the current stability factor for the CB configuration is derived as:

$$S_I = \frac{1}{1 - \alpha_{fb}}$$

Since the value of the denominator is very large, it follows that the current stability factor for the CB configuration is very poor. By way of comparison, a large value of emitter resistance may be utilized in the CE configuration, but the emitter resistance is necessarily zero in the CB configuration.

Next, with respect to the configuration in Figure 2–3(D), the approximate current-stability factor for the CC configuration is derived as:

$$S_I \cong \frac{R_B}{R_E}$$

In other words, the current stability factor of the CC configuration depends on the ratio of base resistance to emitter resistance. As before, it is perceived that the higher the base resistance, the poorer the current-stability factor, and the higher the emitter resistance, the better the current-stability factor.

Base Resistance and Emitter Resistance

Representative combinations of base and emitter resistance values and their effect on collector-current stability are seen in Figure 2–4. Observe that the worst condition occurs when the emitter and base resistances are both zero. The best condition occurs when the emitter resistance is greater than zero (2 kilohms in this example), and the base resistance is zero.

Additional Bias Techniques

With reference to Figure 2–5(A), transformer coupling can provide very low base resistance in the CE configuration; if R_E is large, the collector-current-stability curve is similar to the CC plot in Figure 2–4. If the secondary resistance of T1 is zero (Figure 2–5), the current-stability factor will be zero (ideal).

Fixed base-emitter bias may be provided by means of a voltage-divider network, as shown in Figure 2–5(B). The voltage divider comprises R_F and R_B. From a practical viewpoint, the current-stability factor for this arrangement may be formulated:

$$S = \frac{R_B R_F}{R_B + R_F} \bigg/ R_E$$

Observe that the current-stability factor for this circuit is equal to the ratio of the paralleled resistance value of R_B and R_F to the emitter-resistance value. As before, it is seen that current stability improves with reduction of base ground-return resistance and with increase of emitter resistance.

A more effective design employs negative voltage feedback (voltage feedback) in the voltage-divider network, as shown in Figure 2–5(C). Collector-current stability is thereby improved; if I_C rises, the collector becomes less negative due to the greater IR drop across R_C. In turn, less forward bias is coupled via R_F to the base, which in turn reduces the collector current. Note in passing that R_E provides current feedback in all three circuits.

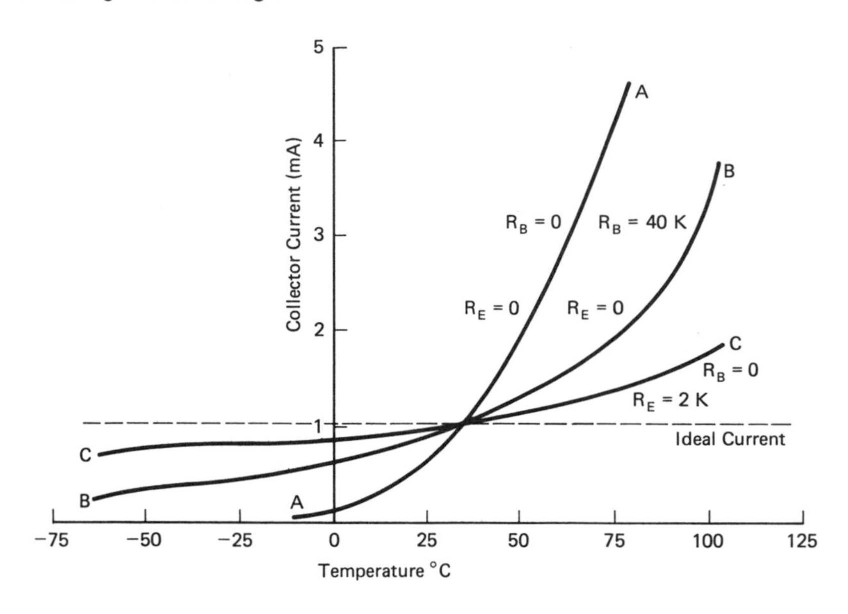

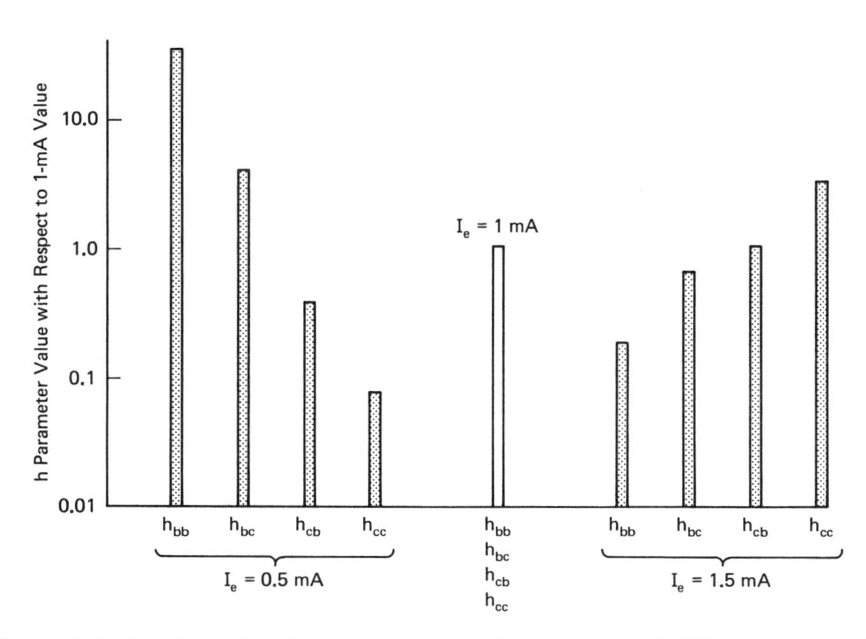

Figure 2–4. Variation of collector current with temperature at different values of emitter and base resistance, and variation of h parameters with emitter current in the CE configuration.

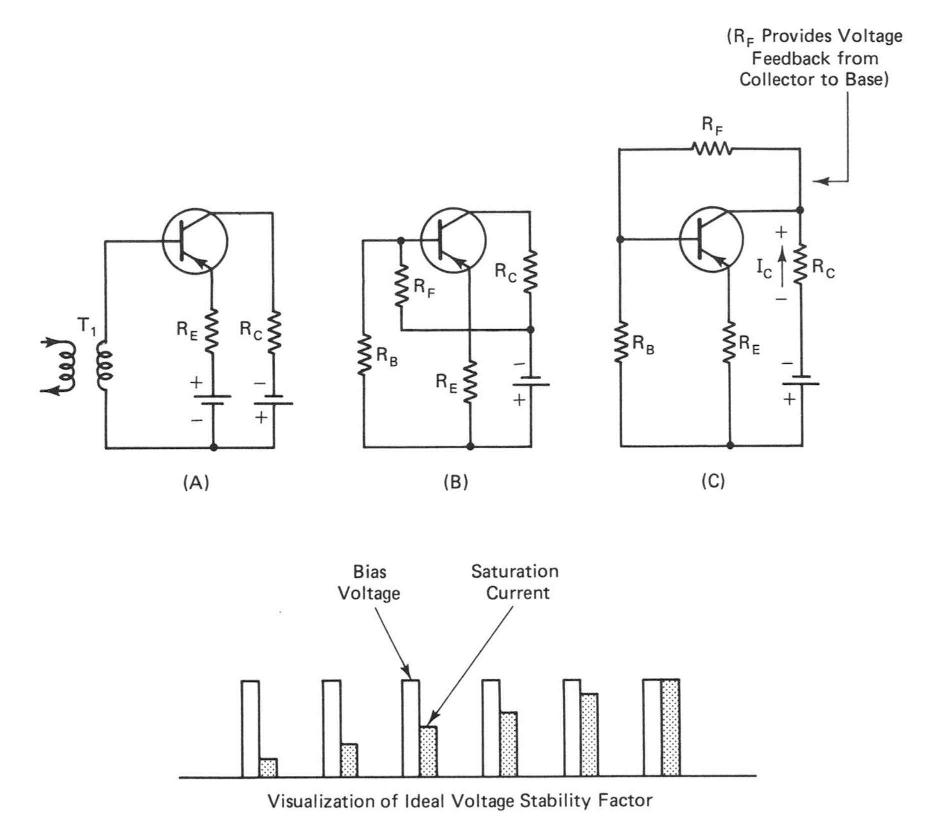

Figure 2–5. Skeleton CE amplifier configurations relating to the basic voltage stability factor formula.

NOTE: Current and voltage stability factors are related as stated by the formulas in Figure 2–3.

VOLTAGE-STABILITY FACTOR

A measure of the collector voltage stability is given by the ratio of an increment in collector voltage to a corresponding increment in saturation current. Stated otherwise, this ratio indicates the effect on the collector voltage of a change in saturation current. It is called the voltage stability factor and is formulated:

$$S_V = \frac{\Delta V_{CB}}{\Delta I_{CBO}}$$

Since the voltage stability factor is the ratio of a voltage to a current, it is expressed in resistance units. Ideally, the voltage stability factor would be zero. This is just another way of saying that ideally, the collector voltage would not be affected by a change in saturation current. With reference to Figure 2–3, the voltage stability factor may be formulated:

$$S_V = -\left[S_I R_E + R_C(1 + \alpha_{fb} S_I)\right]$$

Observe that if the current stability factor is zero, then the voltage stability factor is equal to the collector resistance (R_C). If a transformer with a low-resistance primary winding is utilized, the voltage stability factor can be reduced to near zero.

THERMISTOR STABILIZING CIRCUITS

Emitter-current stabilization versus temperature can be achieved by means of external circuits using temperature-sensitive elements. A thermistor is a familiar example. Suitable thermistors have negative temperature coefficients of resistance. With reference to Figure 2–6, a thermistor is employed to minimize temperature variations in emitter current. Two voltage dividers are utilized: R4-R1, and R2-RT1. The first voltage divider provides application of a portion of V_C between the base and ground (common). The second voltage divider provides application of a portion of V_C between the emitter terminal and ground.

Note that the base is forward-biased and the emitter is reverse-biased in this arrangement. The forward-base bias is greater than the reverse-emitter bias, so that class A operation is obtained. Increase of collector current with temperature is prevented by reduction of the forward-bias value. Thus, R2 and RT1 permit more current flow through the divider as the temperature increases. In turn, reverse-bias voltage is increased and the collector current is reduced.

Base Voltage Control

As exemplified in Figure 2–7, a thermistor may be used to control the base voltage to minimize temperature variations in emitter current. This circuit includes a voltage divider comprising R1 and RT1. A portion of V_C is applied between the base and emitter and

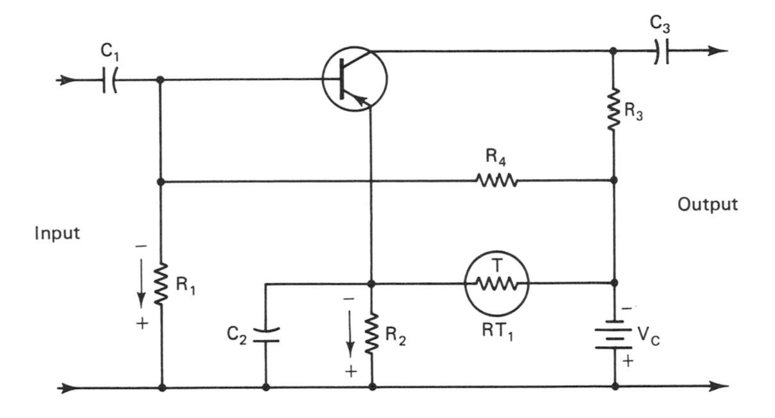

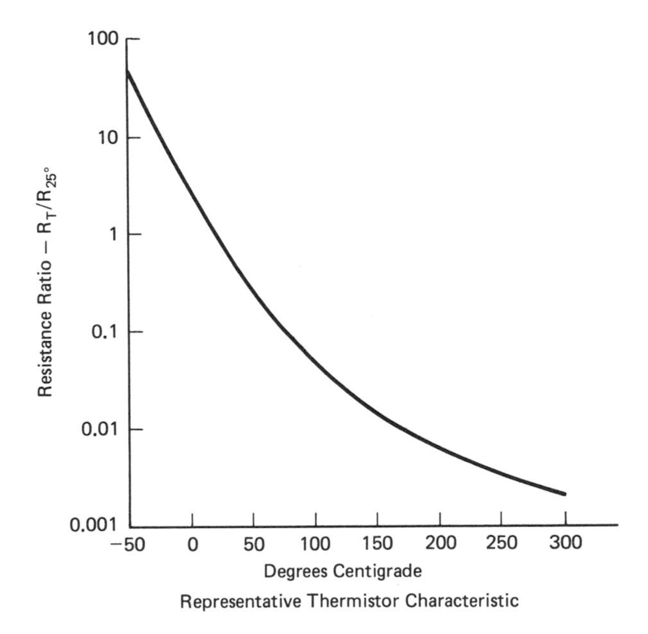

Figure 2–6. CE amplifier configuration with thermistor control of emitter bias voltage.

provides forward bias. If the temperature rises, the resistance of RT1 decreases and more current flows through the divider. In turn, a larger portion of V_C is dropped across R1 and a lesser portion dropped across RT1, with resulting reduction in forward bias and reduction in emitter current.

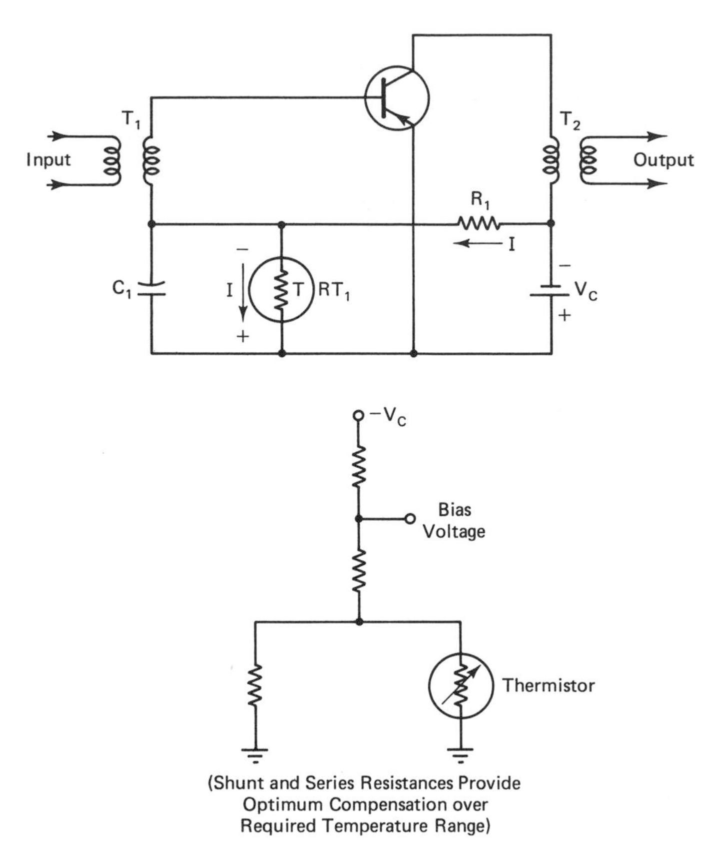

Figure 2–7. CE amplifier configuration with thermistor control of base-bias voltage.

Thermistor Limitations

The ability of a thermistor to limit collector-current variation versus temperature is exemplified in Figure 2–8. Observe that ideal current stabilization occurs at only three points (A,B,C). In other words, the resistance variation provided a typical thermistors does not precisely track the temperature variation in emitter-base junction resistance. Accordingly, designers may prefer some form of diode-current stabilization, as described next.

DIODE STABILIZING CIRCUITS

A diode that is made of the same material as the transistor provides better temperature compensation than a resistor, because the diode

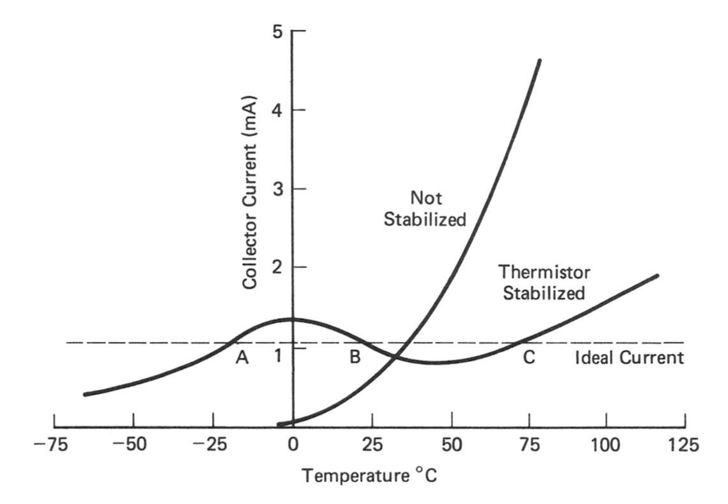

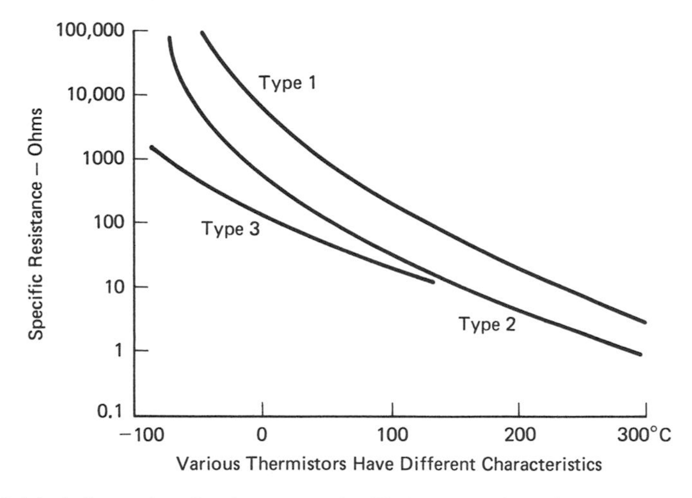

Figure 2–8. Variation of collector current with temperature for thermistor-stabilized and nonstabilized amplifier circuits.

has a matching negative temperature coefficient of resistance. In turn, it provides more precise tracking of the transistor characteristic over a wide range of temperatures.

With reference to Figure 2–9, a forward-biased junction diode CR1 functions as a temperature-sensitive element to compensate for variations in emitter-base junction resistance with temperature. The voltage divider R1-CR1 conducts current I, and the resulting voltage drop across CR1 provides forward bias for the transistor. If

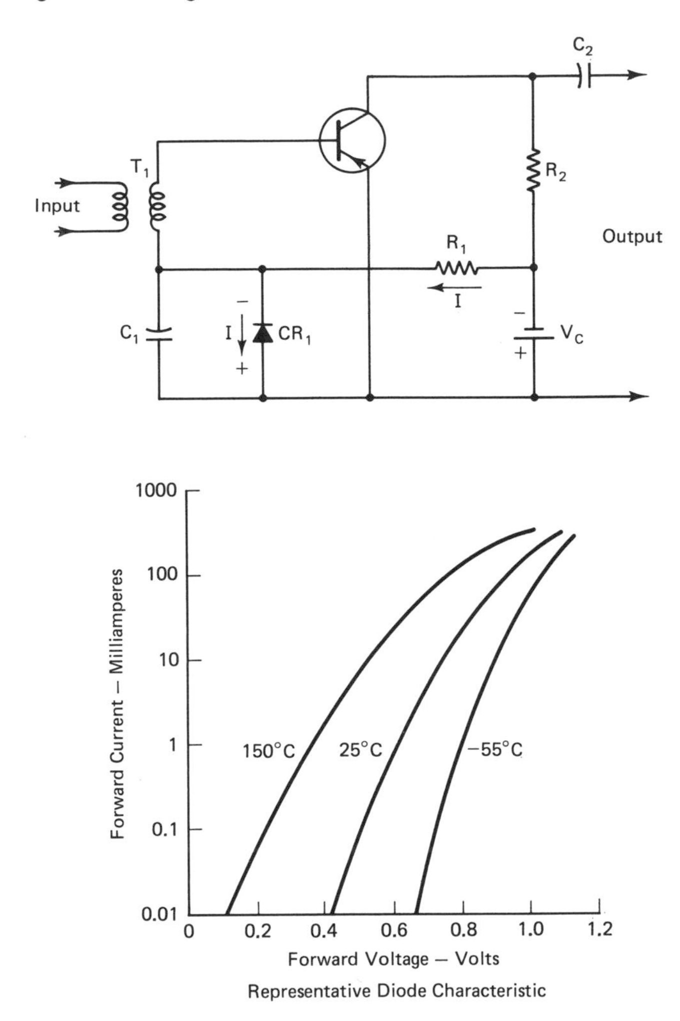

Figure 2–9. CE configuration with single forward-biased diode for temperature compensation of emitter-base junction-resistance variations.

NOTE: The barrier potential of a silicon diode is considered to be 0.7 volt at room temperature; the barrier potential of a germanium diode is considered to be 0.3 volt at room temperature. (Actually, silicon diodes start to conduct when forward-biased over 0.5 volt, and germanium diodes start to conduct when forward-based over 0.1 volt.) As the temperature increases, the barrier potential decreases. A practical rule of thumb states that the barrier potential (voltage) decreases by 0.002 volt for each degree Centigrade increase in temperature.

the temperature increases, the resistance of CR1 decreases, and more current flows through the divider. In turn, there is an increased voltage drop across R1 and a decreased voltage drop across CR1. Thus, the forward bias on the transistor decreases and the collector current is reduced.

Refer to Figure 2–11; the effectiveness of a single diode to stabilize collector current versus temperature is indicated by plot BB. Tracking is comparatively precise below 50°C; however, the sharp increase in collector current (plot BB) at temperatures above 50°C illustrates that the diode does not compensate for saturation-current increase. Therefore, the designer must employ a second diode if saturation current is to be compensated.

DOUBLE DIODE STABILIZATION

In Figure 2–10 a second diode has been included to compensate for temperature variations in saturation current. Resistor R3 and the reverse-biased diode CR2 serve this purpose. At low temperatures, CR2 is essentially an open circuit. CR2 is a selected diode that has a reverse current (I_S) somewhat larger than that of the transistor (I_{CBO}). The reverse current I_S consists of the transistor reverse-bias current I_{CBO} plus a component of current I_1 that is drawn from the battery.

Note that the emitter-base bias voltage in Figure 2–10 is the sum of the opposing voltages across R3 and CR1 (assuming negligible resistance in the secondary of T1). If the temperature increases, I_{CBO}, I_S, and I_1 increase, and the reverse-bias voltage dropped across R3 increases. In turn, the combined circuit action functions to stabilize collector current at high and low temperatures as exemplified by plot CC in Figure 2–11.

REVERSE-BIASED SINGLE-DIODE STABILIZATION

A CE amplifier stage that employs a single reverse-biased diode for temperature compensation of saturation current is shown in Figure 2–12. Observe that this arrangement provides high input resistance. In turn, it may be preferred by the designer if resistance-capacitance coupling is used in the previous stage. The configuration provides two separate paths for the two components of base current.

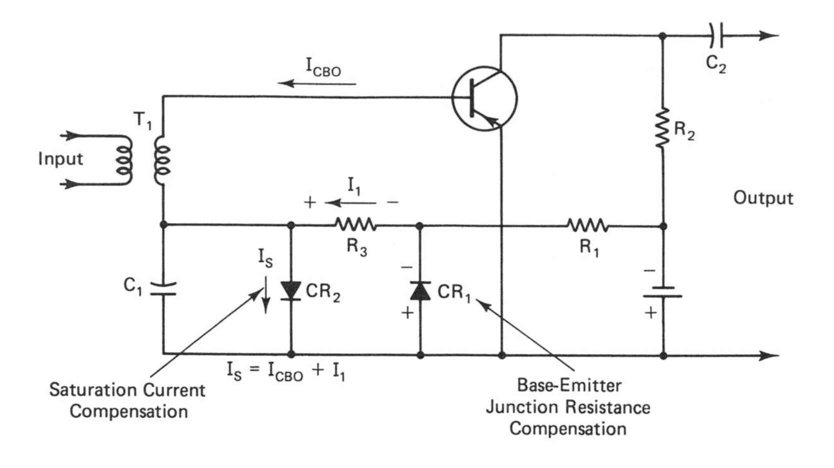

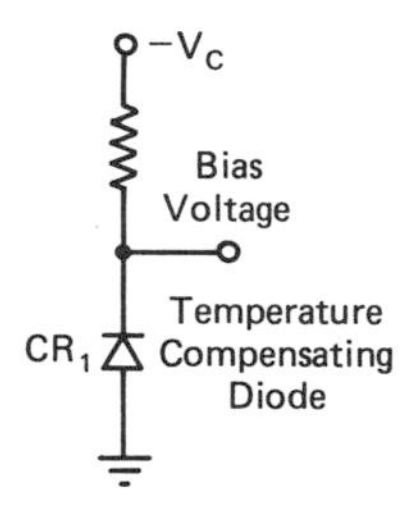

(Temperature Compensating Diode also
Functions to Stabilize the Collector
Current for Variation in Supply Voltage)

Figure 2–10. CE configuration with two diodes for temperature compensation of both base-emitter junction resistance and saturation current.

The base-emitter current ($I_e - \alpha_{fb}I_e$) flows through the base to the emitter via R2, V_C, and R1. However, the saturation current (I_{CBO}) flows from the base lead through CR1, V_B, V_C, R3, and the collector to the base. A selected diode is employed, so that its saturation current matches the transistor over a wide temperature range.

As the operating temperature increases in Figure 2–12, I_{CBO} increases. However, the saturation current of CR1 increases equally, so that emitter current is controlled. Effectively, CR1 functions as a gate that opens wider with an increase in I_{CBO}. Emitter resistor R2 supplements saturation-current compensation by providing emitter-base junction resistance compensation, particularly at lower temperatures.

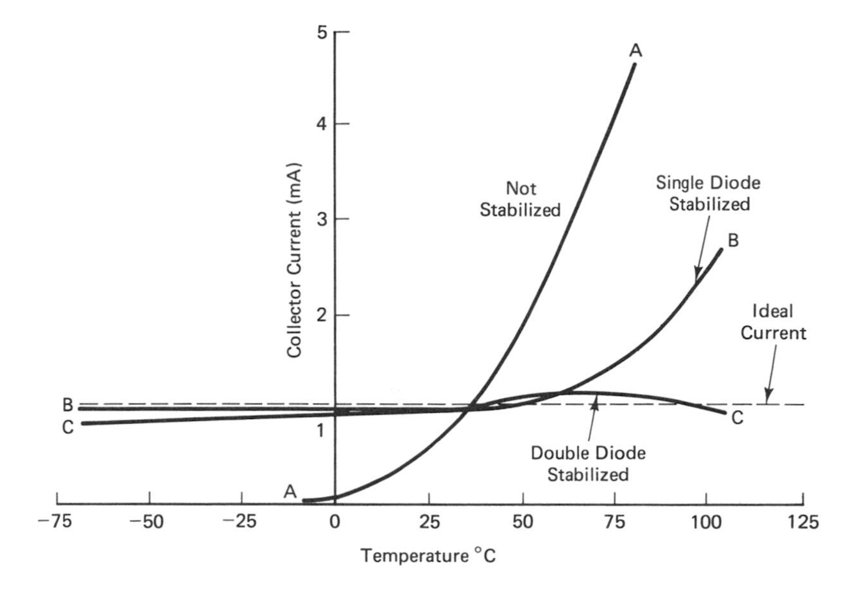

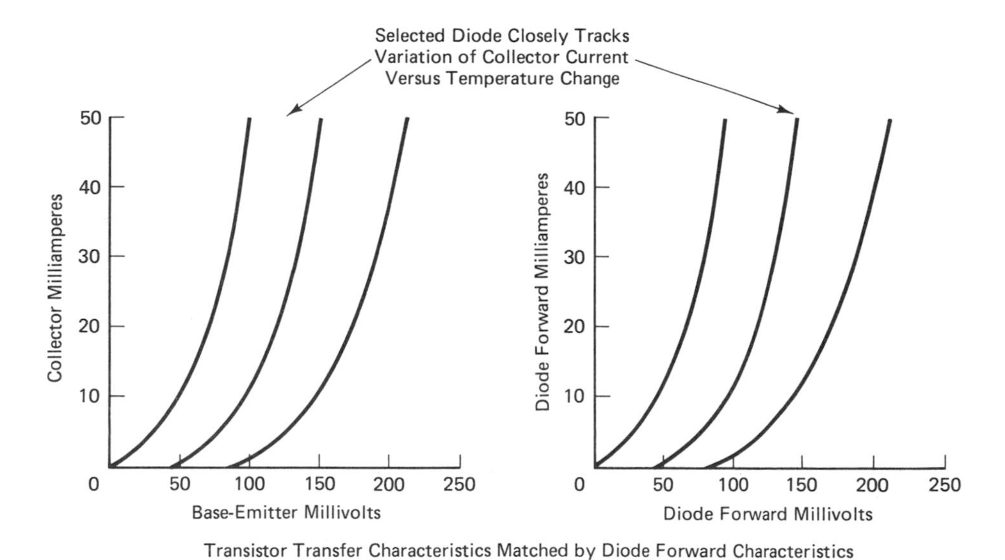

Figure 2–11. Collector-current variation versus temperature for nonstabilized, single-diode stabilized, and double-diode stabilized CE amplifier.

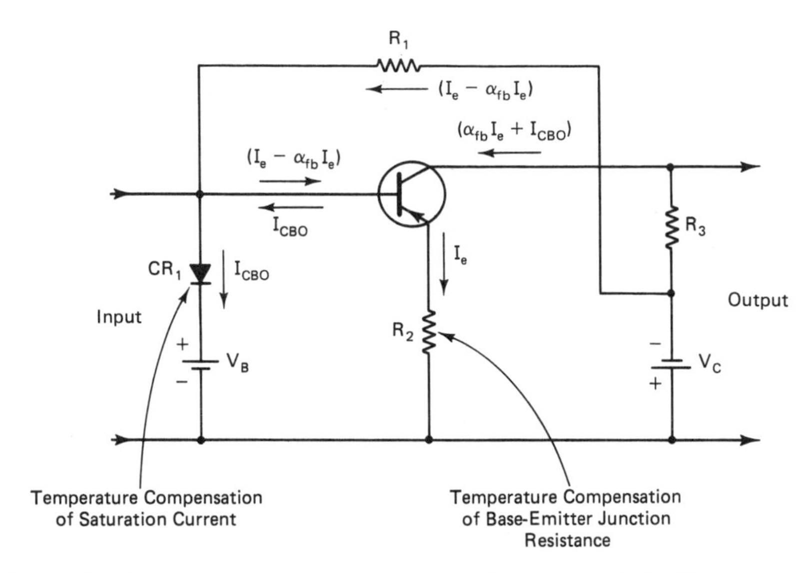

Figure 2–12. Arrangement of single reverse-biased diode in CE amplifier configuration for compensation of saturation current versus temperature.

NOTE: Temperature compensation circuitry functions also to make the amplifier stage relatively independent of tolerances on saturation current of replacement transistors. Note also that conservative ratings of transistor parameters should be assigned to provide for anticipated deterioration in operating characteristics with time.

PROGRAMMED FORMULAS

Current and voltage stability factors are easy to use in design procedures if the short programs shown in Figure 2–13 are typed. These programs are written for the IBM PC Computer, but can be readily converted to run in other types of computers. As previously noted, an ideal current-stability factor would be zero. Similarly, an ideal voltage-stability factor would be zero. The examples are for a simple common-emitter configuration.

It is evident from the voltage-stability formula that the voltage-stability factor increases and approaches the collector-resistance value as the current-stability factor decreases. Accordingly, if the current-stability factor were equal to zero and the collector resistance were equal to zero, then the voltage-stability factor would also be equal to zero.

If desired, the two foregoing programs can be combined into a single current- and voltage-stability-factors program, as shown in

```
5 LPRINT "CURRENT STABILITY FACTOR PROGRAM"
10 LPRINT "Common Emitter Amplifier"
15 INPUT "Re=";A
20 LPRINT "Re=";A
25 INPUT"Rb=";B
30 LPRINT "Rb=";B
35 INPUT "Rf=";C
40 LPRINT "Rf=";C
45 INPUT "Hfb=";D
50 LPRINT "Hfb=";D
60 F=(1/A)/(1/B+1/C+(1-D)/A)
65 LPRINT""
70 A$="###.###"
75 LPRINT "Si="; USING A$; F
80 END

CURRENT STABILITY FACTOR PROGRAM
Common Emitter Amplifier
Re= 58000
Rb= 7300
Rf= 120000
Hfb= .99

Si=  0.119
```

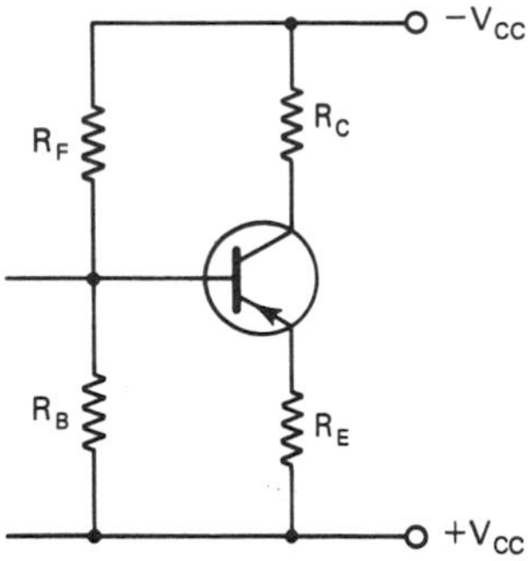

```
5 LPRINT "VOLTAGE STABILITY FACTOR PROGRAM"
10 LPRINT "Common Emitter Amplifier"
15 INPUT "Re=";A
20 LPRINT "Re=";A
25 INPUT"Rb=";B
30 LPRINT "Rb=";B
35 INPUT "Rf=";C
40 LPRINT "Rf=";C
45 INPUT "Hfb=";D
50 LPRINT "Hfb=";D
60 INPUT "Rc=";E
65 LPRINT "Rc=";E
70 INPUT "Si=";F
75 LPRINT "Si="; F:LPRINT ""
80 A$="#######"
85 G=-(F*A+E*(1+D*F))
90 LPRINT "Sv="; USING A$; G
95 END

VOLTAGE STABILITY FACTOR PROGRAM
Common Emitter Amplifier
Re= 58000
Rb= 7300
Rf= 120000
Hfb= .99
Rc= 85000
Si= .118

Sv=-101774
```

Figure 2–13(A). Current and voltage stability factor programs.

NOTE: In this example, a large value (58,000 ohms) has been inputted for R_E in order to clearly illustrate its effect on the stability factor. Stability factors for smaller values of R_E are exemplified in the following program.

```
5 LPRINT "EMITTER RESISTANCE STABILITY FACTOR SURVEY PROGRAM"
10 LPRINT "Common Emitter Amplifier"
15 INPUT"Rb=";B
20 LPRINT "Rb=";B
25 INPUT "Rf=";C
30 LPRINT "Rf=";C
35 INPUT "Hfb=";D
40 LPRINT "Hfb=";D
45 LPRINT ""
50 N=0
55 N=N+100
60 F=(1/N)/(1/B+1/C+(1-D)/N)
65 IF N=100 THEN 75
70 IF N>50 THEN 80
75 A$="#######.#"
80 LPRINT "Re="; USING A$ ; N
85 LPRINT "Si="; USING A$ ; F
90 IF N>2000 THEN 100
95 GOTO 55
100 END
```

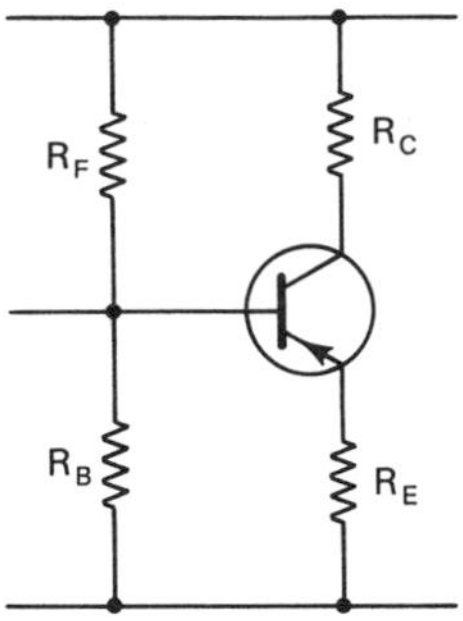

NOTE: This analysis is based on a fixed value for Hfb. Stated otherwise, as the value of Re is increased, the forward-bias value is adjusted as required to maintain Hfb constant.

```
EMITTER RESISTANCE STABILITY FACTOR SURVEY PROGRAM
Common Emitter Amplifier
Rb= 7300
Rf= 120000
Hfb= .99

Re=     100.0        Re=     800.0        Re=    1200.0
Si=      40.8        Si=       7.9        Si=       5.4

Re=     200.0        Re=     900.0        Re=    1300.0
Si=      25.6        Si=       7.1        Si=       5.0

Re=     300.0        Re=    1900.0        Re=    1400.0
Si=      18.7        Si=       3.5        Si=       4.7

Re=     400.0        Re=    2000.0        Re=    1500.0
Si=      14.7        Si=       3.3        Si=       4.4

Re=     500.0        Re=    2100.0        Re=    1600.0
Si=      12.1        Si=       3.2        Si=       4.1

Re=     600.0        Re=    1000.0        Re=    1700.0
Si=      10.3        Si=       6.4        Si=       3.9

Re=     700.0        Re=    1100.0        Re=    1800.0
Si=       9.0        Si=       5.9        Si=       3.7
```

Figure 2-13(B). Emitter-resistance/stability-factor survey program.

```
5 LPRINT "FEEDBACK RESISTANCE/STABILITY FACTOR SURVEY PROGRAM"
10 LPRINT "Common Emitter Amplifier"
15 INPUT"Re=";A
20 LPRINT "Re=";A
25 INPUT "Rb=";B
30 LPRINT "Rb=";B
35 INPUT "Hfb=";D
40 LPRINT "Hfb=";D
45 LPRINT ""
50 N=0
55 N=N+10000
60 F=(1/A)/(1/B+1/N+(1-D)/A)
65 LPRINT ""
70 IF N>150000! THEN 95
75 A$="#######"
80 LPRINT "Rf="; USING A$ ; N
85 LPRINT "Si="; USING A$ ; F
90 LPRINT "":GOTO 55
95 END
```

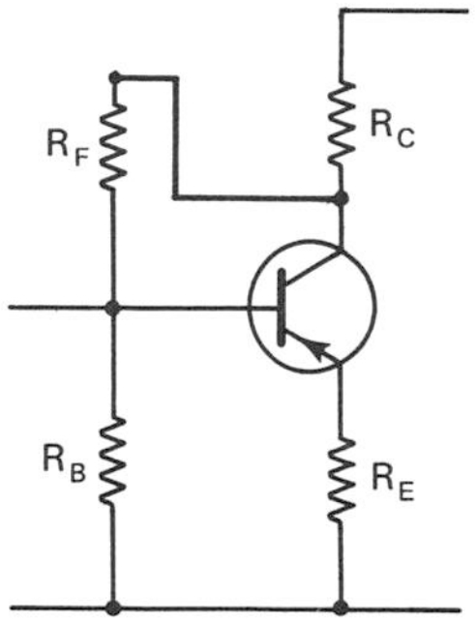

Note: This analysis is based on a fixed value for Hfb. In other words, as the value of Rf is varied, the forward-bias value is adjusted as required to maintain Hfb constant.

```
FEEDBACK RESISTANCE/STABILITY FACTOR SURVEY PROGRAM
Common Emitter Amplifier
Re= 50
Rb= 7300
Hfb= .99

Rf=   10000        Rf=   60000        Rf= 110000
Si=      46        Si=      57        Si=     58

Rf=   20000        Rf=   70000        Rf= 120000
Si=      52        Si=      57        Si=     58

Rf=   30000        Rf=   80000        Rf= 130000
Si=      54        Si=      57        Si=     58

Rf=   40000        Rf=   90000        Rf= 140000
Si=      55        Si=      57        Si=     58

Rf=   50000        Rf=  100000        Rf= 150000
Si=      56        Si=      58        Si=     58
```

Figure 2–13(C). Feedback-resistance/stability-factor survey program.

```
5 LPRINT "EMITTER RESISTANCE/STABILITY FACTOR HEURISTIC ROUTINE"
10 LPRINT "Common Emitter Amplifier"
15 INPUT "Rb=";B
20 LPRINT "Rb=";B
25 INPUT"Rf=";C
30 LPRINT "Rf=";C
35 INPUT "Hfb=";D
40 LPRINT "Hfb=";D
45 INPUT "Maximum Si Limit=";G
50 LPRINT "Maximum Si Limit=";G
55 INPUT "Maximum Re Limit=";H
60 LPRINT "Maximum Re Limit=";H
65 LPRINT ""
70 N=0
75 N=N+50:F=(1/N)/(1/B+1/C+(1-D)/N)
80 IF (F<G)*(N<H) THEN 95
85 IF N>H THEN 110
90 GOTO 75
95 A$="######.##"
100 LPRINT "Re="; USING A$; N: LPRINT "Si="; USING A$; F: END
110 LPRINT "INPUTTED LIMITS ARE OUT OF RANGE; ENTER OTHER LIMITS"
```

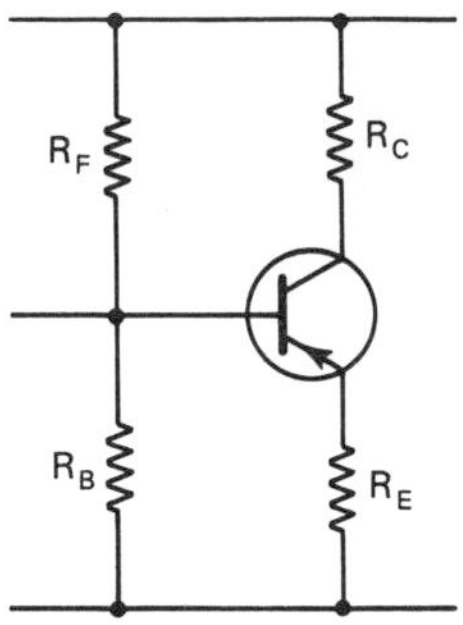

```
EMITTER RESISTANCE/STABILITY FACTOR HEURISTIC ROUTINE
Common Emitter Amplifier
Rb= 7300
Rf= 120000
Hfb= .99
Maximum Si Limit= 15
Maximum Re Limit= 500

Re=    400.00
Si=     14.68
```

Figure 2–13(D). Emitter-resistance/stability-factor heuristic routine.

NOTE: This heuristic routine computes and prints out the values for Re and Si which correspond to the inputted Si and Re limits. However, if the limits are "not in the ballpark," a message is printed out to advise that other limits should be entered. Alternatively, other Rb and/or Rf values may be entered.

Figure 2–14. The exemplified RUN is a repetition of the RUNS in Figure 2–13. There is a difference of −1.71 units in the voltage-stability computations or a difference of 0.07 of 1 percent. This difference results from the rounding-off processes in the computer. See also Chart 2–1.

DESIGN OPTIMIZATION CONSIDERATIONS

Design optimization from the viewpoint of stabilizing circuit calculations requires that device and component tolerances be taken into account in order to determine worst-case conditions. Thus, the current-stability factor is a function of tolerances on the h_{fb} bogie value, on the emitter and base resistor value, and on the bias

```
5 LPRINT "VOLTAGE AND CURRENT STABILITY FACTORS"
10 LPRINT "Common Emitter Amplifier"
15 INPUT "Re=";RE
20 LPRINT "Re=";RE
25 INPUT"Rb=";RB
30 LPRINT "Rb=";RB
35 INPUT  "Rf=";RF
40 LPRINT "Rf=";RF
45 INPUT "Rc=";RC
50 LPRINT "Rc="RC
55 INPUT "Hfb=";FB
60 LPRINT "Fb=";FB
65 SI=(1/RE)/(1/RB+1/RF+(1-FB)/RE)
70 SV=-(SI*RE+RC*(1+FB*SI))
75 A$="###.###"
80 LPRINT "Si=";USING A$; SI
85 A$="#######"
90 LPRINT "Sv="; USING A$; SV

VOLTAGE AND CURRENT STABILITY FACTORS
Common Emitter Amplifier
Re= 58000
Rb= 7300
Rf= 120000
Rc= 85000
Fb= .99
Si=  0.119
Sv=-101845
```

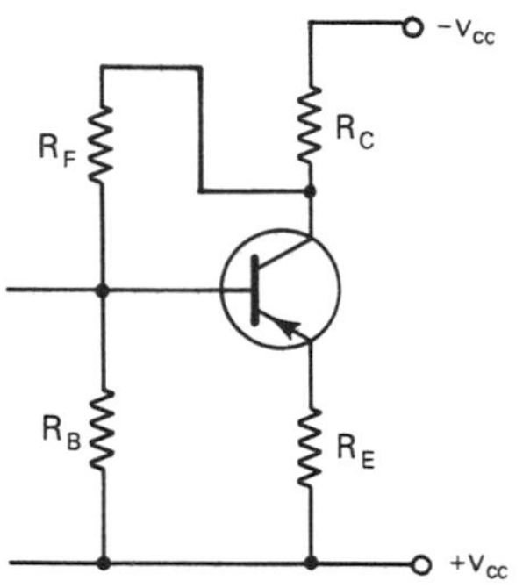

Figure 2–14. Combined current and voltage stability factors program.

resistor value. The voltage-stability factor is also a function of the tolerance on the collector-load-resistor value.

If diodes or thermistors are utilized in a stabilizing circuit, tolerance ratings on the devices should be taken into account. As previously noted, diodes may be selected for optimum stabilizing circuit action. However, a trade-off in production cost is involved. If heat sinks are employed, pertinent tolerances should be observed and the worst-case situation evaluated. Heat-sink tolerances are occasionally associated with ventilation-system tolerances.

After a design has been finalized, a worst-case prototype model should be constructed and its performance verified by appropriate tests and measurements. Transistor and heat-sink temperatures can be easily monitored with thermocouple probes and digital voltmeters. It is occasionally helpful to use recording voltmeters in analysis of prototype operation over a specified range of ambient temperature, associated with a specified range of power-supply voltage variation, and with the pertinent signal sources, such as magnetic tape, disk, FM tuner, or microphone inputs. In each series of tests, worst-case signal-input conditions should be monitored and evaluated.

CHART 2–1

INTEGRATED CIRCUIT BIAS CURRENT, OFFSET VOLTAGE AND CURRENT

The input bias currents for an IC are essentially the currents that must be applied to the inverting and noninverting inputs for proper biasing of the differential input-stage transistors. When an application requires minimum bias current flow, an FET-type IC may be used, or a Darlington-type input section can be employed.

Offset voltage is the deviation from the theoretical zero-volts output level that occurs when the inverting and noninverting inputs are shorted together. Offset current is the difference that occurs between the bias currents to the inverting and noninverting inputs.

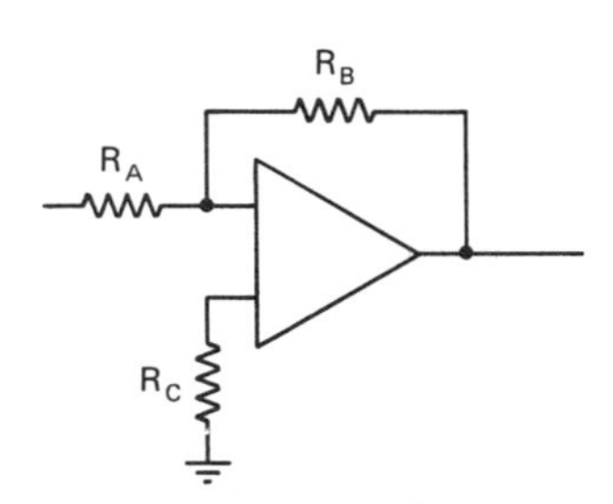

In theory, there would be an output offset of zero when:

$$R_C = \frac{R_A \times R_B}{R_A + R_B}$$

provided that the bias currents flowing into the inputs are equal.

In practice, there will be more or less output voltage offset due to input offset current:

$$V_{os} = I_{os} \times R_B \text{ (approx.)}$$

The designer is concerned with minimizing the offset, particularly in applications such as a cascaded video amplifier. In other words, the input offset of the first stage is amplified by the following stage(s). From the operating viewpoint, offset decreases the dynamic range of the amplifier.

Power-supply stability corresponds to the dependence of offset with respect to power-supply voltage variation. Output offset is dependent on feedback (R_B in the above diagram), and the dependence of offset on power-supply stability is ordinarily referred to the input and is expressed in terms of microvolts/volt.

If the designer employs a precisely regulated power supply, power-supply stability is not a significant factor in amplifier operation. On the other hand, in battery-operated equipment, power-supply stability becomes the dominant design consideration. Note that in a two-supply arrangement, power-supply stability is relaxed to some extent, inasmuch as supply voltage variations generally tend to track reasonably well.

In a single-supply arrangement, the power-supply stability problem is aggravated, and precise regulation is generally required.

3

AUDIO PREAMPLIFIER DESIGN PROCEDURES

AUDIO AMPLIFIER APPLICATIONS AREAS

Audio amplifiers are utilized in high-fidelity systems, public-address systems, radio and television receivers, tape recorders, intercom and telephonic equipment, and many other areas. The input circuit of a bipolar transistor amplifier may draw current either from an input device or from a previous stage. Audio preamplifiers ordinarily operate at power levels measured in micro-microwatts or microwatts.

INPUT RESISTANCE

Low Input Resistance

If a preamplifier is fed from a low-resistance signal source, the designer may employ either the CB or the CE configuration. The CB configuration has a basic input impedance in the range from 30 to 150 ohms.

The CE configuration has a basic input impedance in the range from 500 to 1,500 ohms.

As a rough rule of thumb, a CB configuration will provide an input impedance that is one order in magnitude less than that of a comparable CE stage.

High-Input Resistance

As explained hereafter, it is undesirable to use a high-resistance signal source because of its high noise factor. However, the designer sometimes has no choice in this consideration. If a high-resistance signal source must be used, the preamplifier should have a high-input resistance. Three basic circuits for provision of high-input resistance are shown in Figure 3–1. Although an input transformer may be used for this purpose, this design practice is generally restricted to public-address systems.

As seen in Figure 3–1, an input resistance of 20 kilohms is readily obtainable with the CC or CE configurations. The CE arrangements provide roughly equal input and output resistance, whereas the CC arrangement provides low output resistance. Advantages and disadvantages of these design options are noted in the diagram. As in many design areas, judicious trade-offs are often required.

Variation of voltage gain, current gain, and power gain in the CE configuration is exemplified in Figure 3–2. Observe that the stage gain at the low-frequency end of the frequency range is determined by the values of series coupling capacitors in the circuit. Emitter bypass capacitors also affect stage gain at the low-frequency end. Of course, if direct coupling is employed without emitter bypassing, full gain will be realized down to zero frequency (dc).

Observe also that the stage gain at the high-frequency end of the frequency range is determined by the total shunt capacitance in the circuit, such as junction capacitance and stray capacitance. This shunt capacitance produces the greatest high-frequency attenuation when very high values of load resistances are employed. The frequency characteristics of an amplifier have an associated phase characteristic, as exemplified in the diagram.

Variation of voltage gain, current gain, and power gain in the common-base configuration is exemplified in Figure 3–3. From the standpoint of design perspective, a typical audio preamplifier has a maximum available voltage gain of 69 dB (2,857 times) and a maximum distortion figure of 1 percent, with provision of an output

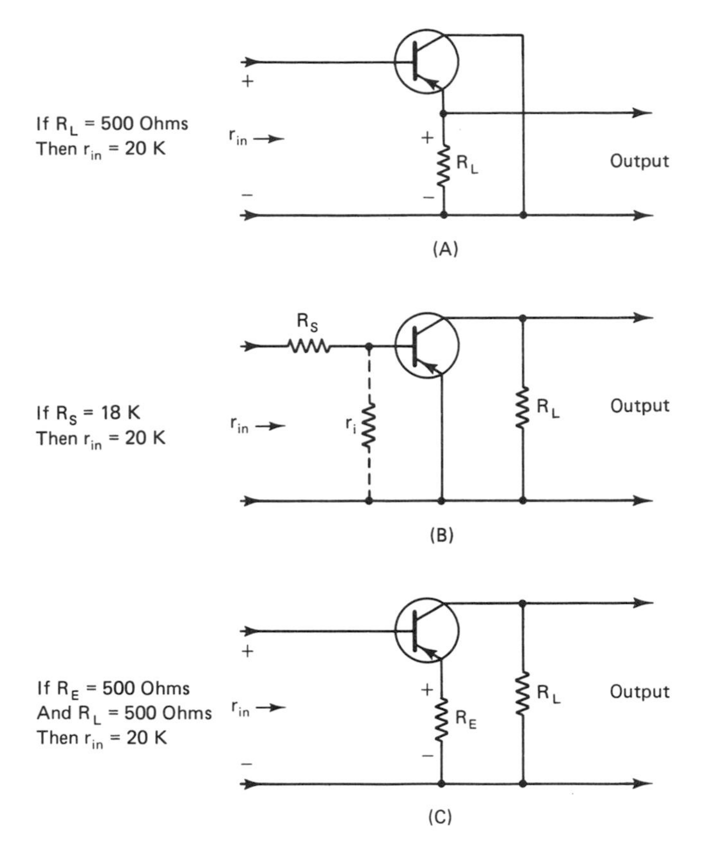

Figure 3–1. Preamplifier configurations for provision of high input resistance.

DISADVANTAGES:

(A) Small variations in current demand by the following stage cause large changes in the value of r_{in}.

(B) Bias stability is poor if the bias is fed to the base via R_S. If bias is fed via a supplementary resistor r_i, the value of r_{in} is reduced accordingly. There is a small loss in current gain, in either case.

ADVANTAGES:

(A) Low output resistance may be an advantage for impedance matching.

(B) The total input resistance remains comparatively constant regardless of variations in current demand by the following stage.

(C) Bias stability is comparatively good.

level of 1 volt rms into 10,000 ohms when operated from a source voltage as low as 350 microvolts. The preamplifier-input impedance is 50 kilohms in this example.

Variation of voltage gain, current gain, and power gain in the common-collector configuration is exemplified in Figure 3–4. At

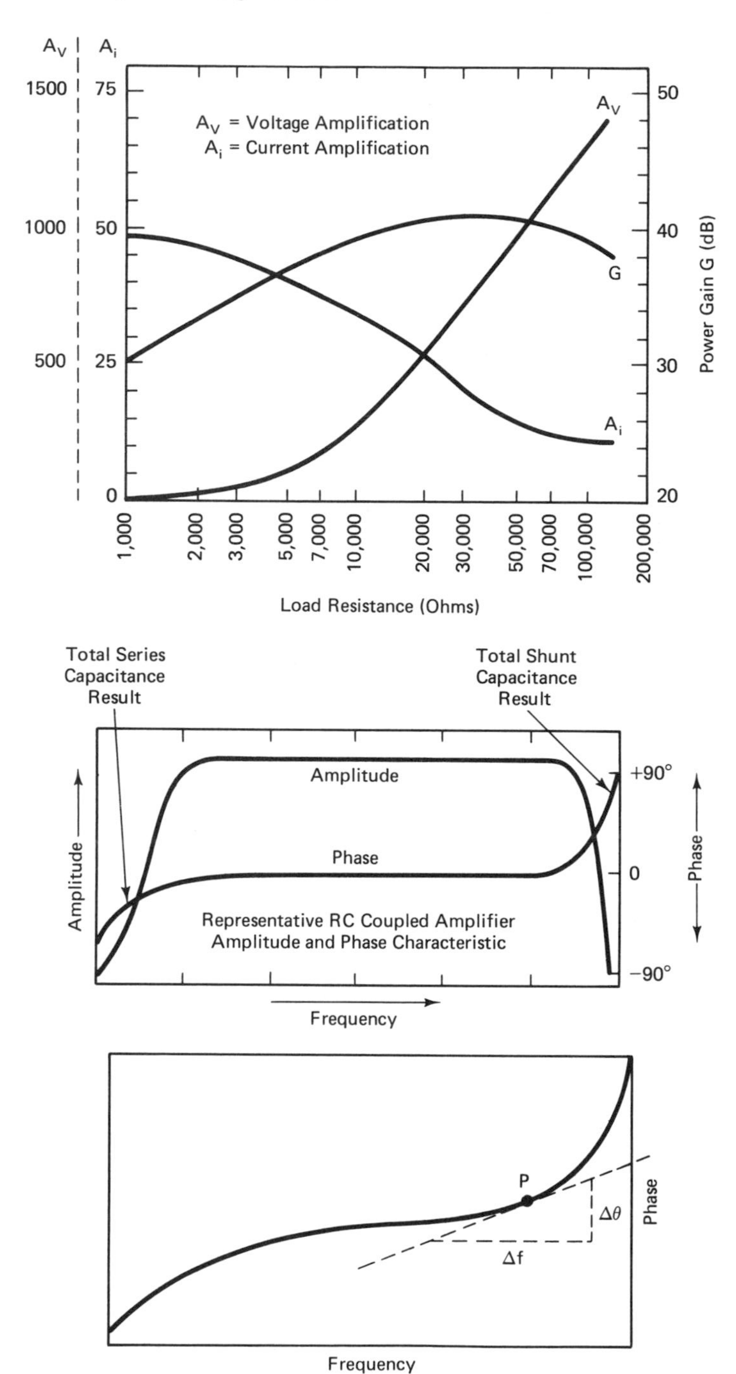

Figure 3–2. Common-emitter voltage, current, and power gain versus load resistance. (Reproduced by special permission of Reston Publishing Company and Derek Cameron from *Handbook of Audio Circuit Design*).

NOTE: The phase characteristic of a network (Figure 3–2) is commonly depicted as a plot of phase-angle value versus frequency. In turn, the slope of the tangent at a point on the phase characteristic is a relative measure of the phase linearity (or nonlinearity). Thus, if a network has a linear phase characteristic, the ratio dθ/df will be constant for all values of f. This is just another way of saying that if a network has a linear-phase characteristic, the phase angle of the output voltage with respect to the input voltage will be proportional to the operating frequency.

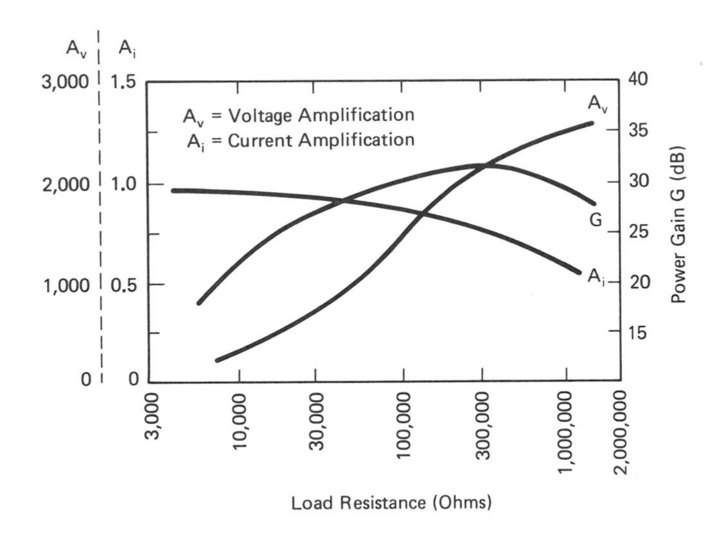

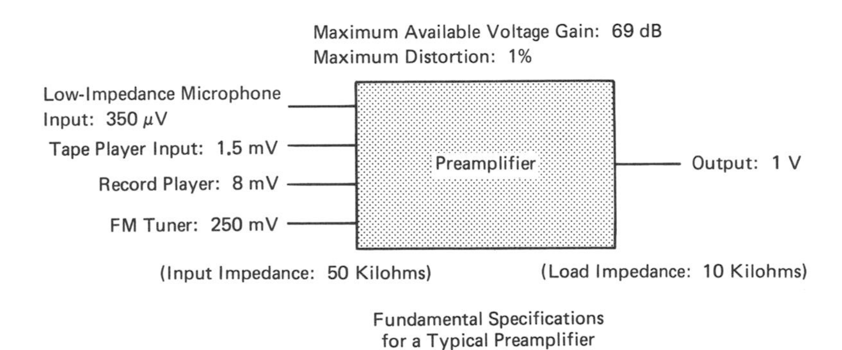

Figure 3–3. Common-base voltage, current, and power gain versus load resistance.

NOTE: The FM tuner channel has high-fidelity response, with a frequency characteristic that is "flat" within ±1 dB from 20 Hz to 20 kHz (typical). On the other hand, intelligible speech can be provided by a frequency response from 300 Hz to 3 kHz.

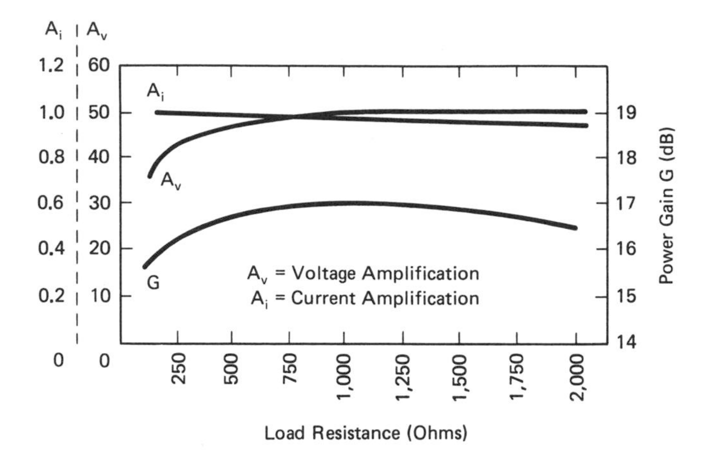

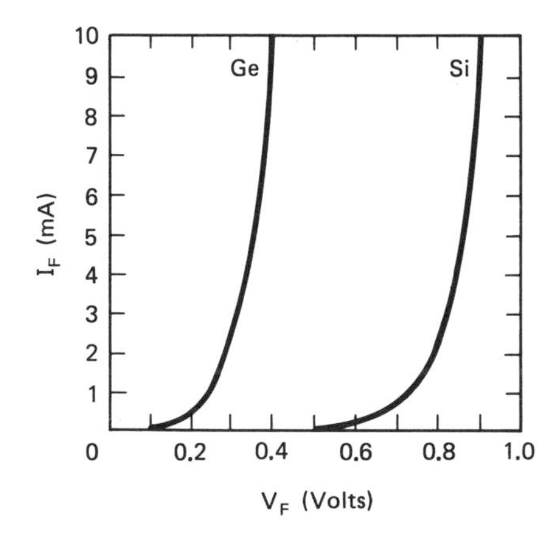

Figure 3–4. Common-collector voltage, current, and power gain versus load resistance.

Note: Silicon and germanium transistors have approximately the same functional characteristics. However, silicon transistors have better performance at higher temperatures and have higher reverse resistance. As seen below, silicon transistors have a comparatively high barrier potential and in turn are operated at a higher forward-bias voltage.

low values of load resistance, voltage gain and power gain decrease appreciably, although current gain is comparatively unaffected. It is helpful to refer back to Chart 1–1 and note that the input resistance to the CC configuration drops rapidly below the 1/Beta value of load resistance.

NOISE FACTOR

The merit of an amplifier with respect to inherent noise is measured in terms of its noise factor (noise figure). This noise figure (F_o) is determined by measurement of signal-to-noise power (S_i/N_i) at the amplifier input and of its ratio to signal-to-noise power (S_o/N_o) at the amplifier output. Thus, the noise factor of the amplifier is equal to the input signal-to-noise ratio divided by the output signal-to-noise ratio:

$$F_o = \frac{S_i/N_i}{S_o/N_o}$$

Stated otherwise, the smaller the value of (F_o), the better is the noise factor of the amplifier. The value of the noise factor is chiefly determined by:

1. Choice of the transistor operating point.
2. Value of the signal source resistance.
3. Value of the signal frequency.

Note that $10 \times \log_{10}$ of the ratio for noise factor gives the value of the noise factor in dB.

Operating Point

The transistor operating point is determined by the zero-signal-base (or emitter) current and by the collector voltage. With reference to Figure 3–5, it is seen that for a given transistor, a low-noise factor can be realized by operating the transistor at an emitter current of less than 1 mA, and at a collector voltage of less than 2 volts. (A high-collector voltage will increase the noise factor more rapidly than will a high-emitter current).

Signal-Source Resistance

A signal-source resistance in the range from 100 to 3,000 ohms assists in minimizing the noise factor, as shown in Figure 3–6. The optimum value occurs at approximately 800 ohms. Inasmuch as the noise factor increases at lower values of signal-source resistance, it

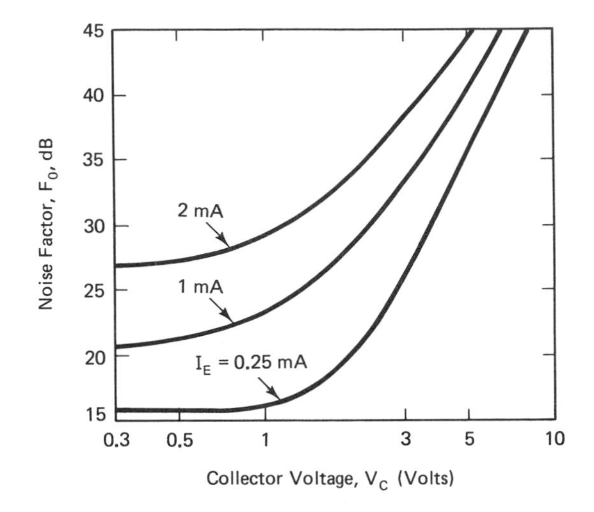

$$dB = 10 \log_{10} (P_1/P_2)$$
$$dB = 20 \log_{10} (V_1/V_2)$$
$$dB = 20 \log_{10} (I_1/I_2)$$

(P_1 and P_2 are Independent of Source and Load Resistance Values, V_1 and V_2 Apply Only to Equal Source and Load Resistance Values)

Figure 3–5. Variation of noise factor with collector voltage and emitter current for a bipolar transistor.

Note: The noise figure (NF) of an amplifier is equal to the ratio of signal power to noise power at the input (S_i/N_i), divided by the ratio of signal power to noise power at the output (S_o/N_o).

Example: *An amplifier with a noise figure of 1 dB impairs (decreases) the signal-to-noise ratio by a factor of 1.26. An NF of 3 dB decreases the signal-to-noise ratio by a factor of 2. An NF of 10 dB decreases the signal-to-noise ratio by a factor of 10. An NF of 20 dB decreases the signal-to-noise ratio by a factor of 100.*

In an ideal design, the lowest value of NF that can be realized is 3 dB (source matched to input resistance).

With all other things being equal, the lowest noise factor will be provided by a transistor that has a beta-cutoff frequency that is no higher than necessary. For example, it is poor design practice to use a transistor with a beta-cutoff frequency of 4 MHz in a 20-kHz audio preamplifier.

The alpha-cutoff frequency of a transistor is defined as the frequency at which its forward-current-transfer ratio in the CB configuration is down 3 dB. The beta-cutoff frequency is defined as the frequency at which its forward-current-transfer ratio in the CE configuration is down 3 dB. The alpha-cutoff frequency is higher than the beta-cutoff frequency. The gain-bandwidth product (f_t) is defined as the frequency at which beta decreases to unity gain (0 dB).

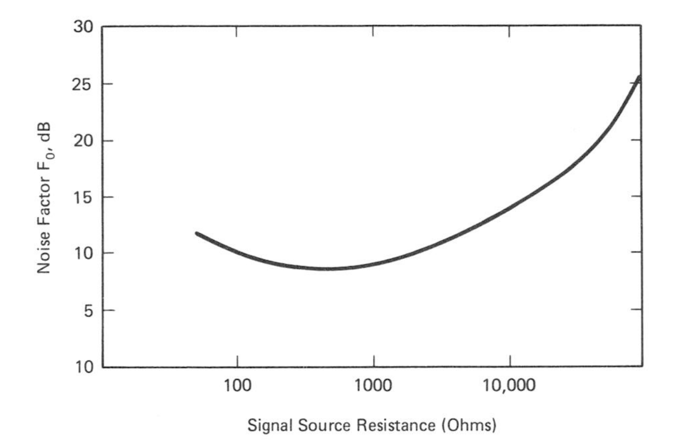

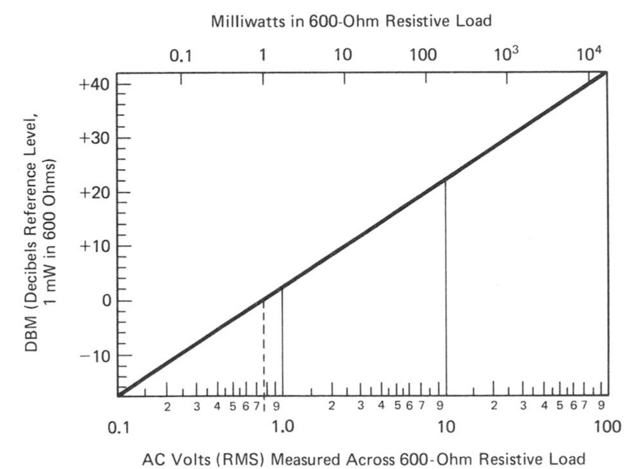

Resistive Load	DBM*
600	0
500	+ 0.8
300	+ 3.0
250	+ 3.8
150	+ 6.0
50	+10.8
15	+16.0
8	+18.8
3.2	+22.7

*DBM Is the Increment to Be Added Algebraically to the DBM Value Read from the Graph

List of dBm Correction Factors

Figure 3–6. Variation of noise factor with signal-source resistance. (Reproduced by special permission of Reston Publishing Company and Miles Ritter-Sanders from *Electronic Meters*).

NOTE: The thermal noise generated by a resistor (or resistive device) is a function of the temperature, the bandwidth of the circuit, and the value of the resistor. This generated thermal noise is formulated:

$$v_n = \sqrt{4kTBR}$$

wherein:

v_n = *rms noise voltage*
k = *Boltzmann's constant* (1.37×10^{-23})
T = *absolute temperature, Celsius plus 273°*
B = *noise bandwidth, Hertz*
R = *resistance ohms*

The noise bandwidth B is approximately equal to the −3 dB bandwidth of the amplifier.

The input noise level to an amplifier is approximated in terms of the noise generated by the Thevenin resistance driving the amplifier.

Example: *The Thevenin source resistance driving an amplifier is 5 kilohms. The amplifier bandwidth (noise bandwidth) is 20 kHz. At an operating temperature of 25°C, the rms noise voltage will be:*

$$v_n = 1.28 \times 10^{-10} \times 10^4$$
$$v_n = 1.28 \text{ microvolts}$$

is advisable for the designer to supplement a very low value of source resistance with sufficient series resistance to provide a total of 800 ohms.

Signal Frequency

As illustrated in Figure 3–7, the noise factor has a minimum value at an operating frequency of approximately 50 kHz. At very low frequencies the noise factor increases substantially. It follows from this relation that low-noise dc amplifiers are difficult to design. The noise factor is generally of concern only in the input stage of an audio system, inasmuch as noise generated by the input stage will be amplified by all subsequent stages. Accordingly, direct coupling can be employed in driver and output stages without regard to their noise factors.

TWO-STAGE AMPLIFIERS

Preamplifiers often comprise two stages. The transistors in a two-stage amplifier can be configured for incidental temperature stabilization, as exemplified in Figure 3–8. The stabilized emitter-collector current of Q1 is utilized by Q2. Observe that the collector current in Q1 is stabilized by the swamping resistor R2. A low value of R1 is employed to assist in bias stabilization.

Next, the stabilized dc collector current of Q1 flows through the emitter-collector circuit of Q2 via direct connection of the Q1 collector to the Q2 emitter. The design advantage of this arrangement is in its elimination of the need for a swamping resistor in the Q2 emitter lead. Although of minor concern in a low-level stage, this feature becomes increasingly attractive in operation at higher power levels.

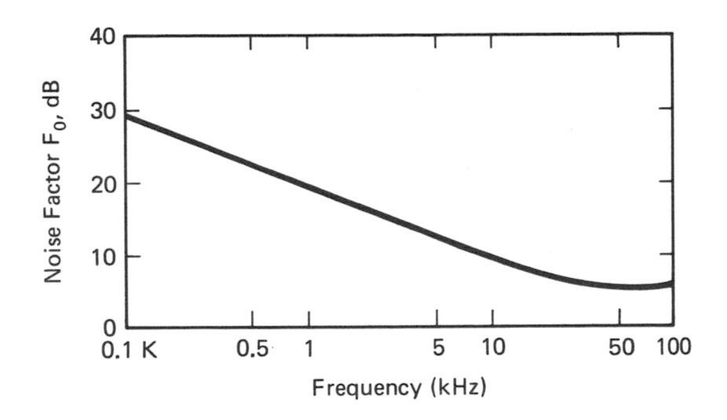

Figure 3–7. Variation of noise factor with frequency.

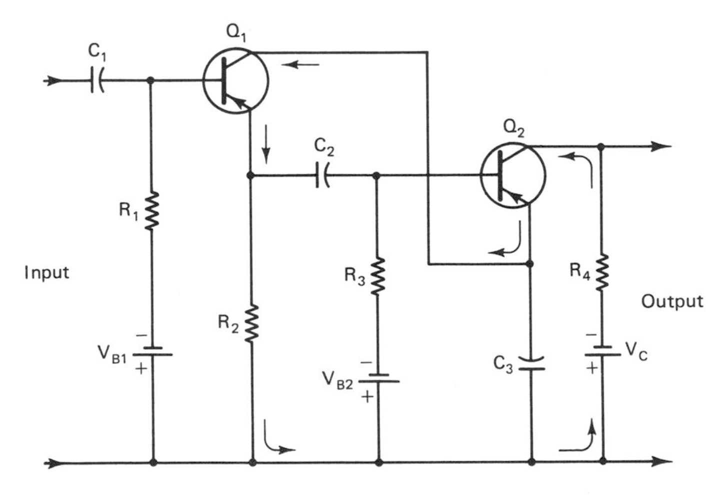

Figure 3–8. Common emitter-collector current is used for temperature stabilization in this cascaded configuration.

NOTE: To calculate the performance of a two-stage amplifier, start with the second stage and calculate its voltage gain, current gain, power gain, input resistance, and output resistance from its h parameter and component values. Then, the load resistance for the first stage is equal to the parallel combination of its resistor load and the input resistance to the second stage. The voltage gain, current gain, power gain, input resistance, and output resistance of the first stage can then be calculated. Finally, the voltage gain of the two-stage amplifier is equal to the product of the voltage gains for the first and second stages. Note that the current gain of the two-stage amplifier is equal to the product of the current gains for the first and second stages and that the dB power gain of the two-stage amplifier is equal to the sum of the dB gains for the first and second stages.

Two-stage and three-stage amplifiers are packaged in integrated circuit form with temperature stabilization provided by differential-amplifier circuitry. However, IC amplifiers are comparatively noisy; designers do not employ them as preamplifiers unless the input signal level is substantial. For example, IC design may be used in telephone-listener applications.

Note in passing that Q1 operates in the CC mode (Figure 3–8), and Q2 operates in the CE mode. In other words, C3 places the collector of Q1 at ac ground potential. Note also that the gain of a two-stage amplifier is equal to the product of individual stage gains. A two-stage amplifier may be evaluated for input resistance, output resistance, voltage gain, current gain, and power gain in the same manner as a single-stage amplifier, if derived hybrid parameters are used, as indicated in Figure 3–8.

Bias Control by Collector Current

Another basic two-stage arrangement is shown in Figure 3–9. This is a direct-coupled configuration wherein an increase in collector current resulting from a temperature rise in Q1 operates to reduce the forward bias on Q2. Observe that Q1 is configured in the CB mode; in turn, its stability factor is ideal. Nevertheless, there remains some variation of its collector current versus temperature. The resulting incremental current flow is depicted by the arrows in the diagram.

A portion of the incremental collector-current flow passes through R3 and develops a voltage drop across R3 as indicated. Another portion of the incremental collector-current flow passes through R2 and develops a voltage drop across R2 as indicated. (These are incremental-voltage polarities, not steady-state voltage polarities.) Observe that the base-emitter bias on Q2 is the sum of the voltages across resistors R3, R2, and battery V_C.

It follows that the indicated voltage drop across R3 aids development of forward bias, whereas the drop across R2 opposes development of forward bias. The designer selects the R2 and R3 values so that the drop across R2 is greater, thus decreasing the forward-bias voltage. In turn, bias control of Q2 by the collector current of Q1 is achieved. Note that R1 functions as an emitter swamping resistor, and R4 functions as a collector-load resistor.

INPUT FREQUENCY CHARACTERISTICS

As exemplified in Figure 3–3, an audio preamplifier may have several inputs. With reference to Figure 3–10, the tuner input frequency characteristic is flat (typically from 20 Hz to 20 kHz). On the other hand, the other input characteristics have high-frequency

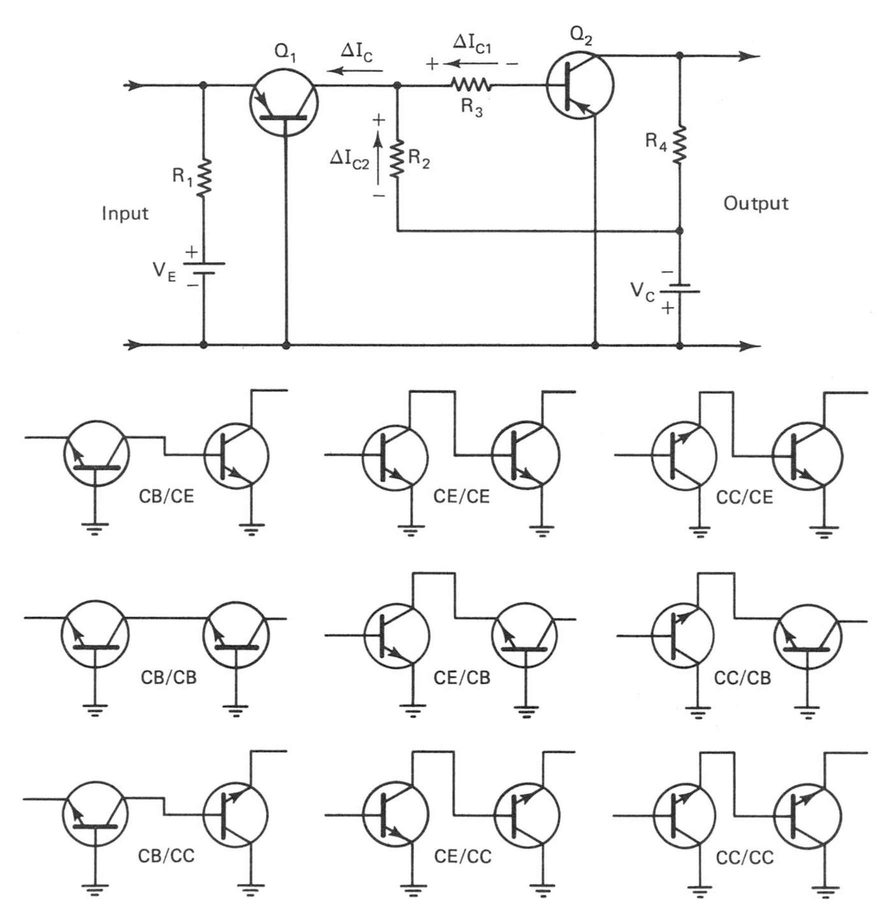

Figure 3–9. Temperature-stabilized dc amplifier arrangement.

Note: Most audio amplifiers are RC-coupled; transformer coupling is used to some extent in utility amplifiers. Direct coupling (as exemplified above) is generally limited to two stages due to temperature instability problems. Op amps are an exception in this regard; an op amp achieves temperature stability over several cascaded direct-coupled stages by design with monolithic fabrication, differential-stage operation, and use of a large amount of negative feedback.

rolloffs. Thus, the RIAA curve has a rolloff of 13 dB per decade; the NAB curve has a rolloff of 20 dB per decade; the MRIA curve has a rolloff of 17 dB per decade. These input-frequency characteristics are obtained from RC coupling circuits, or coupling circuits in combination with negative-feedback loops.

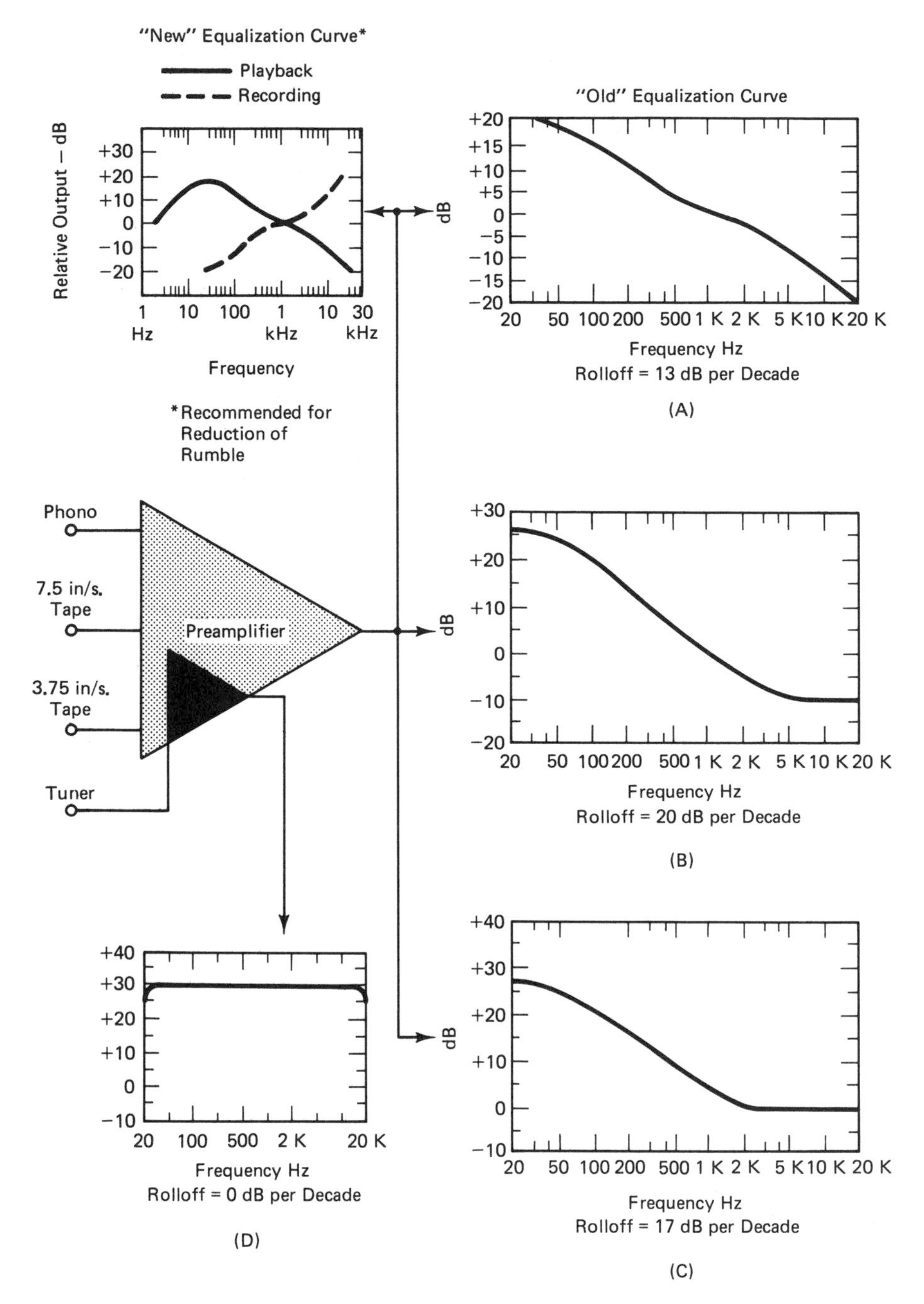

Figure 3–10. Illustration of standardized preamp input-frequency characteristics. (A) RIAA equalization curve for record playback; (B) NAB playback curve for 7.5 in/s. tape; (C) MRIA playback curve for 3.75 in/s. tape; (D) tuner input frequency characteristic is flat. (Reproduced by special permission of Reston Publishing Company and Lloyd Hardin from *Advanced Stereo System Equipment*.)

NOTE: Equalizer network design is discussed in Chapter 11.

Universal frequency-response curves for RC high-pass and low-pass sections are shown in Figure 3–11. In a two-stage preamplifier, two RC sections may be utilized with device isolation as shown in Figure 3–12. Cascaded sections provide more rapid rolloff. Note that design-center values are denoted as bogie values. Audio preamplifiers are usually designed with tone controls. These are basically RC circuits also, as exemplified in Figure 3–13.

Frequency characteristics may be contoured by inserting the RC section(s) between the signal source and the input stage, or the designer may insert an RC frequency-compensating section between stages, as shown in Figure 3–14. A wider range of contouring can be obtained by inserting an RC network in a negative-feedback loop, as also depicted in Figure 3–14.

Audio preamplifiers may include scratch filters and/or rumble filters, as depicted in Figure 3–15. These are usually designed as RC filters that are inserted between the signal source and the input

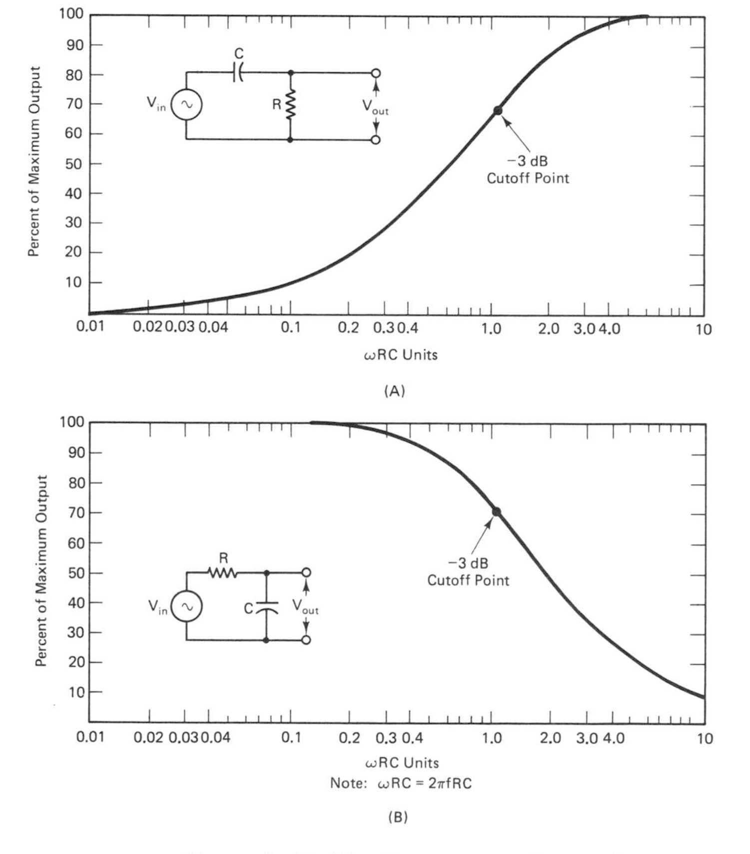

Figure 3–11. (Continues on next page.)

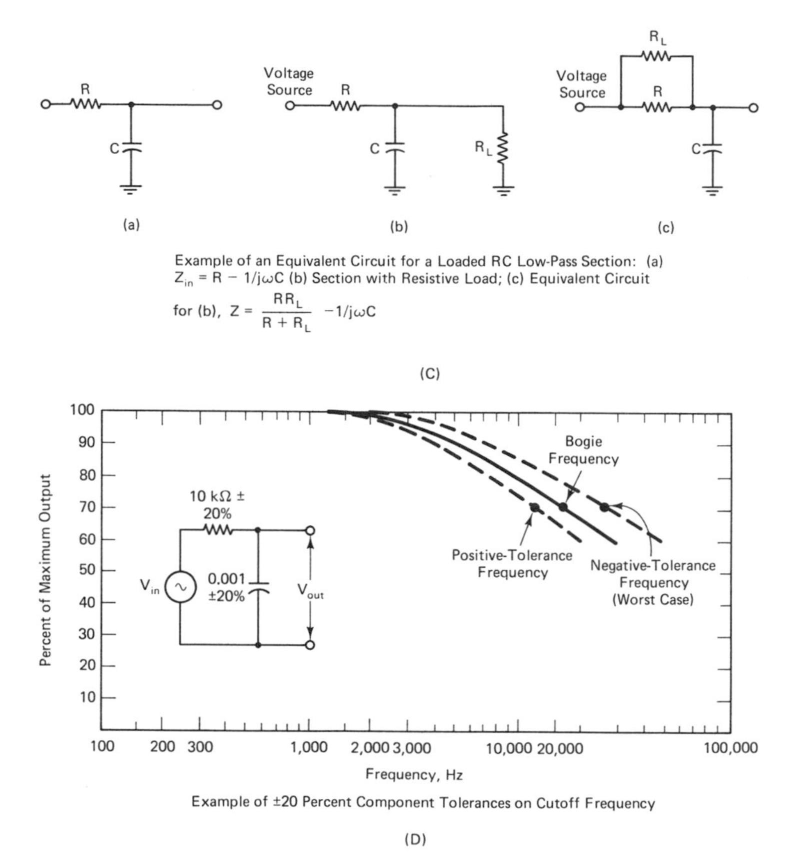

Figure 3–11. Frequency-response curves for RC sections. (A) High pass; (B) low pass; (C) loaded section equation; (D) example of ±20 percent component tolerance on cutoff frequency. (Reproduced by special permission of Reston Publishing Company and Campbell Loudoun from *Handbook for Electronic Circuit Design.*)

NOTE: Analysis of loaded RC high-pass sections is discussed in Chapter 11. Computer programs are provided for convenience.

stage, or between a first and second stage (with or without supplementary negative-feedback action). Also shown in Figure 3–15 is the typical tolerance effect that can be anticipated by the designer on RC coupling-circuit frequency response.

VOLUME CONTROL

Volume controls should be inserted at minimum-noise points in audio circuitry. Typical good design practice is exemplified in Figure 3–16. A volume control should not conduct a substantial

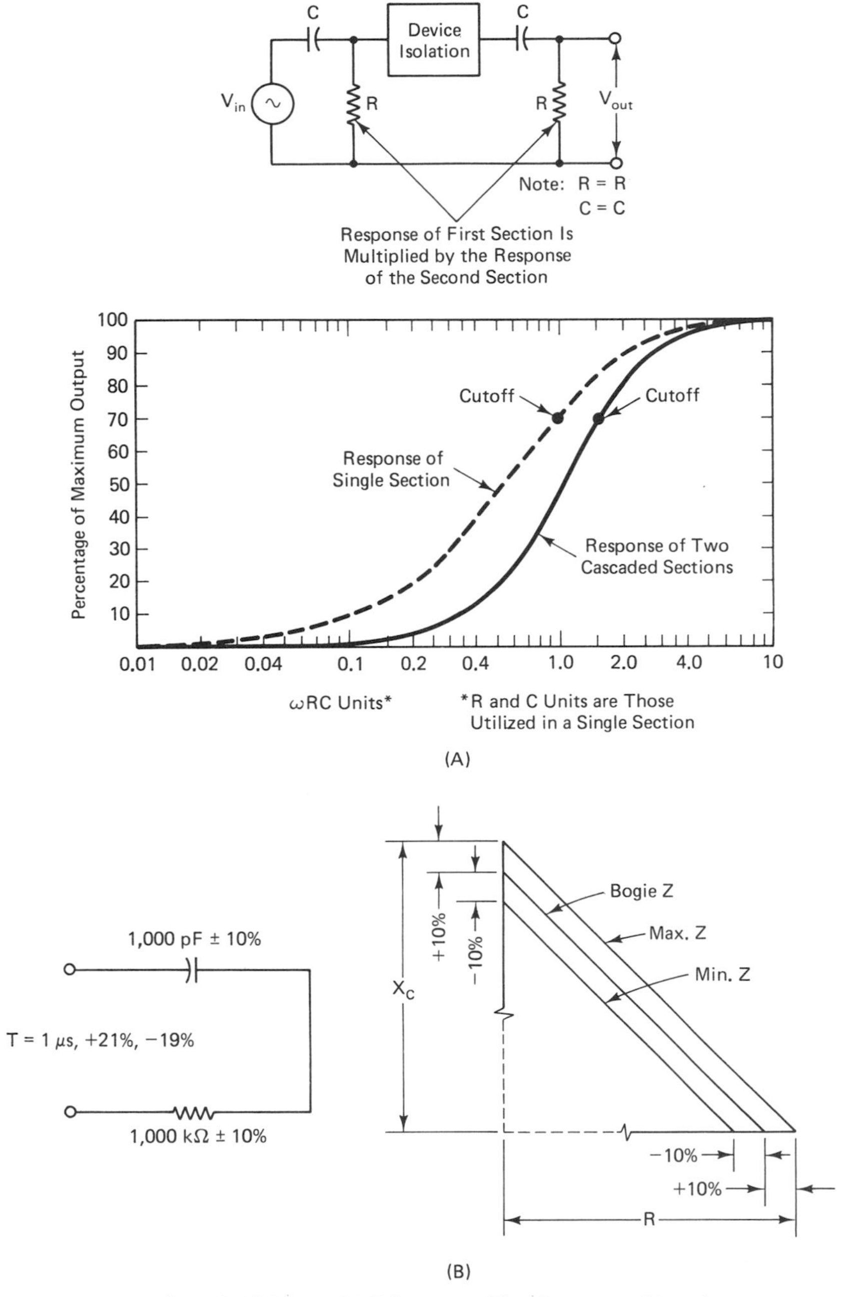

Figure 3–12. Frequency characteristic for two cascaded RC sections with device isolation. (A) Single-section and two-section responses; (B) effect of ± 10 percent component tolerances. (Reproduced by special permission of Reston Publishing Company and Campbell Loudoun from *Handbook for Electronic Circuit Design*.)

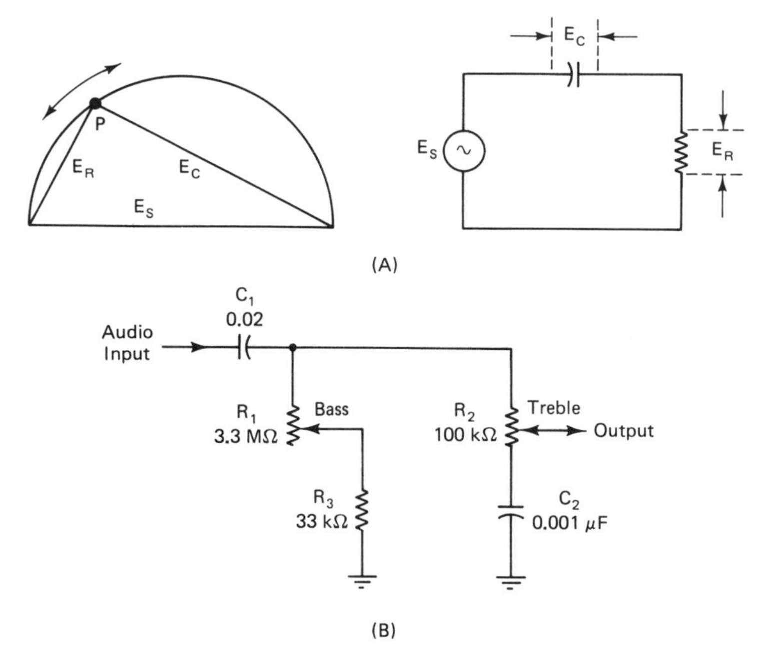

NOTE: A tone control provides a typical bass or treble cut or boost range of ±12dB.

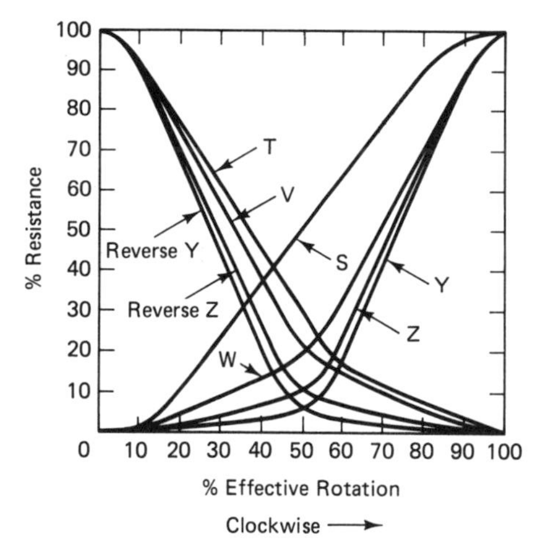

Taper S Straight or Uniform Resistance Change with Rotation
Taper T Right-hand 30% Resistance at 50% of C.C.W. Rotation
Taper V Right-hand 20% Resistance at 50% of C.C.W. Rotation
Taper W Left-Hand 20% Resistance at 50% of C.W. Rotation
Taper Z Left-Hand (Log. Audio) 10% Resistance at 50% C.W. Rotation
Taper Y Left-Hand 5% Resistance at 50% of C.W. Rotation

(C)

Figure 3–13. Basic audio tone-control arrangement. (Continues on next page.)

(A) Locus of voltage variation in an RC circuit; (B) cascaded RC circuits employed in a simple tone-control configuration; (C) typical tone-control taper options. (Reproduced by special permission of Reston Publishing Company and Campbell Loudoun from *Handbook for Electronic Circuit Design*.)

NOTE: Design of various types of tone controls is discussed in Chapter 11. Computer programs are provided for rapid analysis of circuit action and the effect of component tolerances.

amount of dc current and should provide a wide range in signal level. It should also be configured so that all audio frequencies are equally attenuated at any setting of the volume control. Observe in Figure 3–16 that C3 avoids variation in base-bias voltage, and that R2 does not ac-shunt R5 and C2. This arrangement also avoids deterioration of low-frequency response at reduced settings of R2.

PROGRAMMED FORMULAS

Negative feedback is a basic factor in evaluation of the gain, stability, and impedance levels of an amplifier. A simple example of 20 dB negative feedback is shown in Figure 3–17. The voltage gain of the amplifier without negative feedback is:

$$A = V_{out}/V_{in}$$

When the negative-feedback loop is connected into the configuration, the voltage gain is reduced (to 0.1, in this example). Now, the amplifier voltage gain is formulated:

$$A' = V'_{out}/V'_{in}$$

Accordingly, the input voltage to the base is equal to the sum of the original source voltage (V_{in}) and the feedback voltage ($\beta V'_{out}$):

$$V'_{in} = V_{in} + \beta V'_{out}$$

This is just another way of saying that Beta is defined as ratio of feedback voltage to the output voltage:

$$\beta = \text{Feedback Voltage/Output Voltage}$$

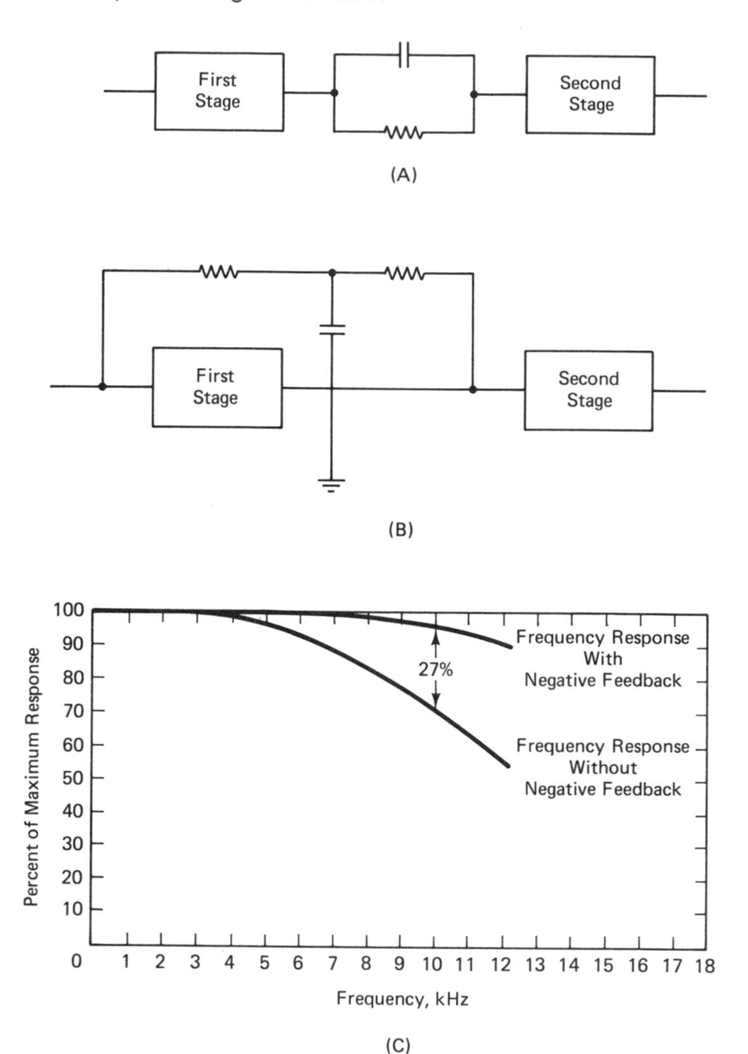

Figure 3–14. Arrangement of an interstage RC-frequency compensation network. (A) Simple coupling circuit; (B) frequency-selective negative-feedback loop; (C) extension of high-frequency response by negative feedback; (D) three-stage preamplifier with low-frequency compensation.

NOTE: *The diagram above shows the effect of 20 dB of negative feedback. Designers call 20 dB (or more) of negative feedback "significant feedback." This means that the feedback is sufficient to usefully reduce distortion. When 20 db of negative feedback is employed, the amplifier gain is reduced to 10 percent of its initial value.*

Example: *If a 10-mV input produces a 2-V output, the initial gain is 200 times. Next, if 90 mV is fed back, a 100-mV input is required to obtain a 2-V output. The stage gain is now 20 times.*

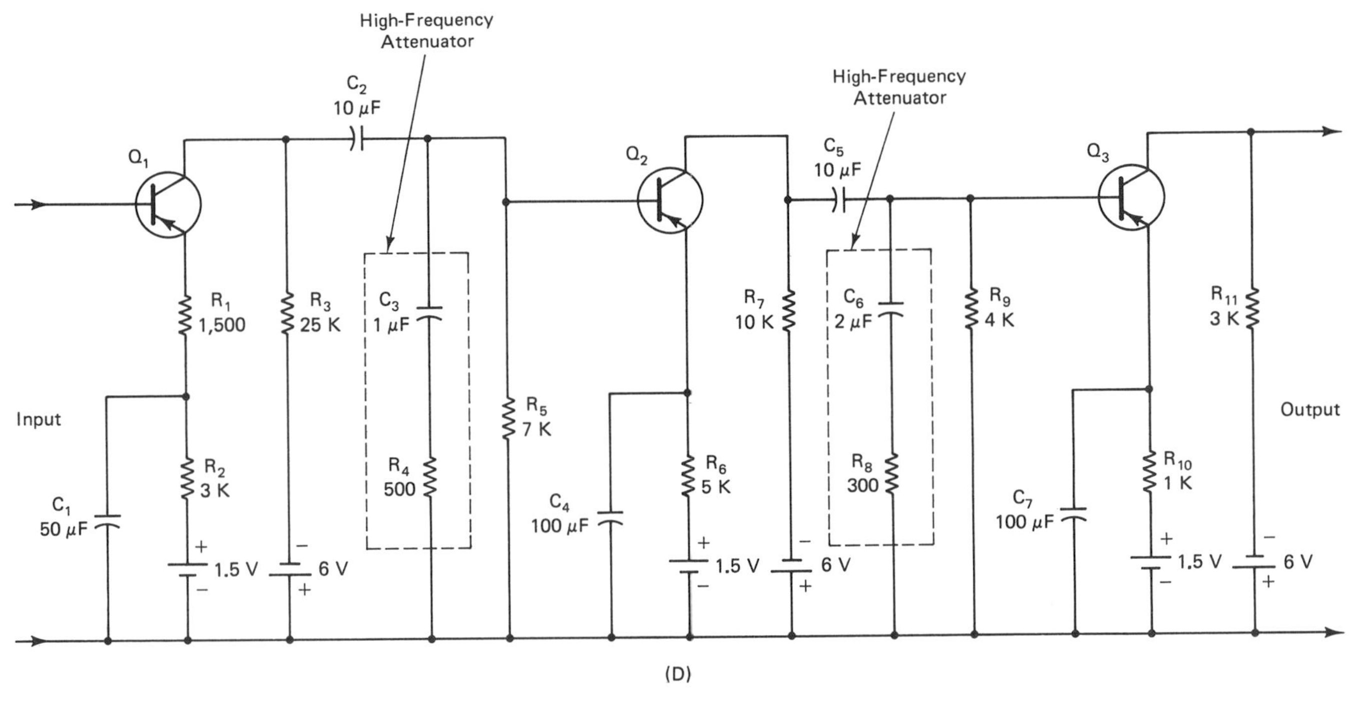

Figure 3–14.

NOTE: This is an example of a well-designed three-stage preamplifier with low-frequency compensation. It compensates for the poor low-frequency output from a transducer. Two high-frequency attenuator sections (low-pass sections) are employed. Q2 and Q3 have an input resistance of approximately 1000 ohms each. Q1 utilizes current feedback via R1, and the input resistance to Q1 is approximately 55,000 ohms. In turn, this preamplifier may be used with a high-impedance transducer. The designer may contour the frequency-compensation characteristic by choice of the values for R4-C3 and R8-C6.

As a practical design consideration, the supply voltages will be obtained from a common power supply. In turn, the internal impedance of the power supply (which is subject to tolerances) can present a source of unsuspected positive feedback. To contend with this problem, the designer should include any necessary power-supply decoupling circuits in the prototype model.

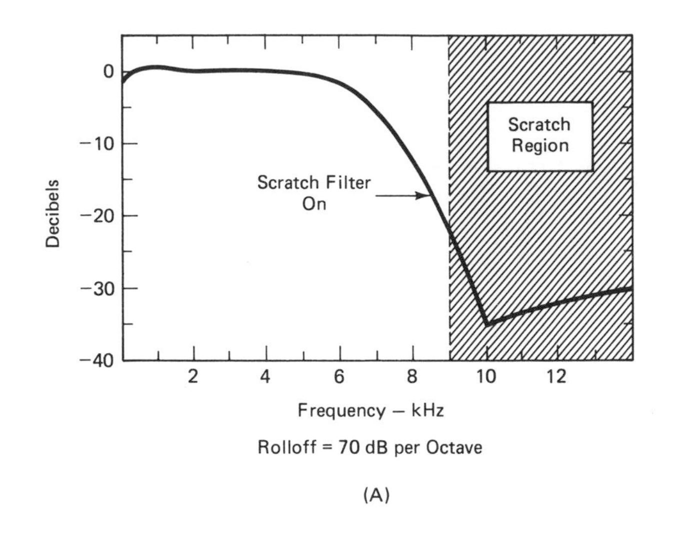

Rolloff = 70 dB per Octave

(A)

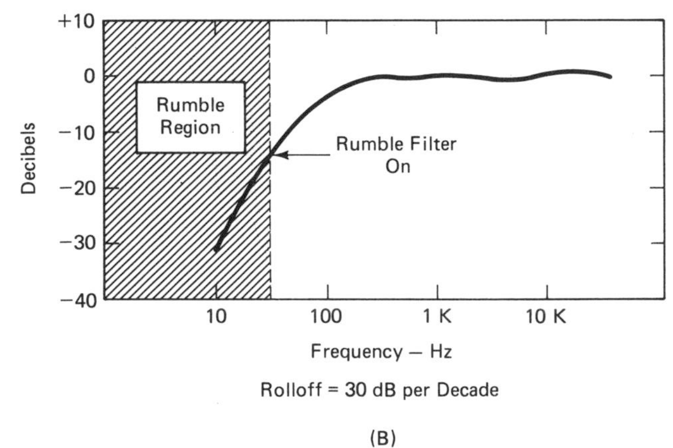

Rolloff = 30 dB per Decade

(B)

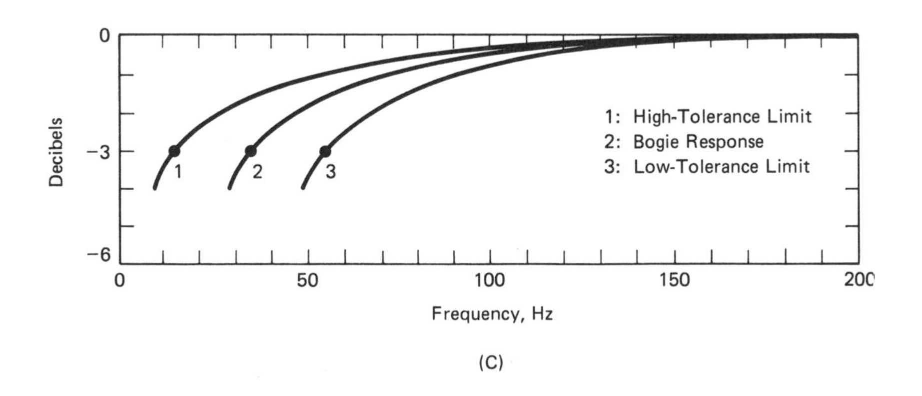

(C)

NOTE: Scratch and rumble filters provide a maximum attenuation of approximately 35 dB.

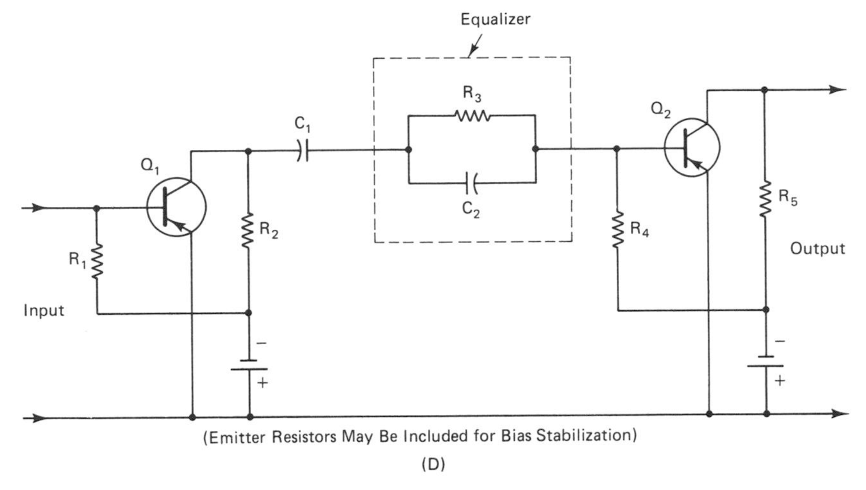

(D)

NOTE: This is an example of a well-designed preamplifier that provides high-frequency compensation for a transducer. It employs a two-stage arrangement with interstage coupling via an equalizer that attenuates the lower signal frequencies. C1 provides some attenuation of low frequencies, and C2 tends to bypass high frequencies around R3. C2 has high reactance at low frequencies, but has low reactance at high frequencies. The designer can contour the refrequency-compensating characteristic by choice of values for C2, R3, and C1.

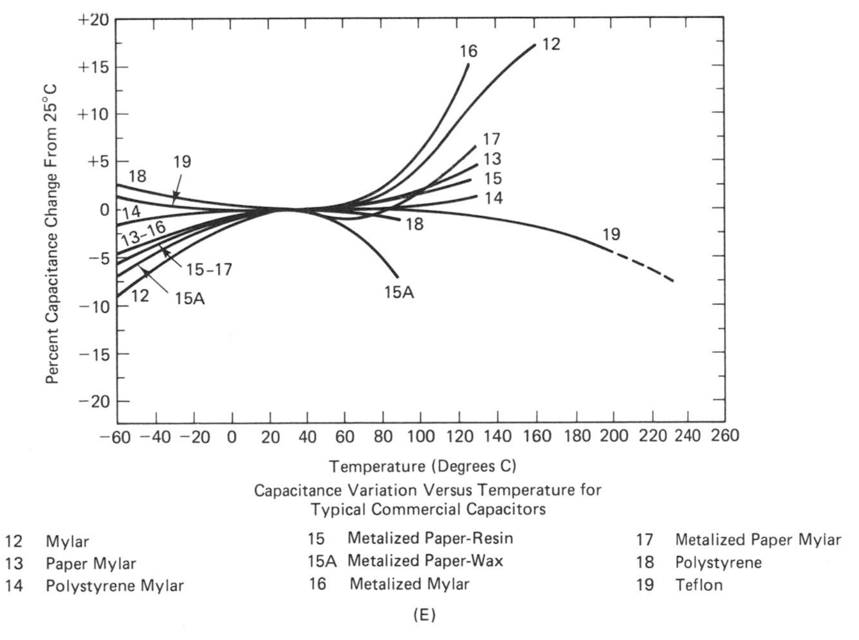

(E)

Figure 3–15. Other RC filter characteristics. (A) Scratch filter; (B) rumble filter; (C) effect of ±20 percent R and C tolerances on frequency response; (D) two-stage preamplifier with high-frequency compensation. (Reproduced by special permission of Reston Publishing Company and Campbell Loudoun, from *Handbook for Electronic Circuit Design*.)

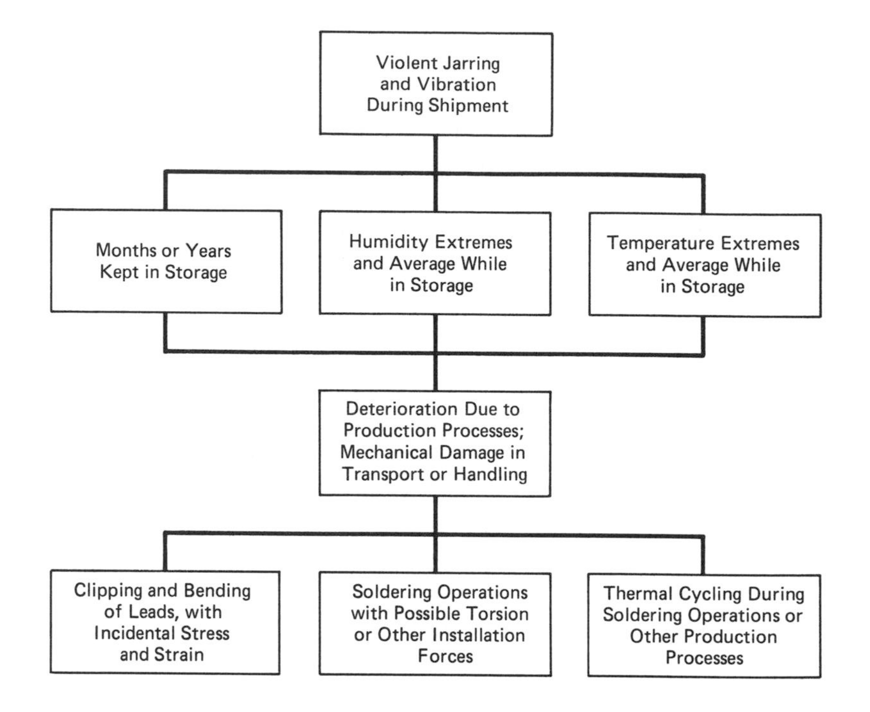

Good design practices recognize that although appropriate consideration has been given to device and component tolerances, rated tolerances tend to drift for reasons given below:

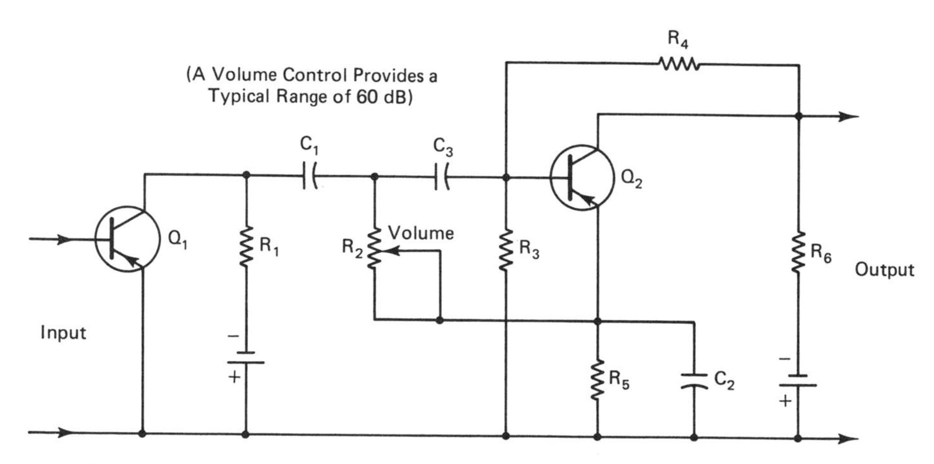

Figure 3–16. Example of good practice in volume-control arrangement.

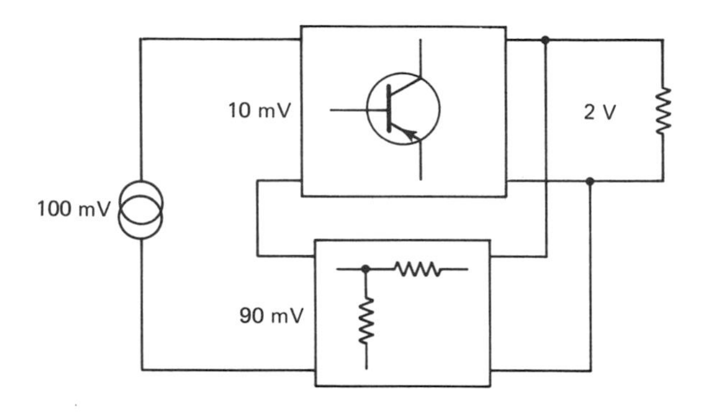

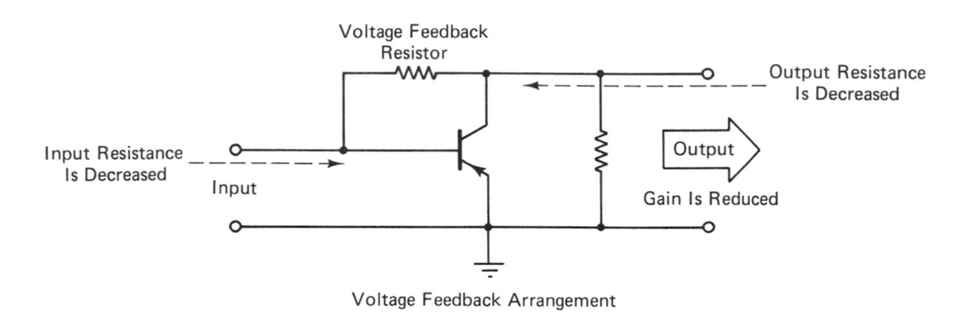

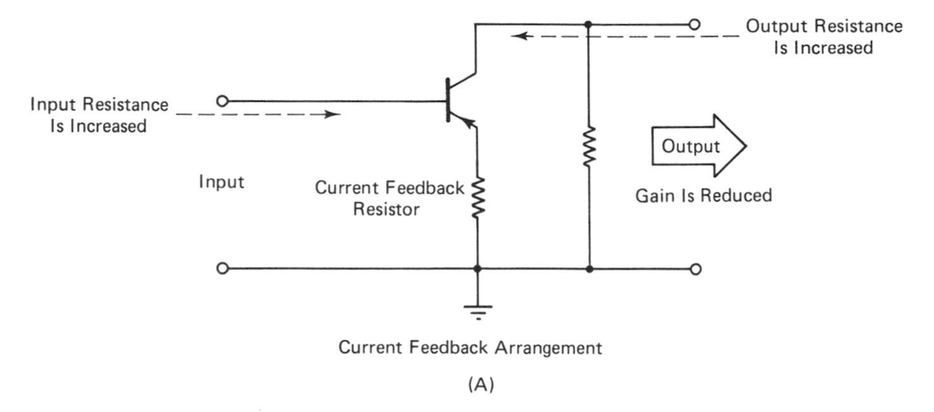

(A)

Figure 3–17(A). Example of a simple negative-feedback arrangement.

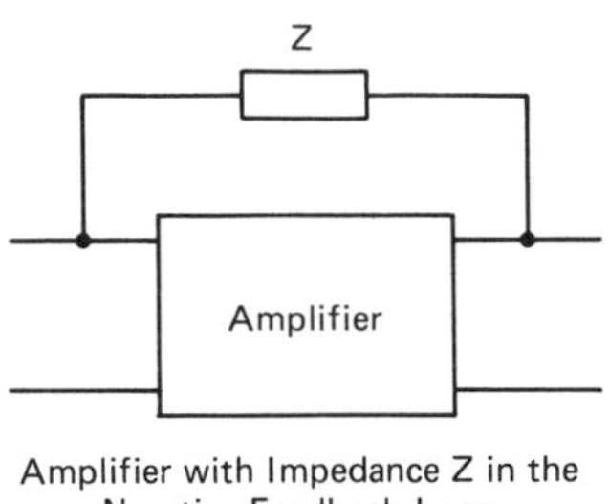

Amplifier with Impedance Z in the Negative Feedback Loop

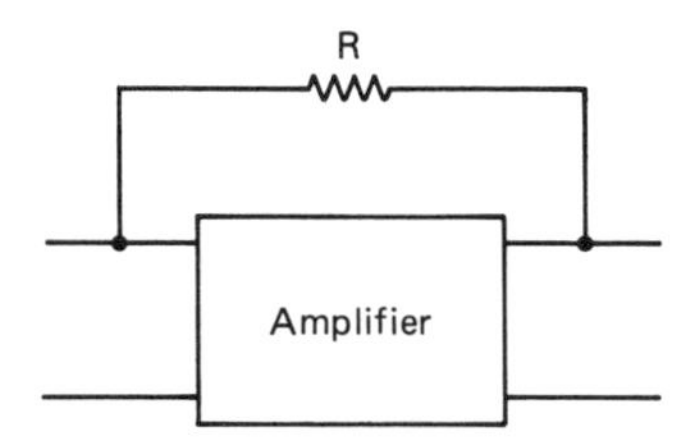

Amplifier with Resistance R in the Negative Feedback Loop (Amplifier Has an Inherent Voltage Gain of A Times)

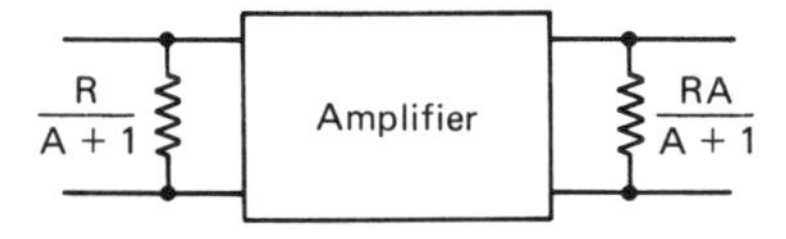

Miller Equivalent Circuit Places Shunt Resistors Across the Input Terminals and the Output Terminals (Shows the Change in Input Resistance and Output Resistance, but Does Not Show the Change in Frequency Response)

(B)

Figure 3–17(B). Miller equivalent circuit for amplifier with voltage feedback.

NOTE: The Miller equivalent circuit for an amplifier with voltage feedback is often helpful to the designer in simplification of an amplifier configuration for analysis. Miller's theorem states that if an amplifier has an impedance Z in series with a voltage negative feedback loop, the feedback loop can be eliminated and an equivalent circuit drawn in which Miller impedances shunt the input terminals and the output terminals. The input Miller impedance is equal to $Z/(A+1)$ *and the output Miller impedance is equal to* $ZA/(A+1)$ *where A is the inherent voltage gain of the amplifier. (It is assumed that the amplifier is linear.)*

The usefulness of the Miller theorem is in its simplification of a feedback configuration; the feedback loop is removed and the feedback impedance is accounted for in terms of shunt input and output resistances.

From a practical viewpoint, if the feedback factor is very large, so that Aβ is much greater than 1, the following approximate formula can be written:

$$A' = -1/\beta \text{ (approx.)}$$

This formula shows that the amplifier gain with negative feedback is practically independent of the original (and highly variable) voltage amplification A. Or, A′ tends to be independent of device and component tolerances, and of supply-voltage variation. With significant negative feedback present, the input resistance of the amplifier is formulated:

$$R'_{in} = (1 - A\beta)R_{in}$$

Suppose next that the original voltage amplification of the circuit deteriorates by a percentage PC. In turn, with negative feedback present, this percentage is reduced by an amount:

$$PC' = PC/(1 - A\beta)$$

In other words, negative feedback improves the gain stability of the amplifier.

And important, the percentage distortion that was inherent in the original amplifier is also reduced by the foregoing amount:

$$D' = D/(1 - A\beta)$$

Thus, if an amplifier has an inherent total harmonic distortion (THD) of 10 percent, and the designer employs 20 dB of negative feedback, with $A = 200$ and $\beta = .045$, THD′ then becomes equal to 1 percent. (Remember that Beta is negative, inasmuch as negative feedback is being utilized.)

These basic relations are easy to apply in design procedures if a short computer program is written, as shown in Figure 3–18. Observe that all the inputted variable assignments in this program are positive (for convenient calculation). The percentage distortion is inputted as hundredths. The example in Figure 3–18 shows that if Beta is equal to 0.1, total harmonic distortion (THD) will be reduced from 15 percent to approximately 3 percent; gain will be reduced from 500 to approximately 10; input resistance will be increased from 10,000 ohms to 510,000 ohms.

```
5 LPRINT "CURRENT FEEDBACK AMPLIFIER PROGRAM"
10 LPRINT "Gain, Distortion, and Input Resistance"
15 LPRINT "Versus Feedback Factor"
20 LPRINT "Common Emitter Configuration"
25 INPUT "Rin=";A
30 LPRINT "": LPRINT "Rin=";A
35 INPUT "Voltage Gain=";B
40 LPRINT "Voltage Gain=";B
45 INPUT "THD/100=";C
50 LPRINT "THD/100=";C
55 INPUT "Bf=";D
60 LPRINT "Bf=";D:LPRINT ""
65 E=B*D
70 A$="###.#"
75 LPRINT "Feedback Factor="; USING A$; E
80 I=1+E:J=I*A
85 B$="#######": LPRINT "Rin With Feedback="; USING B$; J
90 F=B/I
95 C$="###.#"
100 LPRINT"Gain With Feedback="; USING C$;F
105 Z=C/I:D$="##.####"
110 LPRINT "THD/100 With Feedback="; USING D$; Z
115 LPRINT "(Bf = Fraction of Output Voltage Fed Back to the Input)"
```

NOTE: The designer may observe that "gain with feedback" is an indefinite figure unless it is referenced to a stipulated "gain-without-feedback" configuration. This is just another way of saying that a common-emitter configuration may be referenced to a common-base configuration; the common-emitter configuration is then a common-base arrangement with positive feedback. Again, a common-base configuration may be referenced to a common-emitter configuration; the common-base arrangement is then a common-emitter arrangement with negative feedback. In the context of this example, the reference is to a CE stage without external current or voltage feedback.

```
CURRENT FEEDBACK AMPLIFIER PROGRAM
Gain, Distortion, and Input Resistance
Versus Feedback Factor
Common Emitter Configuration

Rin= 10000
Voltage Gain= 500
THD/100= .15
Bf= .1

Feedback Factor= 50.0
Rin With Feedback= 510000
Gain With Feedback=  9.8
THD/100 With Feedback= 0.0029
(Bf = Fraction of Output Voltage Fed Back to the Input)
```

Figure 3–18. A current-feedback amplifier parameter program.

In the event that the original amplifier required an input level of 10 mV rms to develop rated output voltage, the foregoing negative feedback increases the required input level to 510 mV rms. If the gain of the original amplifier varied 20 percent over its rated temperature range, the foregoing negative feedback reduces the gain variation to 0.4 percent. Note that the required input level for development of rated output with negative feedback is approximately equal to the original input level multiplied by the feedback factor. The gain variation with negative feedback is approximately equal to the original gain variation divided by the feedback factor. (See also Figures 3–19, 3–20, and Chart 3–1.)

DESIGN OPTIMIZATION CONSIDERATIONS

Optimization of audio-preamplifier design requires attention to the thermal noise generated by the resistive components of the input circuit. All resistors develop a calculable amount of noise power; however, some types of resistors are "quieter" than others. For example, metal-film resistors generate less noise power than composition resistors. On the other hand, metal-film resistors are comparatively costly. Note that this noise power produced by resistive components of the input circuit is typically 160 dB below one watt for an amplifier with 10 kHz bandwidth.

If the designer has a choice, a transducer may be specified that is predominantly resistive (instead of reactive). As a rough rule of thumb, a transducer source resistance of about 800 ohms and a transistor input impedance of approximately 1200 ohms will result in the minimum noise figure. It is also helpful to specify a transistor type that has a high forward-current transfer ratio. Note that power-line hum and stray signal pickup can be reduced below the noise level by adequate shielding.

If production costs do not prohibit input-transistor selection for minimum noise factor, this can be a viable design option. In other words, some types of transistors are "quieter" than other types, and the noise factor will vary appreciably in a particular production run. In general, bipolar transistors are "quieter" than MOSFETS or integrated circuits. Observe that a bipolar-input transistor may be followed by a MOSFET or IC, inasmuch as the major noise contribution is made by the input circuit.

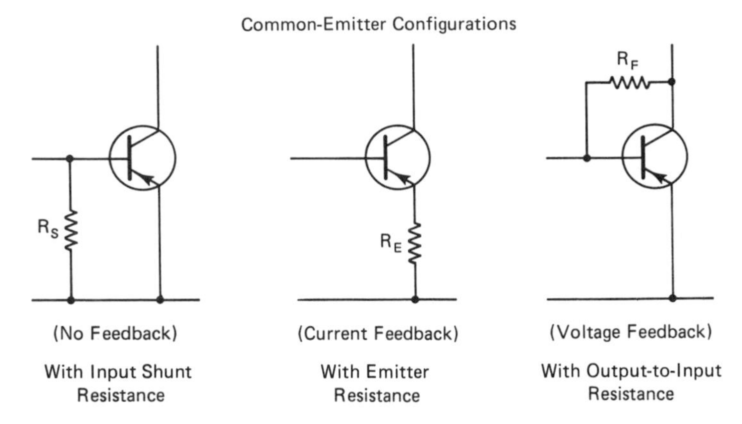

Figure 3–19. Conversion formulas for h parameters to perturbed-h parameters due to input-shunt resistance or negative feedback.

NOTE: (h denotes hybrid parameter for CE configuration without any associated circuit resistance; h' denotes perturbed hybrid parameter due to input shunt resistance; h'' denotes perturbed hybrid parameter due to emitter resistance; h''' denotes perturbed hybrid parameter due to output-to-input resistance.)

$$h'_{ie} = \frac{h_{ie}R_S}{h_{ie} + R_S} \qquad h''_{ie} = h_{ie} + \frac{R_E(1 + h_{fe})}{1 + R_E h_{oe}} \qquad h'''_{ie} = \frac{h_{ie}R_F}{R_F + h_{ie}}$$

$$h'_{fe} = \frac{h_{fe}R_S}{h_{ie} + R_S} \qquad h''_{fe} = \frac{h_{fe} - h_{oe}R_E}{1 + R_E h_{oe}} \qquad h'''_{fe} = \frac{h_{fe}R_F - h_{ie}}{R_F + h_{ie}}$$

$$h'_{re} = \frac{h_{re}R_S}{h_{ie} + R_S} \qquad h''_{re} = \frac{h_{re} + h_{oe}R_E}{1 + R_E h_{oe}} \qquad h'''_{re} = \frac{h_{ie}(1 - h_{re})}{R_F + h_{ie}}$$

$$h'_{oe} = h_{oe} - \frac{h_{fe}h_{re}}{h_{ie} + R_S} \qquad h''_{oe} = \frac{h_{oe}}{1 + R_E h_{oe}} \qquad h'''_{oe} = h_{oe} + \frac{(1 + h_{fe})(1 - h_{re})}{R_F + h_{ie}}$$

NOTE: An example of computer programming for conversion of h parameters to perturbed-h parameters that apply to the CE configuration with voltage feedback is shown next and is followed by a program for computation of characteristics for an amplifier with voltage feedback. Observe that in practical design procedures, computations often involve a combination of input-shunt resistance, with current feedback, with voltage feedback, or with both current and voltage feedback. This requires successive computation of the individual modifications, taken one at a time. It is necessary to first calculate the perturbed-h parameters due to feedback. Then, any further modification of the perturbed-h parameters due to input-shunt resistance may be calculated. Voltage-feedback perturbations may be calculated before current feedback perturbations, or vice versa.

```
5 LPRINT "CONVERSION OF h PARAMETERS INTO"
10 LPRINT "PERTURBED h PARAMETERS"
15 LPRINT "DUE TO VOLTAGE FEEDBACK"
20 LPRINT "(Common Emitter Amplifier)"
25 LPRINT ""
30 INPUT"Hfe=";A
35 LPRINT "Hfe=";A
40 INPUT "Hoe=";B
45 LPRINT "Hoe=";B
50 INPUT "Hie=";C
55 LPRINT "Hie=";C
60 INPUT "Hre=";D
65 LPRINT "Hre=";D
70 INPUT "Rf=";Z
75 LPRINT "Rf=";Z
80 C=C*Z/(Z+C)
85 A=(A*Z-C)/(Z+C)
90 D=(C*(1-D))/(Z+C)
95 B=B+((1+A)*(1-D))/(Z+C)
100 A$="######"
105 LPRINT "Hie (p)="; USING A$; C
110 LPRINT "Hfe (p)="; USING A$; A
115 B$="##.####"
120 LPRINT "Hre (p)="; USING B$; D
125 C$="##.######"
130 LPRINT "Hoe (p)="; USING C$; B
```

R_f

NOTE: Insofar as amplifier bandwidth with negative feedback is concerned, if the amplifier gain is reduced to 10 percent by negative-feedback action, then the bandwidth between half-power points will be multiplied by 10. Again, if the amplifier gain is reduced to 50 percent by negative-feedback action, then the bandwidth between half-power points will be multiplied by 2. (It is assumed that the amplifier rolloff is 20 dB per decade.)

After the perturbed-h parameters have been computed, the modified values can be substituted for the reference values of h parameters in the basic amplifier formulas. An example is shown in (B).

```
CONVERSION OF h PARAMETERS INTO
PERTURBED h PARAMETERS
DUE TO VOLTAGE FEEDBACK
(Common Emitter Amplifier)

Hfe= 50
Hoe= .00002
Hie= 1500
Hre= .0005
Rf= 100000
Hie (p)=  1478
Hfe (p)=    49
Hre (p)= 0.0146
Hoe (p)= 0.000508
```

Figure 3–20(A). Program for converting h parameters into perturbed-h parameters due to voltage feedback.

```
5 LPRINT "COMMON EMITTER AMPLIFIER"
10 LPRINT "Voltage, Current, and Power Gain,"
15 LPRINT "Input Resistance, and Output Resistance"
20 LPRINT "Voltage Feedback Perturbed Parameters"
25 LPRINT ""
30 INPUT"Hfe=";A
35 LPRINT "Hfe=";A
40 INPUT "Hoe=";B
45 LPRINT "Hoe=";B
50 INPUT "Hie=";C
55 LPRINT "Hie=";C
60 INPUT "Hre=";D
65 LPRINT "Hre=";D
70 INPUT "RL=";E
75 LPRINT "RL=";E
80 INPUT "Rg=";F
85 LPRINT "Rg=";F
90 LPRINT ""
95 G=A*E/((C*B-A*D)*E+C)
100 A$="#####.##":LPRINT"Av=";USING A$;G
105 H=A/(B*E+1)
110 LPRINT "Ai=";USING A$;H
115 I=A*A*E/((B*E+1)*((C*B-A*D)*E+C))
120 LPRINT "Gp=";USING A$;I
125 J=(C+(B*C-A*D)*E)/(1+B*E)
130 LPRINT "ri=";USING A$;J
135 K=(C+F)/(B*C-D*A+B*F)
140 LPRINT "ro=";USING A$;K
```

```
COMMON EMITTER AMPLIFIER
Voltage, Current, and Power Gain,
Input Resistance, and Output Resistance
Voltage Feedback Perturbed Parameters

Hfe= 49
Hoe= 5.080001E-04
Hie= 1477
Hre= .0145
RL= 15000
Rg= 1500

Av=  354.35
Ai=     5.68
Gp= 2014.27
ri=  240.63
ro= 3712.82
```

NOTE: If this amplifier is operated without voltage feedback, its parameters then become:

$$Av = 476.19 \qquad ri = 1211.53$$
$$Ai = 38.46 \qquad ro = 85714.28$$
$$Gp = 18315.01$$

Observe that the effect of voltage feedback in this example is to reduce the voltage gain, to reduce the current gain, to reduce the input resistance, and to reduce the output resistance. As previously noted, the effect of voltage feedback is also reduction in distortion and improved bias stability.

Negative feedback also improves the bandwidth of the amplifier. For example, a tradeoff of 20 dB in gain provides a 2.7-dB improvement in frequency response at the −3 dB (half-power) cutoff point.

Figure 3–20(B). Example of amplifier-parameter computation from the foregoing perturbed (voltage-feedback) h parameters.

CHART 3–1

Integrated Circuit Preamplifiers

As previously mentioned, integrated circuits are employed only to a limited extent as preamplifiers, due to the noise factor that is involved. However, if a preamplifier has a comparatively high input level, its noise figure can be disregarded. In turn, a wide-band amplifier, FM detector, and AF preamplifier/driver may be utilized in IC form for sound sections of TV receivers.

A typical IC used for preamplification in this application has the following characteristics:

Input impedance at 4.5 MHz, 11 kilohms and 5 pF

Output impedance at 4.5 MHz, 100 kilohms and 4 pF

IF voltage gain at 4.5 MHz, 67 dB

FM detector output, 250 mV rms

Total harmonic distortion, 3%

Audio voltage gain, 41 dB

AF amplifier output resistance at 1 kHz, 30 kilohms

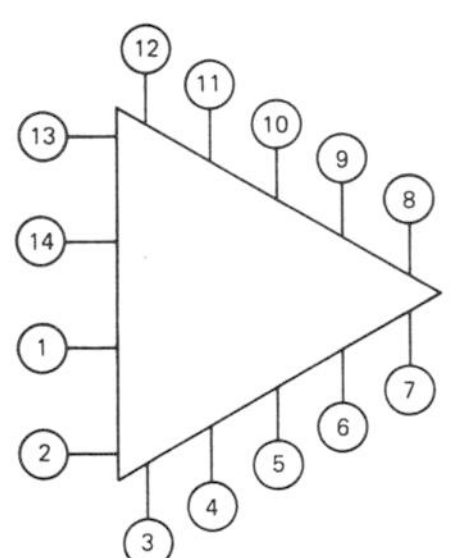

The designer is concerned with avoiding excessive power dissipation under worst-case conditions, and with maintenance of adequate power-supply voltage regulation under all conditions of operation.

In this example, dc power is applied via a series resistor to terminal 14. The required value of series dropping resistor depends upon the power-supply output voltage.

It follows from previous discussion that the tolerance on the value of the series dropping resistor is of initial concern. Observe that if the power-supply output voltage is comparatively high, the required value of series dropping resistor will also be high. In such a case, the IC power dissipation does not vary greatly under various conditions of operation, and a higher bogie value for IC power dissipation is then permissible.

As in transistor circuit design, the worst-case condition includes the maximum specified ambient temperature. Device dissipation derating curves vs. ambient temperature are provided in associated data sheets.

Utility amplifiers are ordinarily designed with comparatively reduced bandwidth. Thus, an amplifier with a frequency response from 300 Hz to 3 kHz will reproduce intelligible speech (although the output does not sound "natural"). Note that 60- or 120-Hz hum is not a problem in a utility amplifier, inasmuch as it has negligible response at hum frequencies. Audio engineers have established that sound output from a utility amplifier is more satisfactory if the frequency response is "trimmed" at both the high end and at the low end. For example, a frequency response from 20 Hz to 3 kHz represents poor design practice. A frequency response from 300 Hz to 20 kHz also represents poor design practice.

Component costs can be reduced by employing direct coupling between stages. However, bias stability becomes impaired, and direct coupling is ordinarily limited to two stages. A four-stage amplifier may be designed with the first and second stages direct coupled and with the third and fourth stages direct coupled, but with RC coupling between the second and third stages. Component costs can also be reduced by utilizing transformerless (ac/dc) power supplies. However, this option is accepted only in design of very low-priced bottom-of-the-line equipment.

Costs can be reduced by relaxing performance specifications—another basic trade-off option. For example, a public-address amplifier may have a frequency response starting at 150 Hz and extending to 10 or 12 kHz with a uniformity of ±3 dB, whereas a high-fidelity amplifier has a typical frequency response from 20 Hz to 20 kHz, with a uniformity of ±1 dB. The rated THD of a high-fidelity amplifier is typically less than 1 percent, whereas the THD of a public-address amplifier is commonly unspecified.

Graphical Approach to Circuit Analysis

Optimization of circuit design is often facilitated by a graphical approach wherein circuit action is determined by drawing vector diagrams on graph paper, instead of employing algebraic analysis or running computer programs. Although graphical analysis lacks the accuracy of either algebraic analysis or computer processing, vector diagrams drawn on graph paper can frequently provide sufficiently precise answers to circuit problems with considerable savings in time and effort for the designer.

With reference to Figure 3–21, a simple graphical construction permits the designer to determine the parallel impedance of a

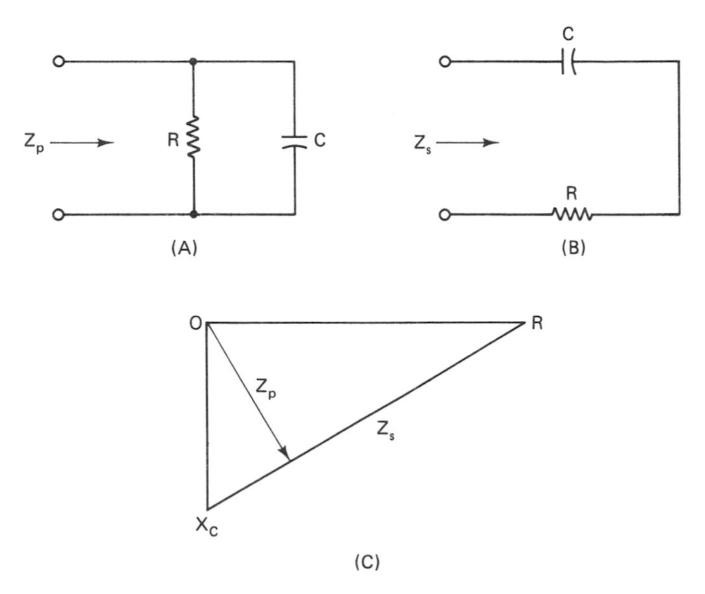

Figure 3–21. Equivalent series and parallel RC sections: (A) Parallel RC section; (B) series RC section; (C) equivalent series and parallel resistance and reactance values. (Reproduced by special permission of Reston Publishing Company and Campbell Loudon from *Handbook for Electronic Circuit Design.*)

NOTE: Graphical analysis sometimes provides a significant advantage over algebraic analysis in that its visual symbolism facilitates variational evaluation, whereby it becomes rapidly apparent how the circuit section will respond to changes in frequency or to increments or decrements in component values.

resistor and capacitor from the vector diagram for the series impedance of the same resistor and capacitor at a given operating frequency. Thus, R and X combine in series to form the impedance Z_s. On the other hand, if these same values of R and X are connected in parallel, their impedance Z_p is given by the altitude of the right triangle.

Conversely, the foregoing construction can be reversed to determine the values of series-connected R and C that will have the same impedance as a given Z_p vector. Stated otherwise, given the magnitude and phase angle of a vector Z_p, an equivalent series circuit is found by constructing a right triangle with Z_p as its altitude, as depicted in Figure 3–22. Thus, the sides of the right triangle give the resistive and reactive values for the equivalent series circuit.

As shown in Figure 3–22, a polar vector Z_p has the rectangular components R comp and X_c comp, which are determined by

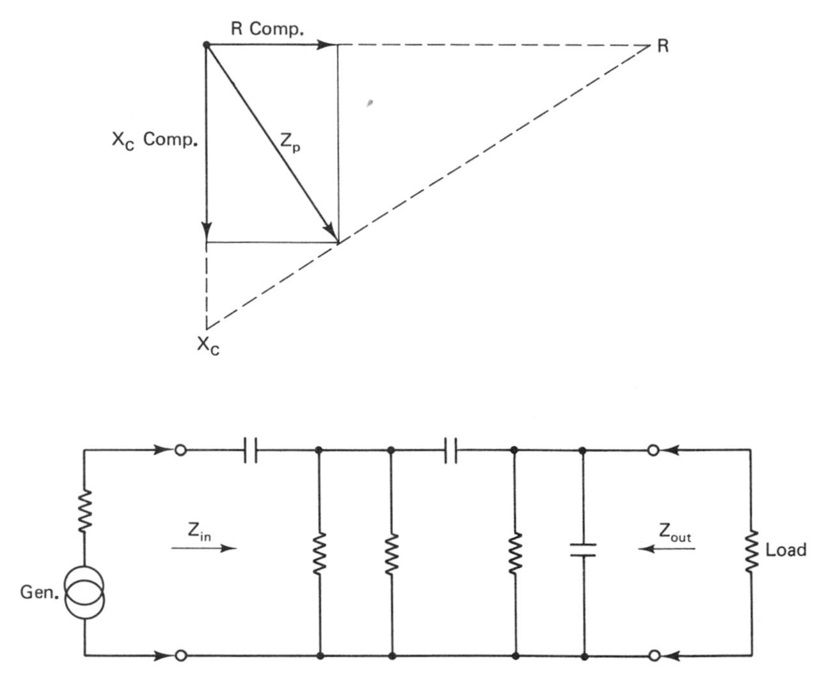

Figure 3–22. Rectangular components of Z_p.

NOTE: If network reduction is started from the output terminals and stepped back to the input terminals, a simple equivalent circuit is obtained that represents the input impedance of the network. On the other hand, if network reduction is started from the input terminals and stepped forward to the output terminals, a simple equivalent circuit is obtained that represents the output impedance of the network.

dropping perpendiculars to the sides of the equivalent series triangle. Observe that equivalent series and parallel RC circuits have the same polar magnitudes and phase angles (same power factor) and dissipate the same amount of power for a given value of applied voltage.

During the course of graphical circuit analysis a simplified vector diagram comprising parallel inductance and capacitance is often encountered. With reference to Figure 3–23, a parallel combination of L and C "looks like" either an inductance or a capacitance, depending on relative reactance values. To analyze the parallel LC section, an X_L axis and X_C axis are drawn at either end of a base line, as shown in the diagram. Then, the X_L and X_C reactance values are located on their axes and lines are drawn from these points through the origins.

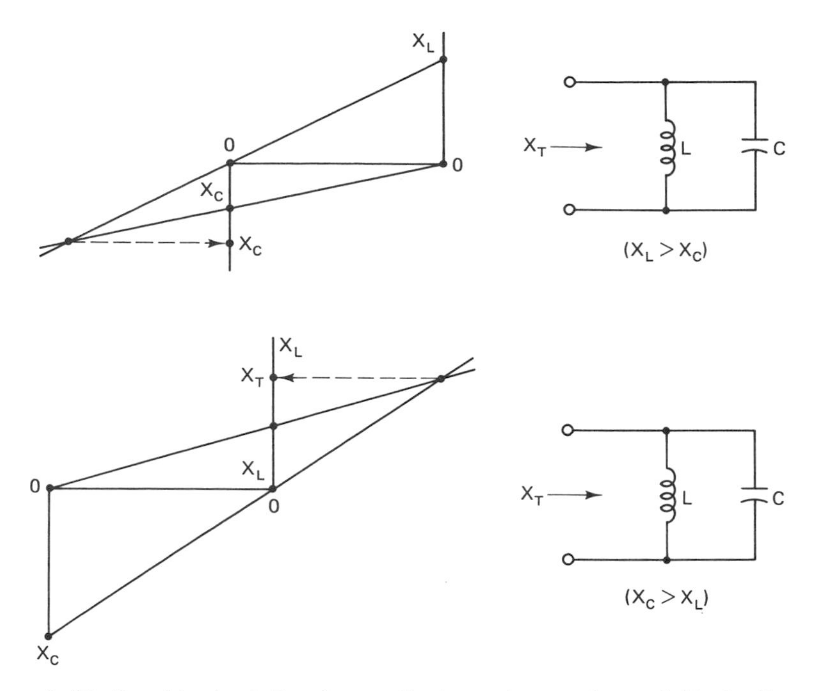

Figure 3–23. Graphical solution for resultant reactance of parallel inductive and capacitance reactances.

NOTE: The base line 0–0 may have any finite length; choose a length that provides a conveniently sized diagram.

The input impedance of any network can be represented at a given frequency by R and C connected in series, or by R and L connected in series. Or, the input impedance can be represented at a given frequency by R and C connected in parallel, or by R and L connected in parallel. Conversely, the output impedance of any network can be similarly represented at a given frequency.

All graphical analysis of networks consists of triangulation steps. When series resistance and series reactance are triangulated into equivalent shunt resistance and shunt reactance, the shunt components are fictitious within the context of the physical network under analysis.

The lines drawn from the X_L and X_C reactance values through the origins intersect at some point to the right or to the left of the axes. In turn, a perpendicular drawn from the point of intersection to the adjacent axis shows the magnitude of the resultant reactance and whether it is inductive or capacitive.

When these basic constructions are employed, starting at the output end of a network, the designer can proceed with considerable rapidity section by section to the input end of the network. The final construction represents the simplest equivalent circuit for the network at the stipulated operating frequency. Since reactance values are employed (often at each step), it is helpful to keep a pocket calculator at hand to quickly "punch out" reactance values.

As a basic illustration of graphical analysis of a network, observe the step-by-step constructions shown in Figure 3–24. The LCR-circuit section comprises series inductance and resistance connected in parallel with capacitance. Analysis is made at a

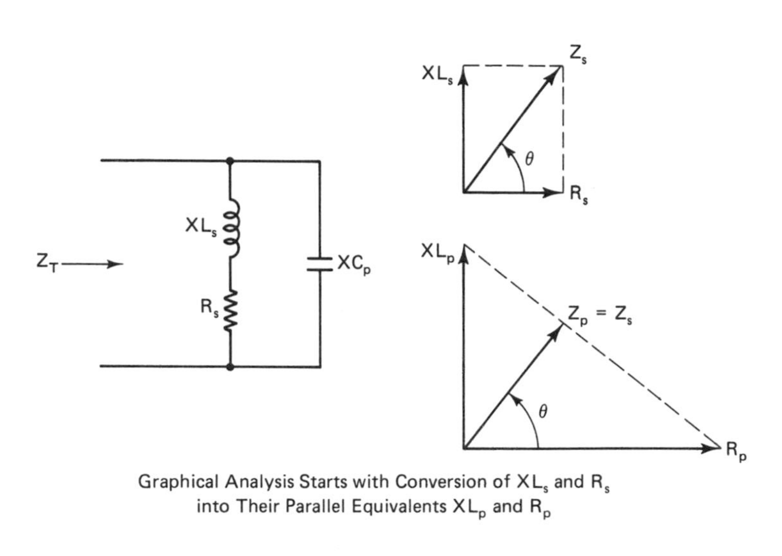

Graphical Analysis Starts with Conversion of XL_s and R_s into Their Parallel Equivalents XL_p and R_p

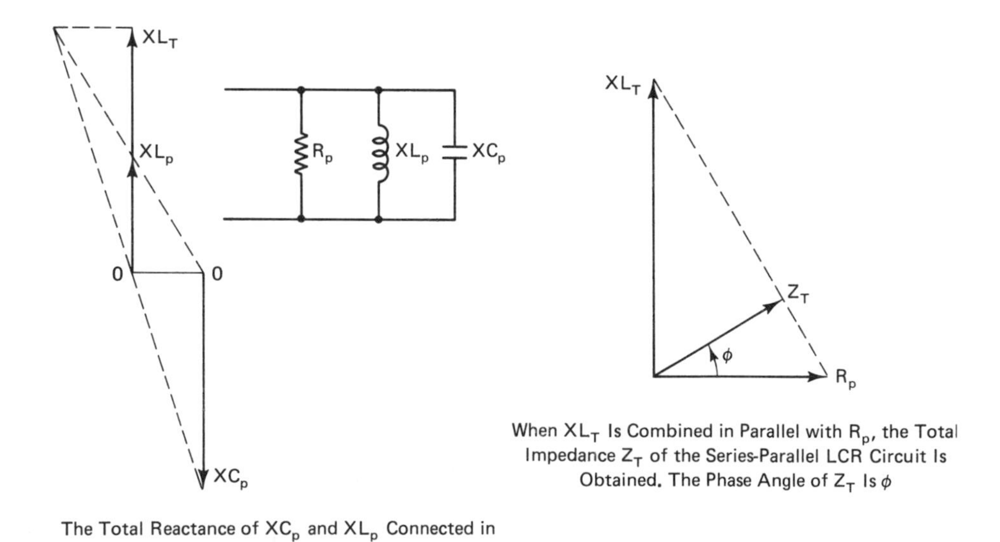

The Total Reactance of XC_p and XL_p Connected in Parallel is Equal to XL_T

When XL_T Is Combined in Parallel with R_p, the Total Impedance Z_T of the Series-Parallel LCR Circuit Is Obtained. The Phase Angle of Z_T Is ϕ

Figure 3–24. Example of graphical analysis of a series-parallel LCR-circuit section.

specified frequency. In turn, the inductive reactance and series resistance form an impedance Z_s with a phase angle θ. Next, the equivalent parallel components of Z_s are constructed. The phase angle θ remains the same.

Then, the equivalent parallel inductive reactance XL_p is combined with the parallel capacitive reactance XC_p. In this example, the resultant reactance XL_T is inductive (XL_p draws more current than XC_p). Finally, the resultant reactance XL_T is combined with the equivalent parallel resistance R_p to obtain the total impedance Z_T of the original LCR series parallel-circuit section at the specified frequency. The phase angle of Z_T is ϕ. Note that like parameters combine graphically in series and in parallel as shown in Figure 3–25.

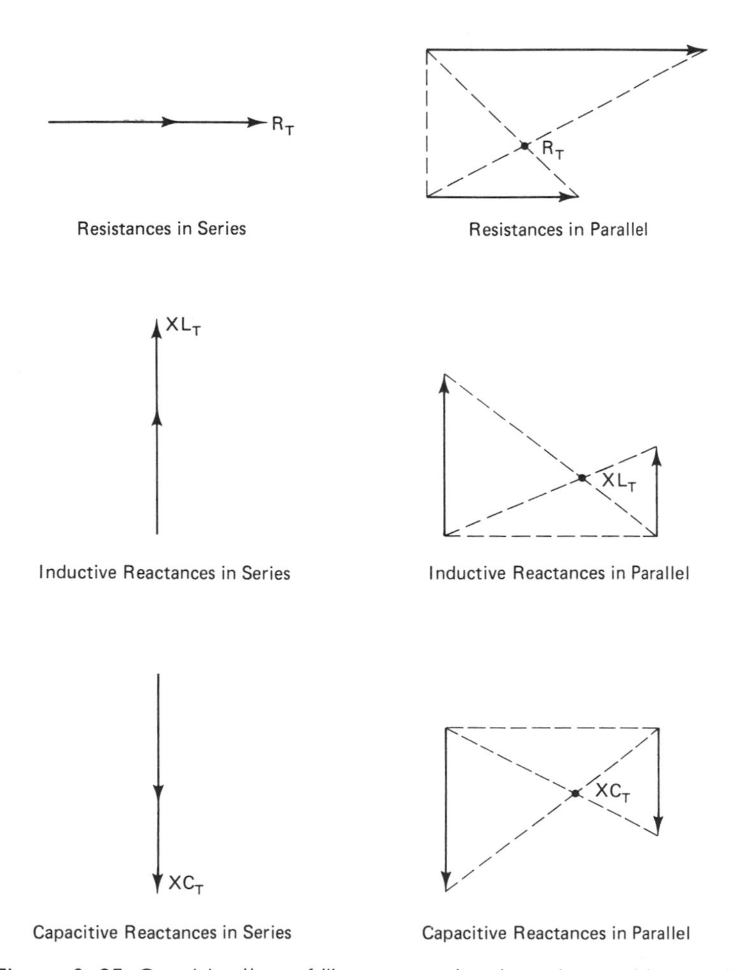

Figure 3–25. Combination of like parameters in series and in parallel.

As shown in Figure 3–26, graphical analysis procedures are facilitated by drawing the vectors on graph paper. In other words, graph paper eliminates the necessity for measuring line lengths and ensures that reactance vectors are drawn at 90° angles to resistance vectors. Observe also in Figure 3–26 that the length of the base line in the diagram for determination of XL_T is longer than its corresponding base line in Figure 3–24. This is an example of the

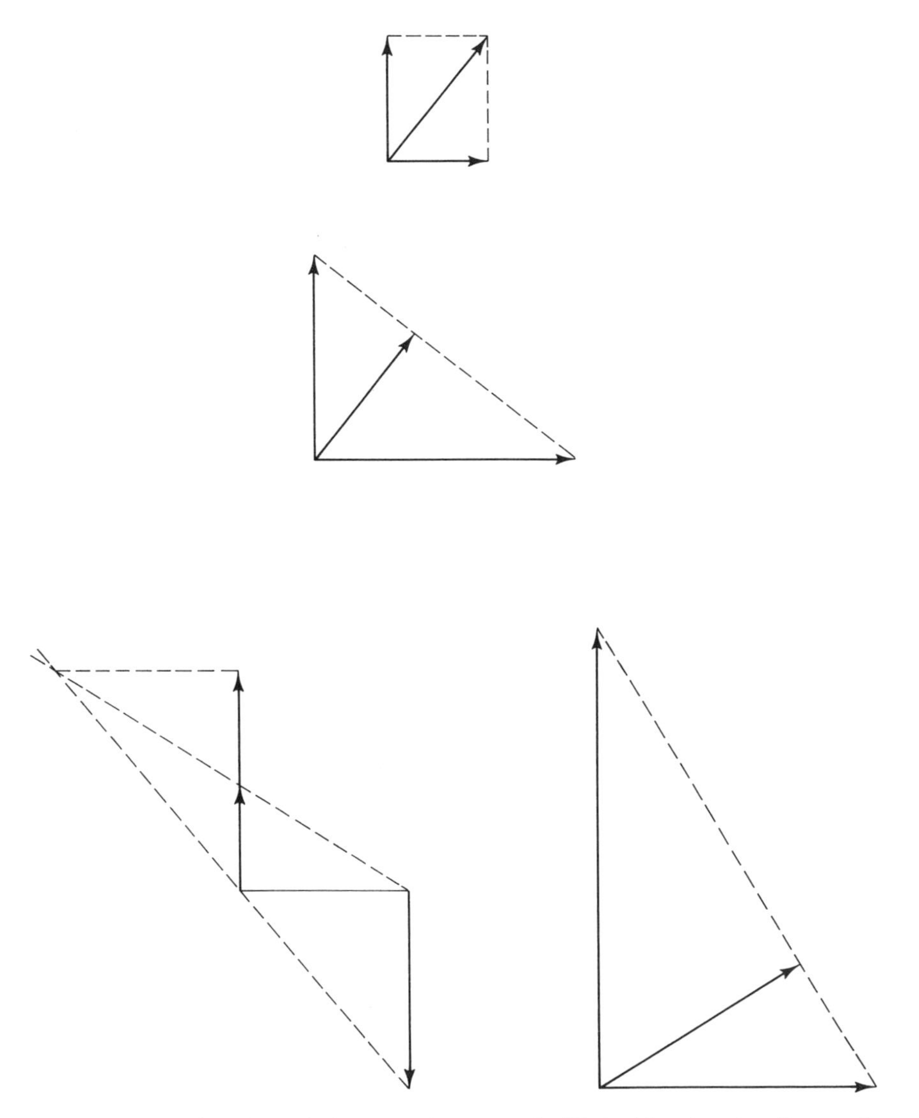

Figure 3–26. Graphical-analysis procedures are facilitated by drawing the vectors on graph paper.

circumstance that the base-line length is arbitrary and may be chosen from the viewpoint of convenience.

A geometrical model of the network under analysis can be derived by chaining the graphical procedures in step-by-step reduction of the network as exemplified in Figure 3–27. Geometrical models often provide an advantage over algebraic models in that the former present a helpful "picture" of network interrelations and permit the designer to easily evaluate the practical significance of component tolerances.

When the designer needs to quickly and easily derive the optimized algebraic equations for an arbitrary network, a geometrical model can be sketched promptly. Since the sketch involves only successive triangulations, this chained graphical analysis points directly to the simplest possible algebraic model for the network. Thus, the geometrical model of a network is a powerful derivational tool within the framework of abstract mathematics.

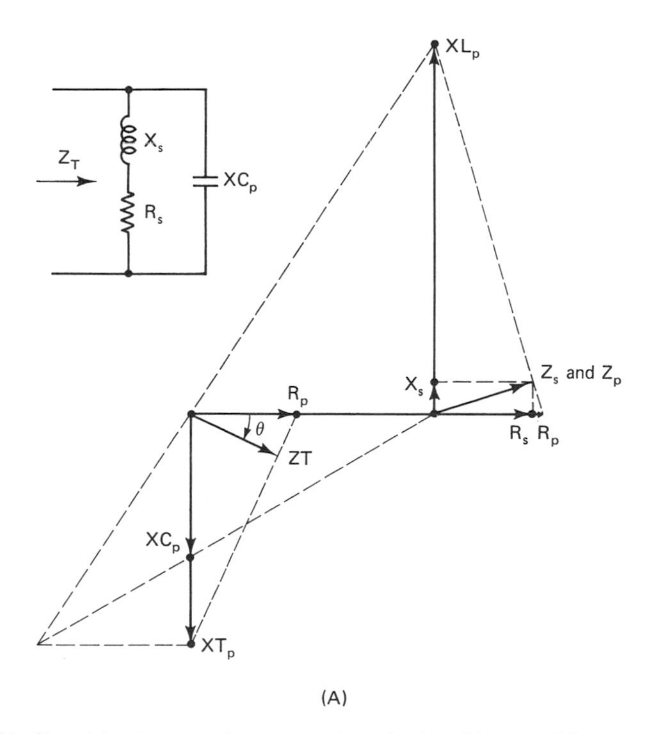

Figure 3–27(A). Graphical procedures may be chained to provide a geometrical model of the network under analysis.

NOTE: At resonance, with the current in phase with the line voltage, ZT becomes equal to R_p.

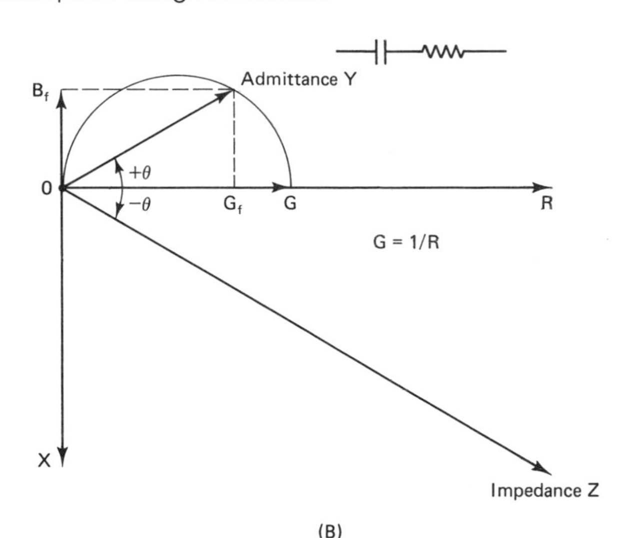

Figure 3–27(B). Construction of the admittance vector Y corresponding to the impedance vector Z.

Note: Admittance vectors provide simplified analysis of networks that have shunt branches. Shunt admittances add in the same manner that series impedances add.

Given the resistance R and the reactance X, the resultant impedance vector is Z and the phase angle of the impedance is $-\theta$. In turn, G is equal to 1/R, and G is the diameter of a semicircle, as shown. The angle of the admittance vector is $+\theta$, which determines the location of the admittance vector Y on the semicircle. This construction provides a rapid and easy procedure for deriving an admittance vector from an impedance vector. Note that Y has the components G_f and B_f.

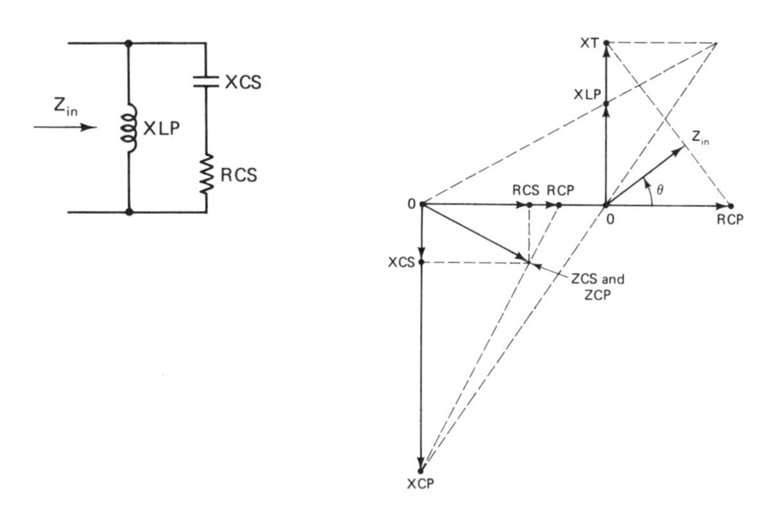

Figure 3–27(C). Chained graphical analysis of series capacitance and resistance in parallel with pure inductance.

Note: This is a basic chained graphical procedure in the same category as that shown in Figure 3–27(A). It is a basic step in the analysis of elaborate RCL networks.

Note that when progressively elaborate networks are graphically analyzed to obtain a chained geometric model, the diagram becomes comparatively complex. To avoid confusion in "reading" a complex diagram, the designer may sectionalize the "architecture" and delineate auxiliary diagrams in the same general manner that auxiliary views are delineated in conventional mechanical drawings. In other words, each auxiliary diagram depicts a separate section of the network, with its relation to previous and following sections indicated by means of a dotted projection lines.

Sectionalization of geometric models does not impair their utility from the functional viewpoints of tolerance evaluations, derivation of optimized algebraic equations, or presentation of potentially viable developmental variations. Note that the first section can be terminated at any convenient point, and projection lines drawn to an auxiliary section. Stated otherwise, the resultant vector in the first section is thereby projected into the start of the auxiliary section. (See Figure 3–28.)

Similarly, this auxiliary section can be terminated at any convenient point and projection lines drawn to a second auxiliary

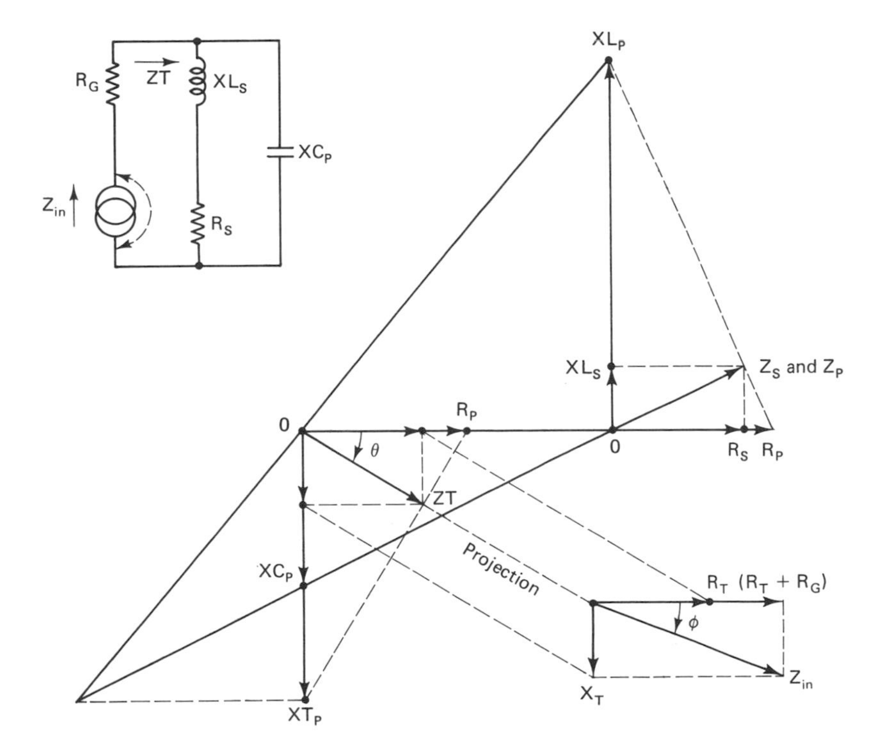

Figure 3–28. Example of projected view of a vector for a following step in graphical analysis.

section. Thus, a chained geometrical model is obtained that is easy to "read," compared with an elaborate and complex model without projections to auxiliary diagrams.

Note that positive resistance and negative resistance combine in parallel as shown in Figure 3–29. If the negative resistance has a smaller absolute value than the positive resistance, their resultant is negative. On the other hand, if the positive resistance has a smaller value than the absolute value of the negative resistance, their resultant is positive. If the resultant is positive, the circuit will be stable. However, if their resultant is negative, the circuit will be unstable.

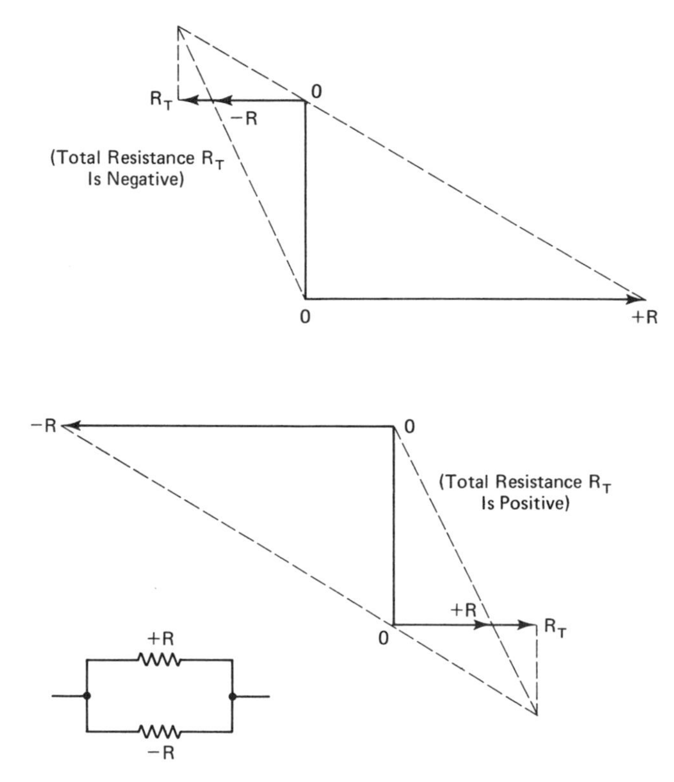

Figure 3–29. Combination of positive resistance in parallel with negative resistance.

4

DRIVER AND POWER-AMPLIFIER REQUIREMENTS

CHARACTERISTIC CURVES

Calculation of current, voltage, and power gain for a transistor in the CE configuration can be accomplished with respect to the output static characteristic curves for the device, as shown in Figure 4–1. The output characteristic curves are plots of collector current versus collector voltage with the base current as the running parameter. Note in passing that manufacturer's characteristic curves are bogie (design-center) specifications, and rated tolerances must be taken into account in practical design procedures.

With reference to Figure 4–1, the known parameters for the exemplified amplifier are:

1. Collector supply voltage is 10 volts.
2. Load resistor R2 has a value of 1500 ohms.
3. Emitter-base input resistance (r_i) is 500 ohms.
4. Input current has a value of 20μA p-p.
5. Operating point (X) is at 25μA base current and 4.8 volts collector potential.

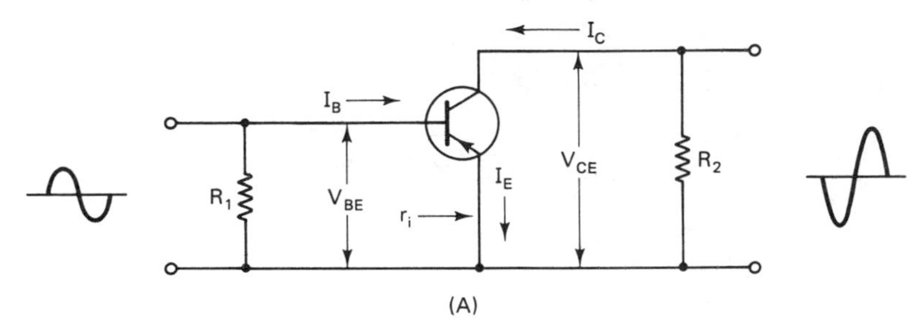

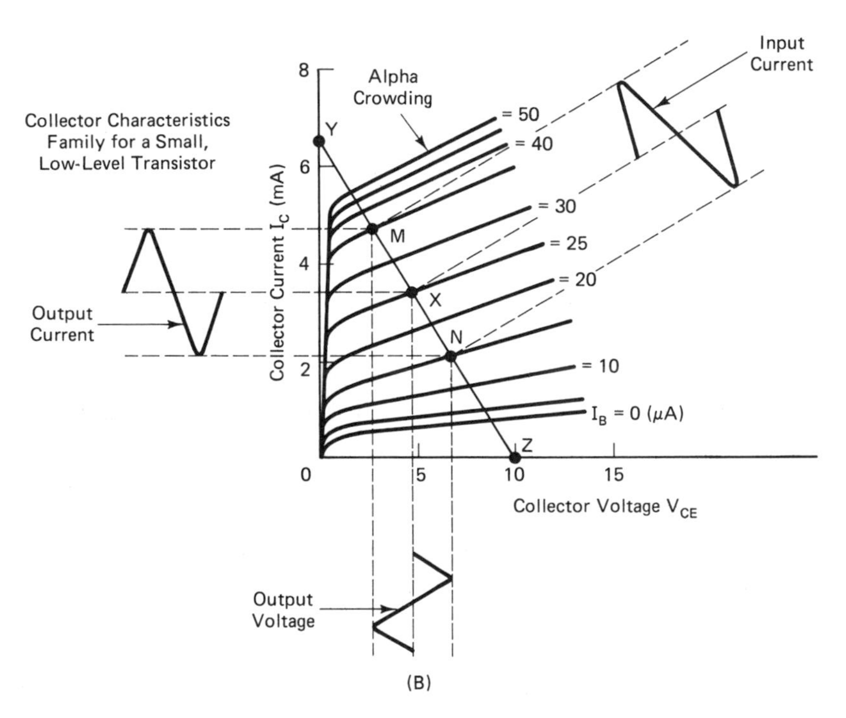

Figure 4–1. Skeleton CE amplifier diagram and example of output characteristic curves with a load line.

NOTE: The collector family of curves provides a helpful "picture" of transistor operation. A typical audio-amplifier inverter and driver transistor has a forward-current transfer ratio (h_{FE}) in the range from 70 to 350. Its maximum collector-current rating is 700 mA, and its maximum base-current rating is 200 mA. Its maximum V_{CE} rating is 70 volts. A typical audio-amplifier small-signal low-level transistor has an h_{FE} rating in the range from 100 to 250. Its maximum collector-current rating is 50 mA. It may be operated with V_{CE} voltages up to 35 volts.

The 1500-ohm load line is established by points Y and Z, and the operating point is denoted by X. In turn, the input-current, output-voltage, and output-current relations (for this example) are as indicated. Current gain is equal to the ratio of collector-current change to the corresponding base-current change. In this example, the ratio is seen to be 2.6/0,02, or 130.

Voltage gain is equal to the ratio of collector-voltage change to the corresponding base-voltage change. The base-voltage change is equal to the input current change multiplied by the input-resistance value (500 ohms in this example). Thus, the voltage gain is equal to $(6.7-2.7)/0.01$, or 400.

Power gain is equal to the product of voltage gain and current gain, or $400 \times 130 = 52{,}000$. A power gain of 52,000 times is equal to a dB gain of $10 \log_{10} 52{,}000 = 47$ dB. Note in passing that inherent nonlinearity in transistor action becomes evidenced as alpha crowding wherein the current gain decreases as the base current increases. Percentage distortion due to alpha crowding can be approximated by graphical analysis. However, it is customary design procedure to "breadboard" the configuration and to directly measure percentage distortion with a harmonic distortion meter (or intermodulation analyzer).

DYNAMIC-TRANSFER CHARACTERISTIC

Amplifier stage action is visualized to best advantage by means of the dynamic-transfer characteristic, as exemplified in Figure 4–2. A 1500-ohm load line is utilized, and its relation to the transfer characteristic is indicated by the P-Z plots for 2.5 and 10 volts collector potential. Observe that the dynamic-transfer characteristic is nonlinear and that more or less distortion will occur in amplifier operation. This distortion can be largely corrected by significant feedback, which has the effect of predistorting the input signal to compensate for transistor nonlinearity.

PHASE INVERTER/DRIVER STAGE

With reference to Figure 4–3, driver stages precede the power output stage. Most power-output stages are push-pull configurations that require two driving signals, each 180° out of phase with

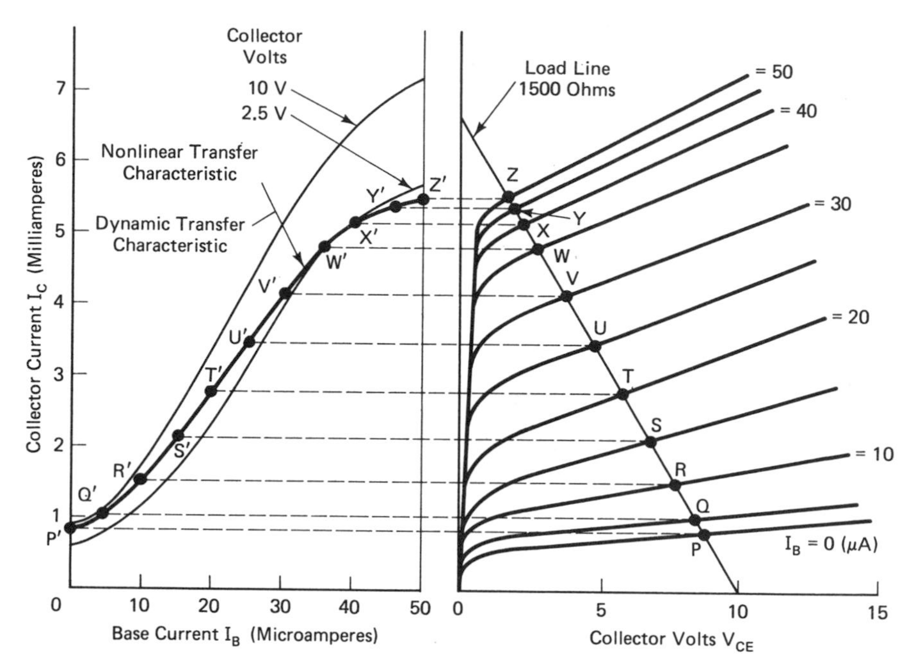

Figure 4–2. Dynamic-transfer characteristic is constructed from forward-transfer and output curves with load line.

NOTE: A transistor-transfer characteristic curve is not directly employed in design procedures. On the other hand, the transfer characteristic provides a meaningful overview of stage distortion problems. Observe that this transfer characteristic assumes a linear load of 1500 ohms. In a worst-case situation when an inverter transistor is used to drive a class-B output stage, the load line may become highly nonlinear, thereby introducing another source of inverter distortion. This distortion can be reduced (at the cost of driving power) by inserting series resistance between the driver and the output stage. A large amount of negative feedback is commonly employed from the power-output stage to the inverter input for minimization of distortion.

the other. In utility-amplifier designs, transformer coupling (with a tapped secondary) may be used for this purpose. RC coupling provides better fidelity (and may be less costly). When RC coupling is employed, the driver must provide inverter action such as exemplified in Figure 4–3.

Observe that the voltage gains are indicated as dB values in Figure 4–3. A word of caution to the beginner is in order here: "amplifier men" often designate voltage gain values in dB units, as if the input voltage and the output voltage were dropped across

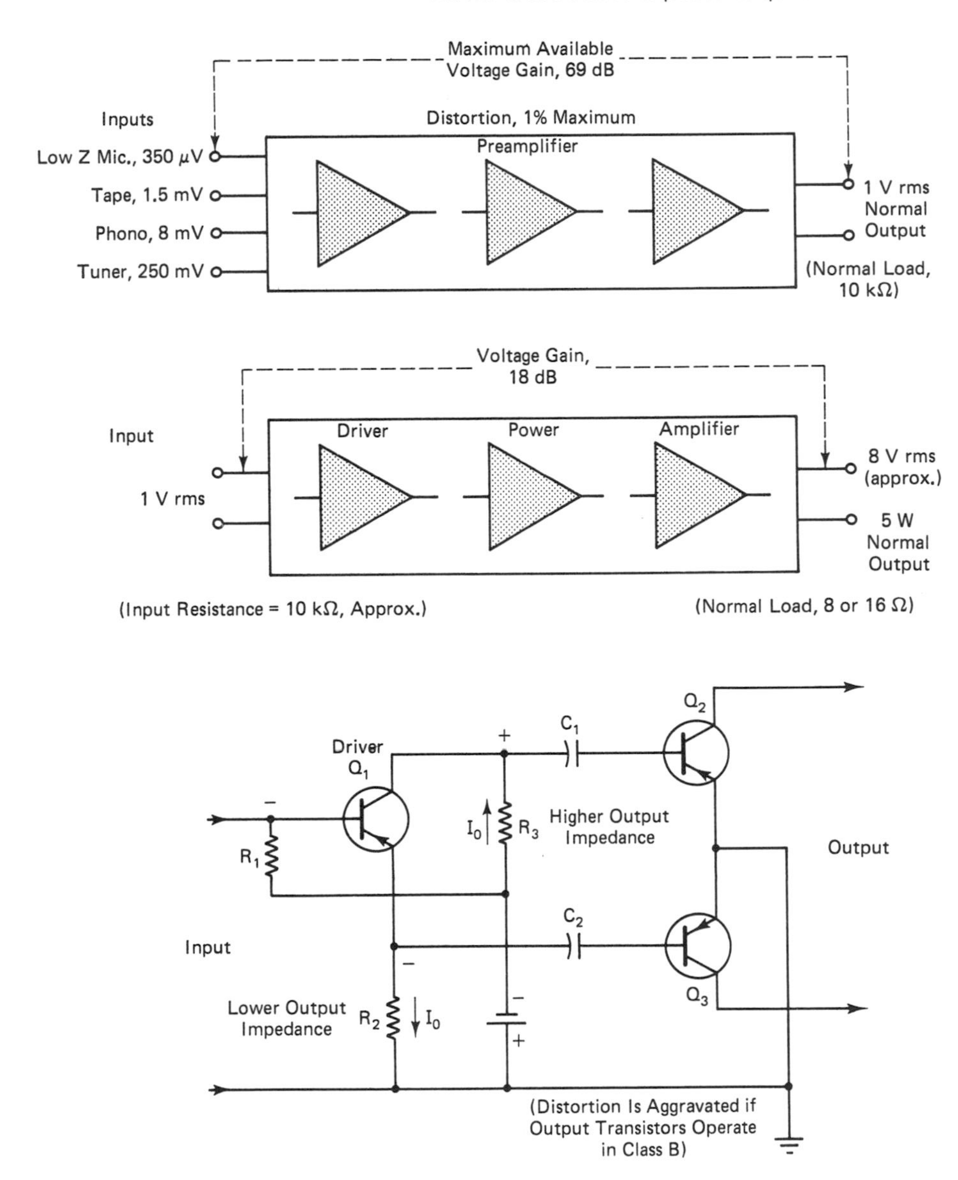

Figure 4–3. Example of a one-stage phase inverter driving a push-pull output stage.

equal resistance values (which is seldom the fact). In turn, the dB gain figure has no meaning apart from the particular amplifier that is under discussion. (Power gain values in dB units are always true decibel figures.)

Beginning designers may also be cautioned against neglect of possible reactive factors when evaluating dB values. For example, an amplifier may be designed to work into an 8-ohm load. At 400 Hz, a 50μF capacitor has a reactance of approximately 8 ohms. If the capacitor is connected across the amplifier output terminals, an 8-ohm load is provided (at 400 Hz). However, the dB voltage-gain figure is completely meaningless inasmuch as no real power is outputted by the amplifier. Stated otherwise, the dB power-gain figure is zero. The bottom line is that when an audio signal voltage is measured across an impedance, only the in-phase component of the voltage "counts" when calculating dB values.

In the example of Figure 4–3, equal voltage outputs result if R2 = R3. On the other hand, unequal (unbalanced) impedances are inherent in the configuration. In other words, the emitter output impedance of Q1 is less than the collector output impedance. In turn, progressive distortion occurs with increase in signal level. Refer next to Figure 4–4. Impedance unbalance is corrected in this arrangement by means of series resistor R4. The designer chooses values for R2 and R4 such that the signal source impedance for Q2 is equal to the signal source impedance for Q3. Resistor R2 has a higher value than R3 to compensate for the IR drop across R4. (See also Chart 4-1.)

TWO-STAGE PHASE INVERTER

A large amount of negative feedback is introduced by R2 in Figure 4–4. Accordingly, a comparatively large signal input is required to drive the inverter, and the designer may need to utilize a two-stage phase inverter such as shown in Figure 4–5. Note that the two-stage phase inverter not only operates with comparatively low-level signal input, but also develops more power output than the single-stage phase inverter.

With reference to Figure 4–5, Q1 operates in the CE mode, and Q2 operates in the CB mode. Output transistor Q3 is driven by Q1, and Q4 is driven by Q2. Q1 also develops a small signal-voltage drop across R2 to drive Q2. The base of Q2 is grounded for ac via C1. Although some current feedback is provided by R2, the input resistance to Q1 is not substantially increased due to the low impedance shunted across R2 by the input resistance of Q2.

A two-stage phase inverter/driver configuration may employ two CE circuit sections, as exemplified in Figure 4–6. This arrangement provides much more driving power than the one-stage configuration shown in Figure 4–4, and it also presents equal source

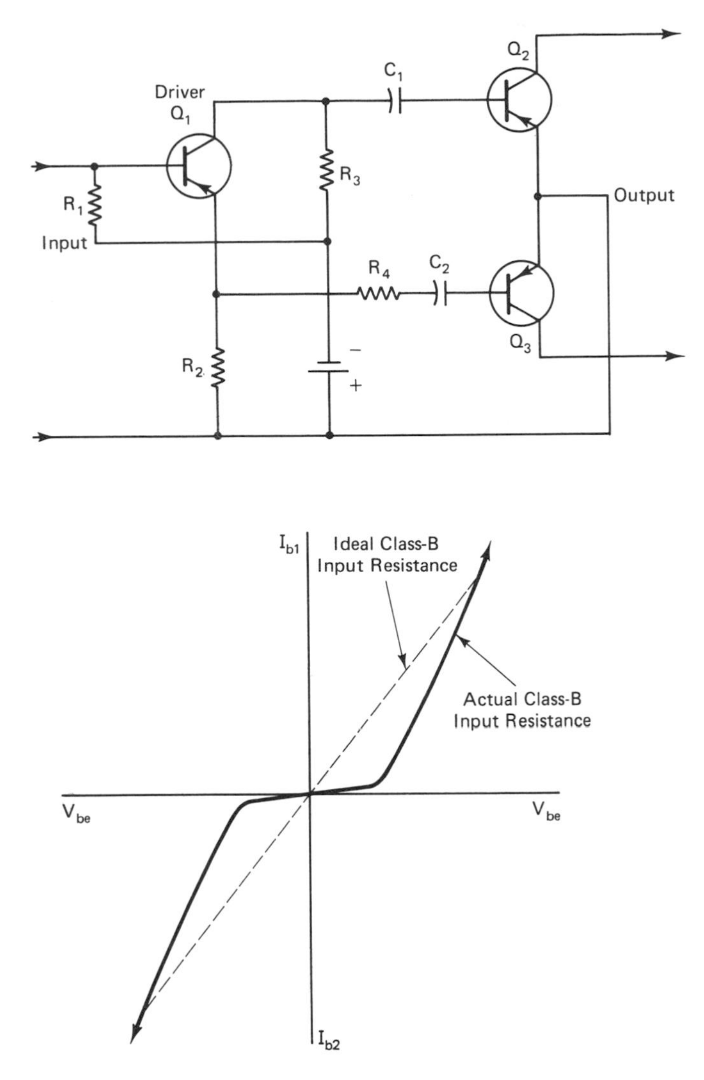

Figure 4–4. One-stage phase-inverter configuration with equalized output impedances.

NOTE: A zero-bias class-B stage develops crossover distortion, which is predominantly third-harmonic distortion. During the distortion interval h_{fe} suddenly drops to zero.

impedances for the two input sections of the following push-pull output stage. Negative feedback (voltage feedback) is provided by R1 and R5; this feedback assists in linearizing the inverter. Bias stabilization is provided by R2 and R7. The output from Q2 is made equal to the output from Q1 by suitable choice of resistance value for R4. (See also Chart 4-1.)

CHART 4–1

Operational Amplifier Characteristics

An operational amplifier is basically a dc-coupled high-gain differential amplifier in IC form. It is characterized by extremely high open-loop gain, with extremely high input impedance and with very low output impedance. An op amp is ordinarily used with a large amount of external feedback whereby its input terminal becomes a virtual ground, and its gain is essentially equal to the ratio of feedback resistance to input resistance.

A typical general-purpose op amp has the following parameters:

Open-loop gain: 200,000 times
Open-loop input impedance: 5 megohms
Input offset voltage: 5 mV
Common-mode rejection ratio: 90 dB
Drift versus temperature: 5 mV/C°
Bias current: 50 nA
Frequency response: Full power to 10 kHz, unity gain to 1 MHz
Full power response: Undistorted output of ±10 volts into a 1000 ohm load
Slew rate: 0.5V/μs at unity gain

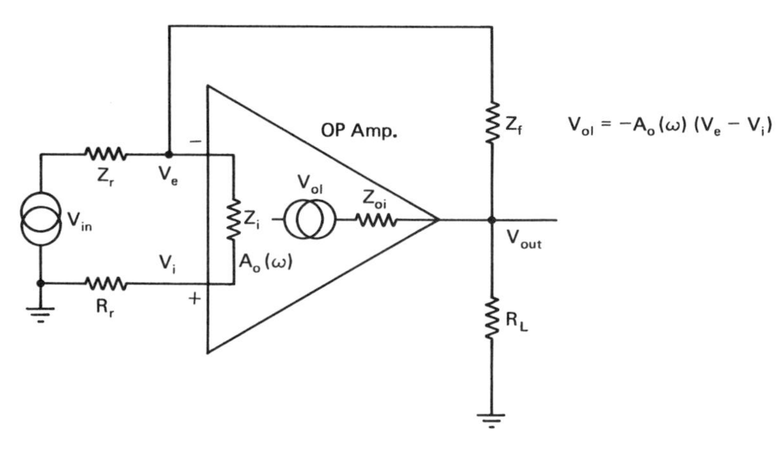

$$V_e = \frac{V_{in}(Z_f + Z_{oi})/\!/(Z_i + R_r)}{Z_r + (Z_f + Z_{oi})/\!/(Z_i + R_r)} + \frac{V_{oi}Z_r/\!/(Z_i + R_r)}{Z_f + Z_{oi} + Z_r/\!/(Z_i + R_r)}$$

$$V_i = V_e R_r/(Z_i + R_r)$$

$$\frac{V_{out}}{V_i} = \frac{Z_{oi}(Z_i + R_r) - A_o(\omega)\,Z_i Z_f}{(Z_f + Z_{oi})(Z_i + R_r) + (Z_r(Z_f + Z_{oi} + Z_i + R_r) + A_o(\omega)\,Z_i Z_r)}$$

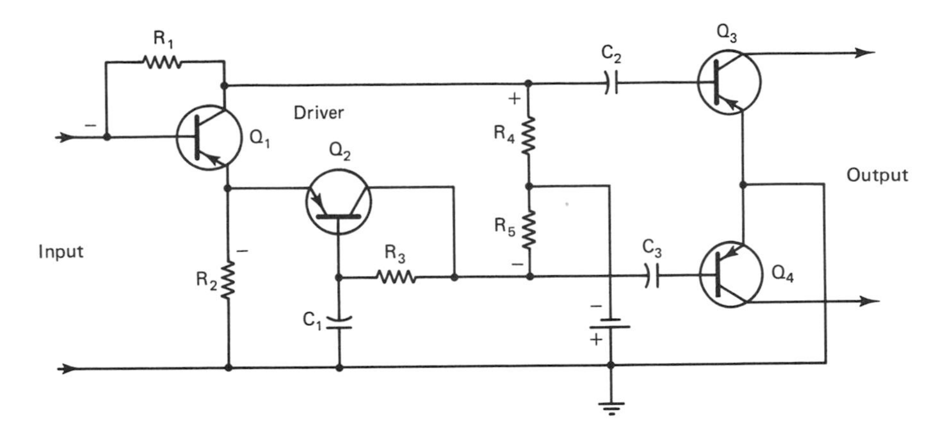

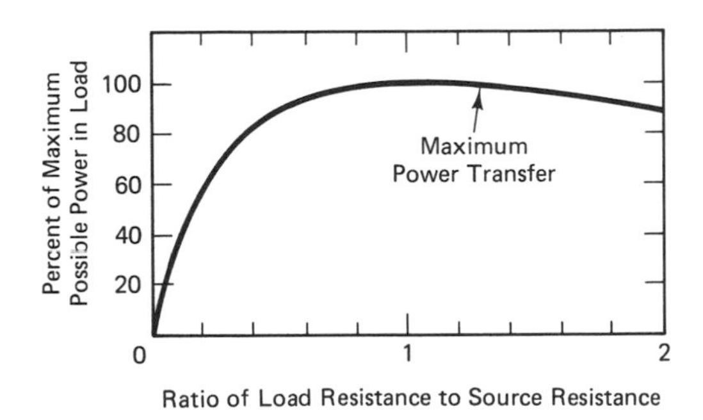

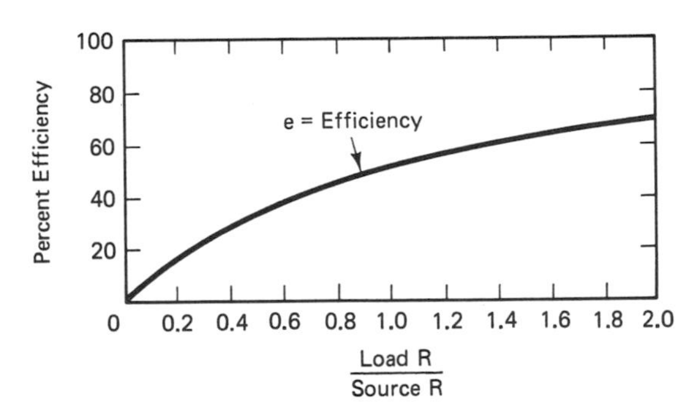

Figure 4–5. A typical two-stage phase inverter/driver arrangement.

NOTE: The output stage is the load on the driver stage. It is desirable to transfer as much power as possible from the driver stage to the output stage. Efficiency is a secondary consideration. The worst-case situation occurs when the load resistance is substantially smaller than the source resistance.

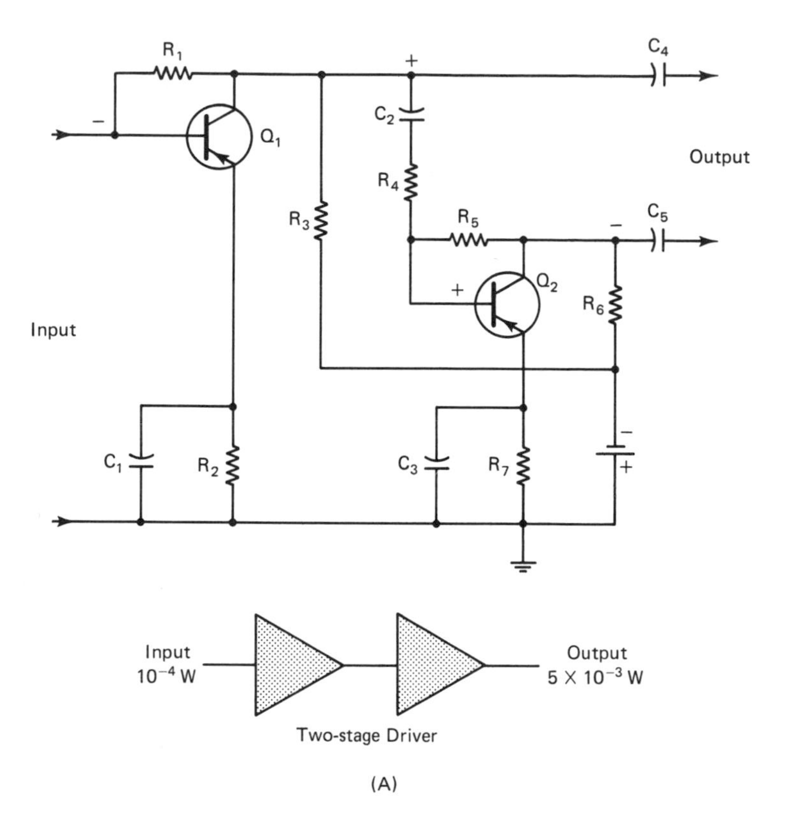

Figure 4–6(A). Example of a two-stage phase inverter/driver with both transistors operating in the CE mode.

NOTE: If a driver stage (or stages) were operated between a typical preamplifier and a typical output amplifier, it would provide a power gain of 50 times (17 dB):

In this example, the driver is preceded by a preamplifier that supplies a signal level of 10^{-4} watt. In turn, the driver is followed by a power amplifier that requires an input signal level of $5x10^{-3}$ watt. Accordingly, the driver must develop a power gain of 50 times, or 17 dB.

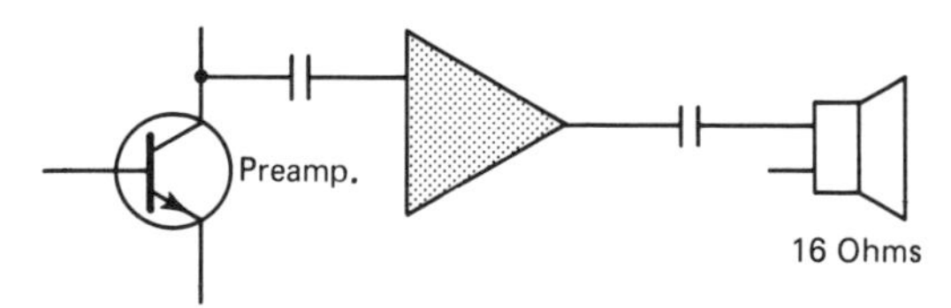

Figure 4–6(B). Example of a transistor preamplifier followed by a low-power IC amplifier. (The ICs are essentially operational amplifiers that provide low output impedance. Designers may refer to "Encyclopedia of Integrated Circuits," by Walter H. Buchsbaum, Sc.D.)

NOTE: Figure 4–6(B) is a typical example of an IC driver/output arrangement with a bipolar transistor preamplifier. The transistor preamp assists in reduction of the noise level. The amplifier system provides 1 mV input sensitivity for 200 mW output and a voltage gain of 1,780 times, in this example. The frequency response is from 100 to 10,000 Hz.

The dynamic range of an amplifier denotes its signal power capability. The dynamic range is specified in dB as the ratio of the maximum usable output signal level (ordinarily for a signal-to-noise ratio of approximately 20 dB). Designers regard a dynamic range of 40 dB as acceptable; a dynamic range of 70 dB is exceptional for any audio system.

For essential ready-reference data on analog integrated circuits, see Encyclopedia of Integrated Circuits, *by Walter H. Buchsbaum, Sc.D.*

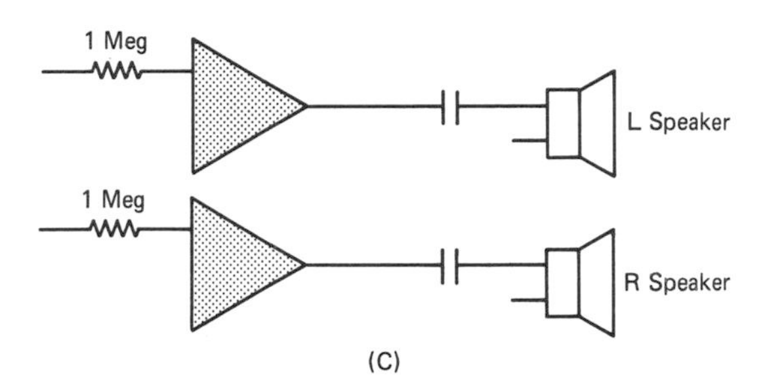

Figure 4–6(C). Example of a stereo phono amplifier implemented with integrated circuits. (The ICs are essentially operational amplifiers that provide low output impedance. Designers may refer to *Encyclopedia of Integrated Circuits* by Walter H. Buchsbaum, Sc.D.)

NOTE: This is an example of a stereo phono amplifier that uses integrated circuits for preamp and driver/output operation. A transistor preamp stage is not required in this application because the phono cartridge provides substantial input voltage so that noise is not objectionable. Observe that the phono cartridge feeds into the ICs via 1-megohm series resistors. These resistors function as low-pass RC filters in combination with the input capacitance of the ICs. Thereby, the RIAA equalization curve is approximated.

Integrated circuits can often offer some cost advantages in preamp and driver applications. As an overview, a typical IC package such as the CA3052 four-amplifier array has a noise figure of 2 dB at 1 kHz, an open-loop voltage gain of 58 dB at 10 kHz, and an open-loop input impedance of 90 kilohms at 1 kHz. (See Chart 4–1.)

POWER-AMPLIFIER DESIGN

Audio-frequency power amplifiers, unlike small-signal amplifiers, are seldom designed for maximized power gain. Power amplifiers are usually designed for provision of maximum output with minimum distortion and with maximum efficiency (minimum dc power demand). Although these are predominant design goals, judicious trade-offs are ordinarily required from the viewpoints of device and component costs, production costs, sales appeal, product compatibility, and so on.

Single-ended class-A and push-pull class-A power amplifiers are used only in small equipment in which poor power efficiency can be accepted and in which minimum distortion is a prime requirement. With reference to Figure 4–7, a matching transformer is ordinarily provided in a class-A amplifier between the output transistor and the speaker. The transformer is comparatively costly, and distortion is a problem ("iron third harmonic") due to the quiescent collector-current flow. Consequently, an "overdesigned" transformer must be utilized for minimum distortion.

Push-pull class-A power amplifiers employ transformers with center-tapped windings, whereby opposing quiescent current flows cancel each other out, thereby avoiding magnetic-core saturation. Although distortion can be minimized thereby, an output trans-

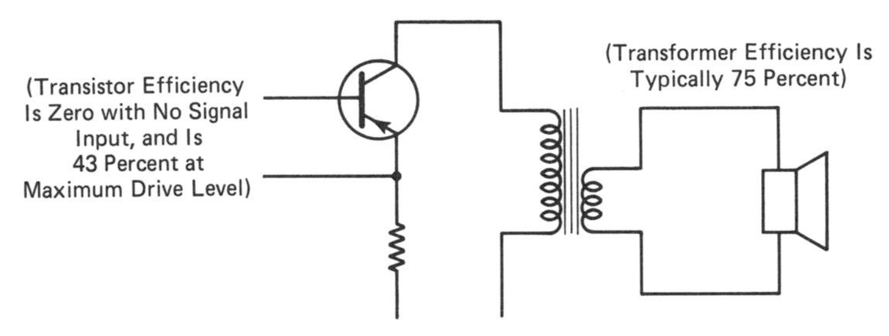

Figure 4–7. Example of a simple but inefficient output stage configuration.

NOTE: *From the standpoint of design perspective, it is helpful to observe the simple class-A audio-output configuration shown above. An output transformer is required to obtain reasonable power transfer to the speaker (a disadvantage on both cost and fidelity counts). Typically, a 5-mW rms base drive will develop a 4-W rms output; the base input resistance is on the order of 100 ohms. This arrangement may also be driven from a push-pull source.*

Practical Example: *A 2N301 power transistor may be operated in class A from a 14-V supply to develop 5 watts rms output at a maximum distortion of 5 percent THD. An efficiency of 45 percent is obtained at maximum output. With a forward bias of 0.35 volt, the quiescent current drain is 900 mA. A signal source resistance of approximately 10 ohms is appropriate. The stage provides a power gain of 38 dB. Note that with no input signal, the collector dissipation is 11 watts, in this example. For this reason (and others) the elementary class-A output-stage arrangement represents poor design practice.*

Observe that the output transformer cannot be eliminated by connecting the voice coil of the speaker directly in series with the emitter, for example, because of the dc offset current that would flow through the voice coil. However, dc offset current can be avoided by use of the complementary symmetry configuration, as subsequently described.

former with high-fidelity frequency response is nevertheless an expensive component. As a practical design note, negative feedback is seldom feasible for reduction of output-transformer distortion. This is just another way of saying that if negative feedback is employed from the speaker back to the input of the amplifier, instability becomes a major problem. The phase-shift versus frequency characteristic of the transformer usually results in development of positive feedback at some critical frequency, with the result that the amplifier "takes off" and oscillates uncontrollably.

Incidentally, the designer must also beware of reliance upon negative feedback to reduce distortion in the event that the amplifier is driven into its cutoff or saturation region. In other words, whenever h_{fe} becomes unduly diminished, or is reduced to zero, the amplifier then has little or no gain, and negative feedback action disappears. This is commonly stated as "negative feedback cannot correct clipping action."

With reference to Figure 4–8, operating efficiency can be realized by employing a push-pull class-B configuration. The basic class-B arrangement operates the transistors at zero bias. Essentially, the bias is less than zero due to the barrier potential of the transistors, and an effective reverse bias of approximately 0.5 volt is present. Consequently, crossover distortion occurs. As shown in Figure 4–8, crossover distortion can be minimized by operating the transistors in class AB, using sufficient forward bias to cancel out the barrier potential.

Capacitance-Diode Input Coupling

An inspection of Figure 4–9(A) shows that RC input coupling to a class-B (or class-AB) push-pull amplifier is accompanied by rectification of the drive voltage via the base-emitter circuits. Unless R3 and R4 have sufficiently low values, it is evident that drive-signal rectification can reverse-bias Q1 and Q2. Thus, although the cost of an input push-pull transformer (and its inherent limited frequency response) have been avoided, a new source of serious distortion is confronted.

To avoid reverse-biasing the transistors by the drive signal, while maintaining high input resistance to their bases, capacitance-diode coupling may be employed, as shown in Figure 4–9(B). As the drive signal rises on Q1, C1 "sees" the reverse resistance of CR1, or C1

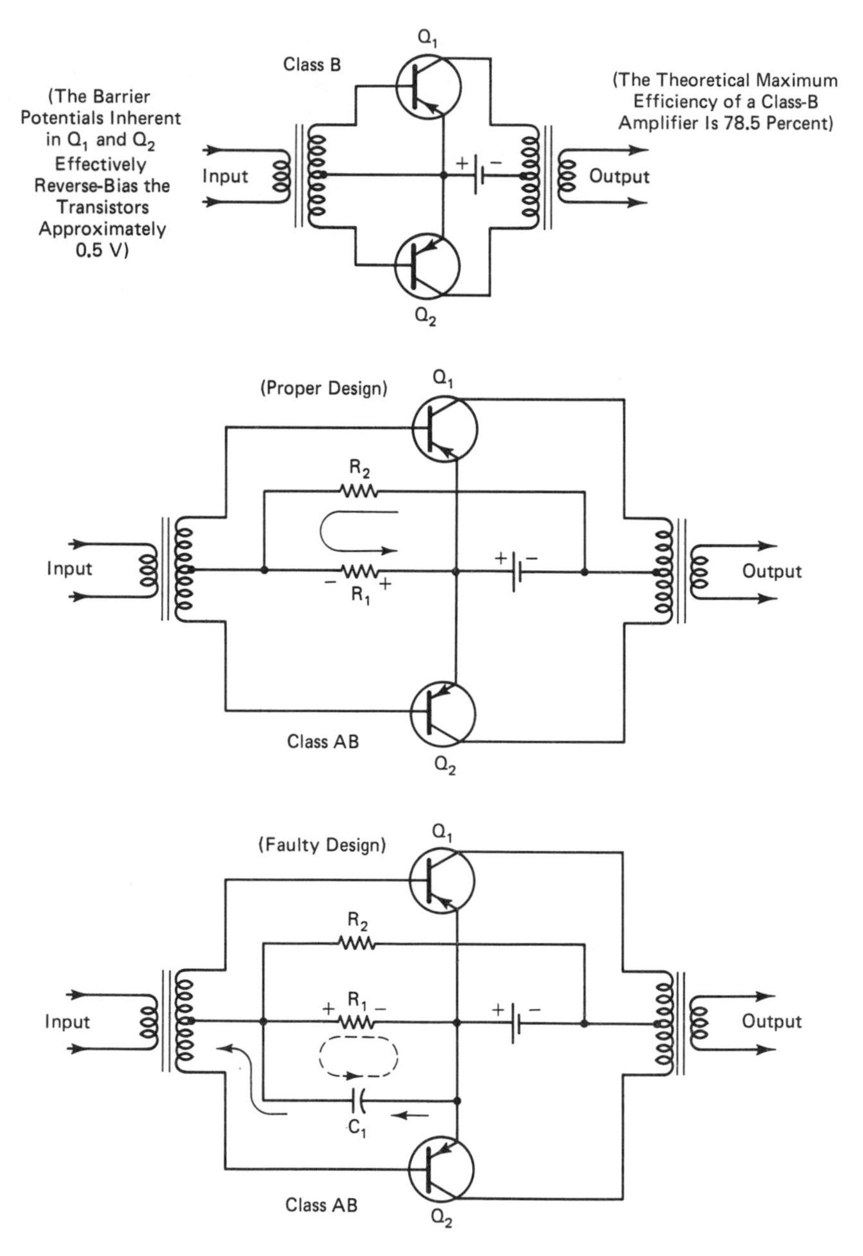

Figure 4–8. Transformer-coupled class-B push-pull output-stage configuration and examples of class-AB arrangements.

NOTE: A bypass capacitor C1 cannot be employed in the class-AB push-pull configuration because it will store a charge on signal peaks and upset the bias circuit, thereby introducing distortion. Observe that the class-AB arrangement can be designed to eliminate crossover distortion; on the other hand, the class-B arrangement has inherent crossover distortion.

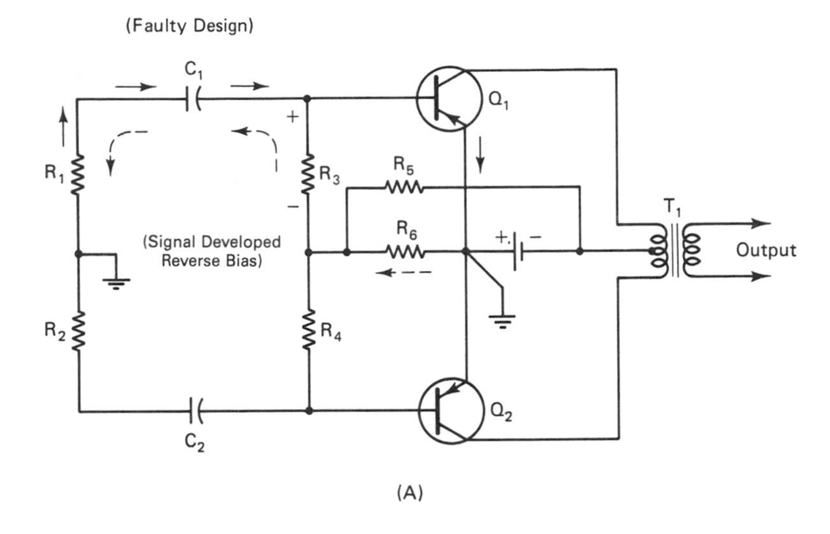

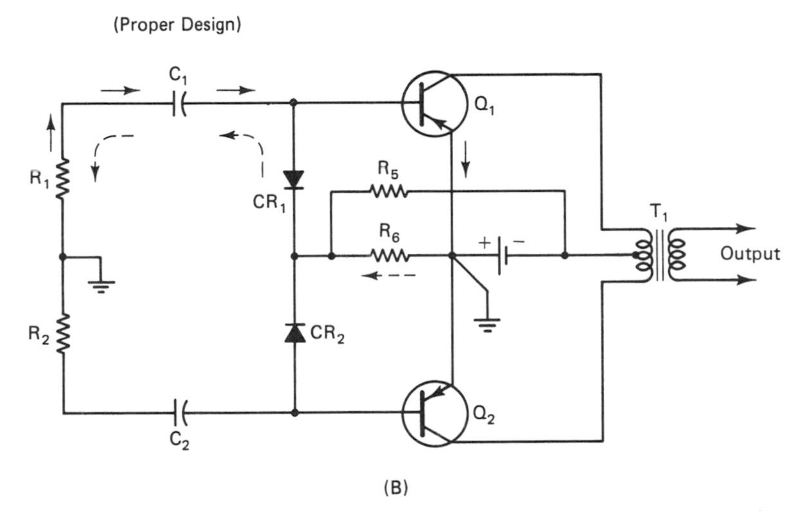

Figure 4–9. Example of class-AB push-pull configuration with RC input coupling and capacitance-diode input coupling.

NOTE: As the drive signal to Q1 (below) becomes more negative, base current flows. After the drive signal peaks, C1 would be left with a positive charge except for CR1. In other words, CR1 drains the positive charge from C1 as rapidly as it tends to develop. Thus, Q1 does not become reverse biased by the rectified drive signal.

"sees" a very high input resistance to Q1. As the drive signal falls, it does not leave a charge on C1 because C1 discharges via CR1 as rapidly as the drive signal falls. The same principle applies to Q2, C2, and CR2.

Observe that the transistors are slightly forward-biased in both Figures 4–9(A) and (B). This forward-bias voltage is applied via R5 to CR1 and CR2, with the result that the diodes conduct slightly, as do Q1 and Q2. Accordingly, a small forward-bias current flows in the base-emitter circuits of both configurations when no drive signal is applied. Application of a drive signal results in a greater current flow through the diodes.

COMPLEMENTARY SYMMETRY OPERATION

Various power-amplifier design difficulties are eased by employing some form of complementary symmetry circuit, such as depicted in Figure 4–10. Observe that the need for a phase-inverter drive stage is avoided—single-ended drive is applied to Q1 and Q2. Similarly, an input transformer is not utilized. Since C1 drives an NPN transistor and a PNP transistor, discharge diodes are not needed in the input circuit.

Observe also that there is no net dc offset current flow in the emitter circuit, which permits direct connection of the voice coil in series with the emitter return lead. Stated otherwise, this is a practical output-transformerless amplifier configuration. The basic circuitry shown in Figure 4–10 operates in class B; it can be alternatively configured for class-A or for class-AB operation. A simplified equivalent circuit is shown in Figure 4–10(B). With no signal input, both contact arms are in their "off" positions. Application of an input signal results in one or the other contact arm moving along R1 or R2.

The complementary symmetry power-amplifier circuit exemplified in Figure 4–10 has substantial negative feedback (current feedback). However, crossover distortion is present; stated otherwise, the stage has zero gain during the crossover-distortion interval, and there is no negative-feedback action during this interval. To minimize or eliminate crossover distortion, class-AB amplification may be used by addition of resistors R1, R2, and R3 depicted in Figure 4–11. The value of R1 is very small, and it has negligible unbalancing effect on the drive signal to Q2.

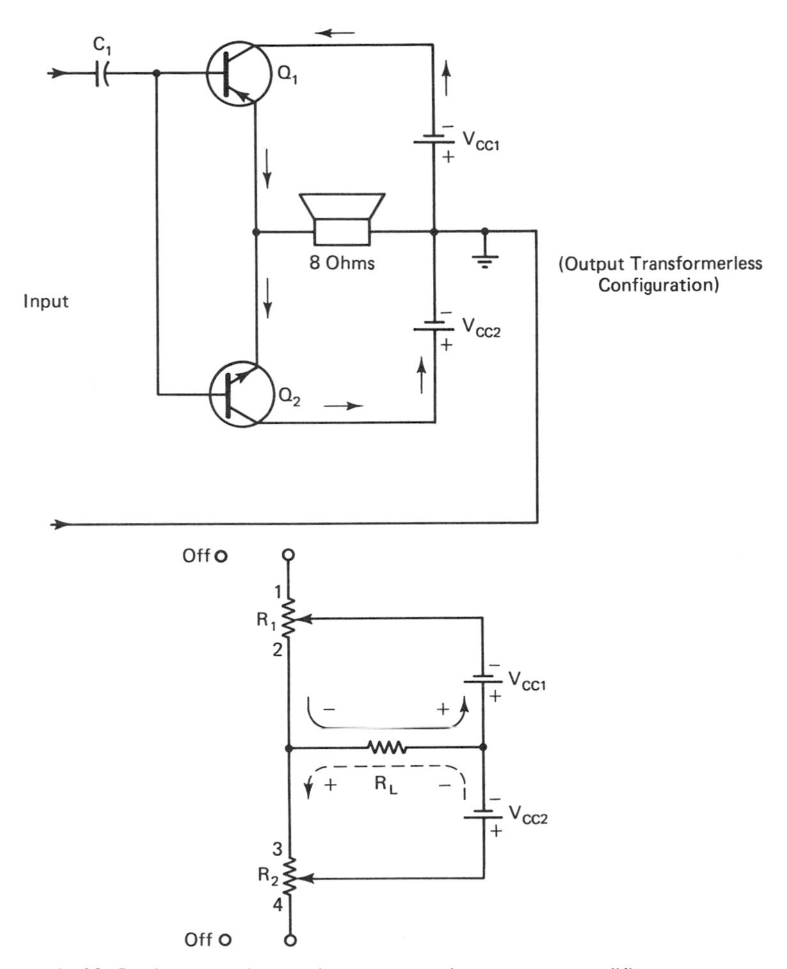

Figure 4–10. Basic complementary symmetry power-amplifier arrangement.

NOTE: Typical transistors used in a medium-power configuration have the following characteristics:

h_{ie} = *2000 ohms*
h_{fe} = *50*
h_{re} = 600×10^{-6}
h_{oe} = *25 microsiemens*

Since the transistors conduct alternately, the basic hybrid-parameter formulas may be used to calculate the combined performance of Q1 and Q2. Thus:

A_v = *0.26*
A_i = *51*
G_p = *13.3*
r_i = *1600 ohms*

Although the power gain is not impressive, the input resistance is high and the stage is easy to drive. Moreover, the simplicity of the circuit translates into low production costs.

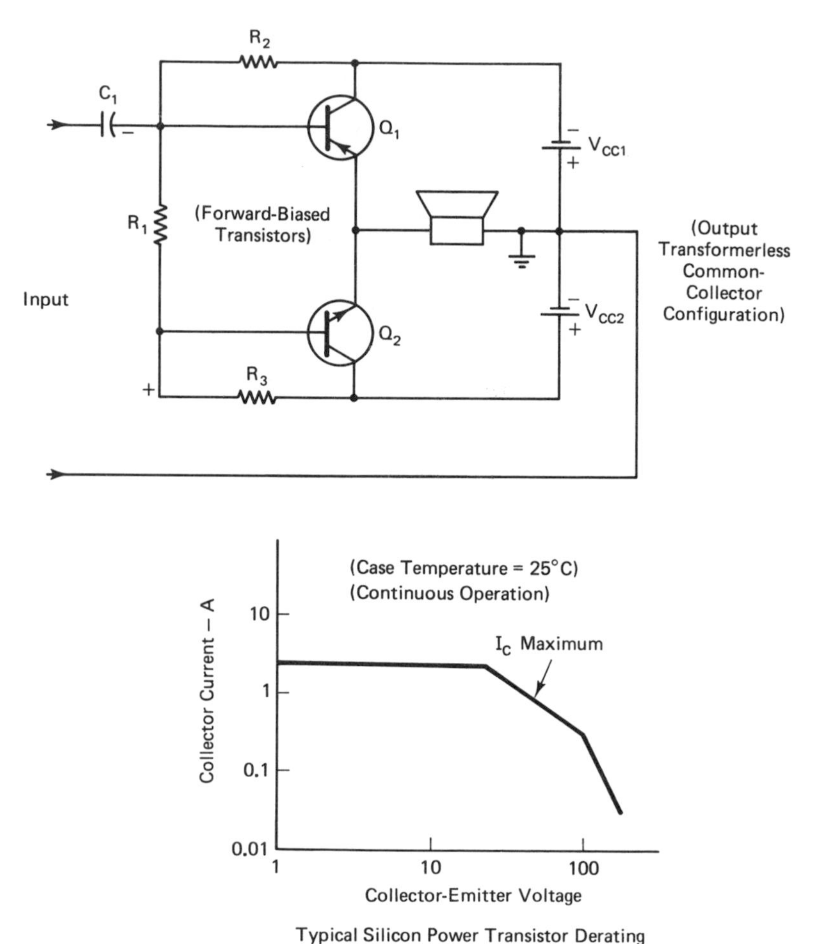

Figure 4–11. Class-AB complementary symmetry power-amplifier configuration.

Note: The boundary limits for safe operation of a power transistor at a specified temperature are determined by the collector-current flow and the collector-emitter voltage. Conservative design restricts the collector-current flow to the continuous-operation region. However, brief pulses of current that exceed the I_C maximum boundary for continuous operation will not damage the transistor.

Compound Connection in Complementary Symmetry

The compound connection in complementary symmetry is exemplified in Figure 4–12. Q1A and Q1B are compound-operated, and Q2A and Q2B are compound-operated. The chief advantage of this arrangement is its high power gain, compared with two-

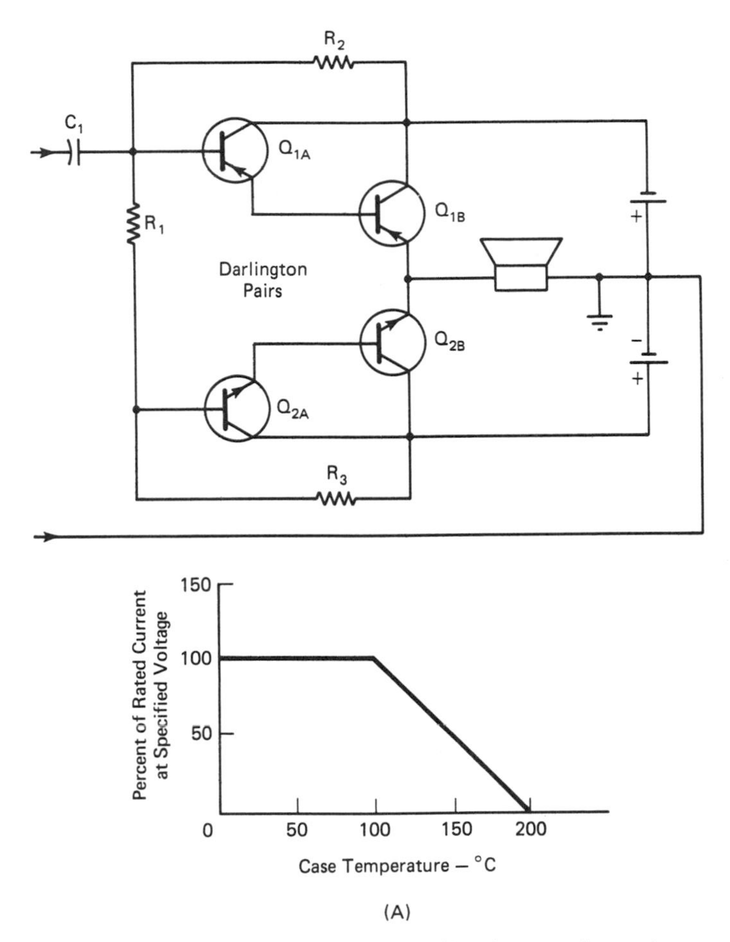

Figure 4–12(A). The basic compound connection in complementary symmetry.

NOTE: The designer should observe rated temperature derating curves, as exemplified above. This thermal (dissipation limited) derating curve is typical; it should be used only as a guide. In the case of a prototype model, the specific temperature derating curve should be followed for the particular transistor that is utilized.

transistor complementary symmetry configuration. Q1A and Q2A operate in the common-collector mode, with the result that the stage has comparatively high input resistance and is easy to drive.

Since PNP power transistors are more costly than NPN power transistors, the designer often prefers the quasi-complementary arrangement shown in Figure 4–13. A low-power PNP transistor is directly coupled to a high-current NPN transistor for simulation of a high-current PNP transistor. Observe that the driver Q1 operates

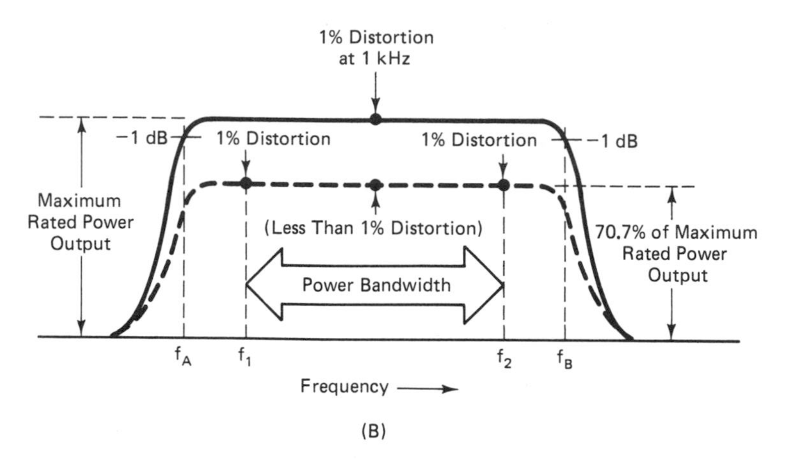

Figure 4–12(B). Distinction between maximum-rated power output and power bandwidth. (Reproduced by special permission of Reston Publishing Company and Robert Russell from *Electronic Troubleshooting with the Oscilloscope.*)

NOTE: This diagram illustrates the distinction between maximum-rated power output and the rated-power bandwidth of an amplifier. In other words, the maximum rated-power output is defined as the power level at which the distortion is 1 percent at 1 kHz. (The distortion will be greater as the low-frequency cutoff and the high-frequency cutoff regions are approached.) At maximum rated power output, the amplifier bandwidth is defined as the frequency span between the −1 dB points on the frequency response curve (f_A and f_B). Next, at a power-output level equal to 70.7 percent of the maximum rated power output, the distortion at 1 kHz will be less than 1 percent. Now, the 1 percent distortion limits occur at frequencies f_1 and f_2. In turn, the power bandwidth of the amplifier is defined as the frequency span between f_1 and f_2. The power bandwidth rating of an amplifier is directed to its realistic limits of high-fidelity performance.

in class A; the other transistors operate in class AB or class B. Q2 conducts on negative half cycles and Q3 conducts on positive half cycles of the drive signal. The outputs from Q2 and Q3 are further amplified by Q4 and Q5. Half-cycle outputs are combined into full-cycle outputs in the speaker.

From the functional viewpoint, half of the supply voltage appears at point $V_{cc}/2$. Bias current is established by Q1. Diodes D1 and D2 set and maintain the idling current in the output circuit with respect to temperature changes. Capacitor C2 functions in a positive-feedback bootstrap circuit to equalize the half cycles of drive voltages. At higher levels of drive signal, the bias on Q3 and Q4 tends to shift from class-AB to class-B operation. This shift is minimized by the large amount of negative feedback provided around the circuit (a loop from the speaker to the base of Q1, not shown in the diagram). Capacitor C1 can be eliminated if additional circuitry is included to cancel dc current flow.

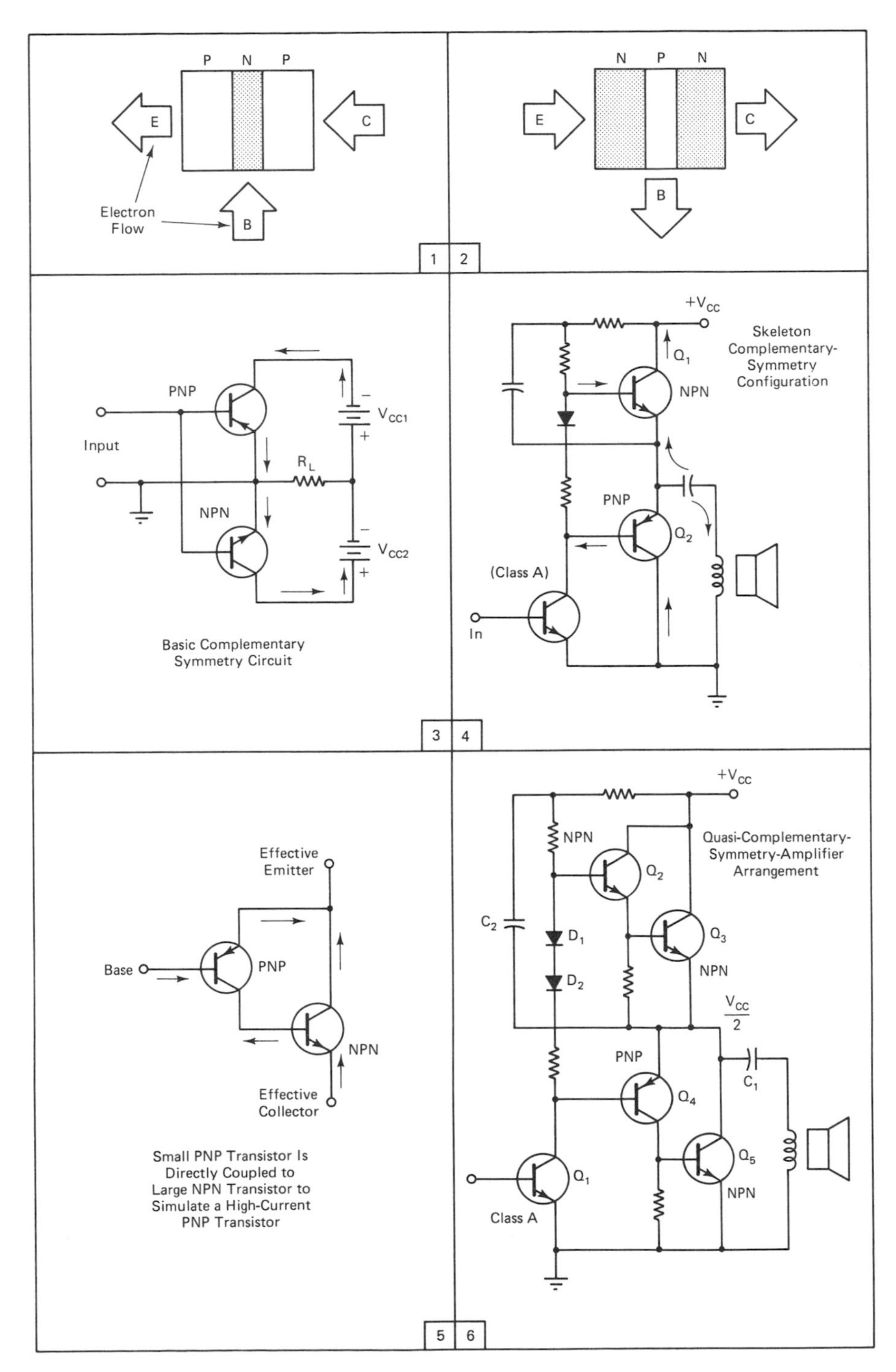

Figure 4–13. Plan of a quasi-complementary output configuration.

HEATSINK REQUIREMENTS

If the maximum rated junction temperature (T_j) is exceeded, a power transistor will be damaged or destroyed. Dissipation of substantial heat requires that the transistor be mounted on a thermally conductive metal plate (heatsink). There are three components in the heat-flow system. The total resistance to heat flow is called the thermal resistance (R_{th}), which is related to the power dissipation of the transistor, and to the difference in temperature between the junction and the environmental (ambient) temperature. This relation is stated by the thermal law:

$$\Delta T = PR_{th}(j-a)$$

where

ΔT_j = $T_j - T_a$ (junction temperature minus ambient temperature Centigrade)

P = maximum power dissipation of the transistor in watts

$R_{th}(j-a)$ = thermal resistance from junction to ambience in °C/W

The parameter ΔT_j in the thermal law is known, being stated in the data sheet for the transistor, and the ambient temperature is readily measurable. P denotes the actual maximum power dissipation of the transistor, and can be evaluated from the circuit arrangement. In a practical situation, $R_{th}(j-a)$ comprises the sum of three thermal resistances from junction to ambience:

$$R_{th_{(j-a)}} = R_{th_{(j-c)}} + R_{th_{(c-s)}} + R_{th_{(s-a)}}$$

where

$R_{th_{(j-c)}}$ is the thermal resistance from the junction to the case

$R_{th_{(c-s)}}$ is the thermal resistance from the case to the heatsink

$R_{th_{(s-a)}}$ is the thermal resistance from the heatsink to ambiance

Next, the material of which the heatsink is fabricated is noted, and the required size of the heatsink is calculated for adequate dissipation of the generated heat. Copper has high thermal conductivity

EXAMPLE: A 2N4347 power transistor operates at a maximum dissipation of 10 watts, and the ambient temperature will not exceed 50°C. The data sheet states that the maximum permissible junction temperature is 200°C. The thermal resistance from junction to case is listed as 1.5°C/W. The thermal resistance from case to heatsink must be assumed; a reasonable value is 0.5°C/W. In turn, the total thermal resistance from junction to ambience is:

$$R_{th(j-a)} = (200°C - 50°C)/10W = 15°C/W$$

Subtracting the known components of the thermal resistance path, the heatsink resistance is:

$$R_{th(s-a)} = 15°C/W - 2°C/W = 13°C/W$$

(R_{th} = 1.38°C/W for SWG 18), compared with aluminum (R_{th} = 1.86°C/W). However, aluminum is less costly and is generally employed. Note that steel has considerably less thermal conductivity (R_{th} = 3.97°C/W). The emissivity of a black anodized heatsink is much better than that of a bright metallic surface. Vertical mounting is more effective than horizontal mounting. Forced-air cooling is used when necessary to supplement convection cooling.

PROGRAMMED FORMULAS

The designer is concerned with the impedance and phase angle of R and C connected in series, or connected in parallel. It can be helpful to have a short computer program available for "punching out" these values, as shown in Figure 4–14. Note that if there is device isolation between RC coupling circuits, their phase shifts are additive.

Note on Integrated Circuits

Integrated circuits are comparatively limited in regard to the audio power input that can be provided. However, when used with adequate heatsinks, suitably designed integrated circuits can supply up to 5 watts rms of audio output. This type of IC is a ruggedized version of the IC described in Chart 3–1 and has a typical rated maximum power dissipation of 3.9 W. The rated total-system harmonic distortion from the 4.5-MHz IF input to the audio output is 1.5 percent in this example.

```
5 LPRINT"RC SERIES AND PARALLEL IMPEDANCE AND PHASE ANGLE"
10 LPRINT "Series Z and Parallel Z Polar and Rectangular"
15 LPRINT"..............................................."
20 INPUT"C (Mfd)=";C
25 LPRINT"C=";C
30 INPUT"R (Ohms)";R
35 LPRINT"R=";R
40 INPUT"f (Hz)=";F
45 LPRINT"f=";F:LPRINT"":LPRINT"(Series Parameters)"
50 LPRINT"RS=";R
55 XC=1/(6.283*F*C*10^-6)
60 A$="#########.#":LPRINT"-jXC=";USING A$;-XC
65 ZS=(R*R+XC*XC)^.5
70 LPRINT "ZS=";USING A$;ZS
75 TH=-ATN(XC/R):PH=-ATN(R/XC)
80 LPRINT "Series Phase Angle=";USING A$;TH*360/6.283
85 LPRINT"":LPRINT"(Parallel Parameters)"
90 ZP=(R*XC)/((R*R+XC*XC)^.5)
95 RP=ZP*COS(PH):XP=ZP*SIN(PH)
100 LPRINT"RP=";USING A$;RP
105 LPRINT"-jXP=";USING A$;XP
110 LPRINT"ZP=";USING A$;ZP
115 LPRINT"Parallel Phase Angle=";USING A$;PH*360/6.283
```

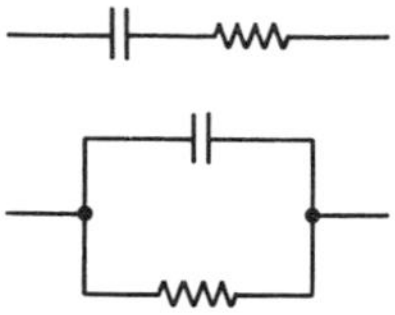

```
RC SERIES AND PARALLEL IMPEDANCE AND PHASE ANGLE
Series Z and Parallel Z Polar and Rectangular
...............................................
C= 5
R= 500
f= 400

(Series Parameters)                          (Parallel Parameters)
RS= 500                                      RP=          12.4
-jXC=         -79.6                          -jXP=        -77.6
ZS=         506.3                            ZP=          78.6
Series Phase Angle=          -9.0            Parallel Phase Angle=          -81.0
```

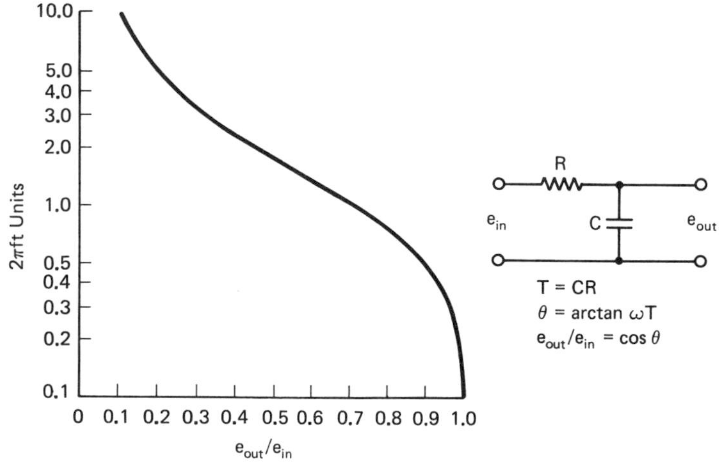

R
e_{in}
C
e_{out}
T = CR
θ = arctan ωT
e_{out}/e_{in} = cos θ

2πft Units
10.0
5.0
4.0
3.0
2.0
1.0
0.5
0.4
0.3
0.2
0.1
0 0.1 0.2 0.3 0.4 0.5 0.6 0.7 0.8 0.9 1.0
e_{out}/e_{in}

(A)

Figure 4–14(A). An R-C series and parallel impedance and phase angle program.

```
5 LPRINT"HALF-POWER LOW AND HIGH CUTOFF FREQUENCIES"
10 LPRINT "C in Series With Source and Sink":LPRINT "C in Shunt to Source and Si
nk":LPRINT"..............................."
15 LPRINT""
20 INPUT"C (Mfd)=";A
25 LPRINT"C (Mfd)=";A
30 INPUT"Load R (Ohms)=";B
35 LPRINT"Load R (Ohms)=";B
40 INPUT"Gen R (Ohms)=";C
45 LPRINT"Gen R (Ohms)=";C:LPRINT ""
50 LPRINT"(For R and C in Series)":LPRINT ""
55 D=1/(6.283*A*10^-6*(B+C))
60 A$="##########.#":LPRINT"Low P/2 f (Hz)=";USING A$;D
65 LPRINT "Phase Angle=45 Degrees":LPRINT ""
70 LPRINT "(For R and C in Parallel)":LPRINT "":PRINT"(Enter RUN 75)":END
75 INPUT "C (Mfd)=";E
80 LPRINT "C (Mfd)=";E
85 INPUT"Load R (Ohms)=";F
90 LPRINT "Load R (Ohms)=";F
95 INPUT "Gen R (Ohms)=";G
100 LPRINT"Gen R (Ohms)=";G
105 H=(F+G)/(6.283*E*10^-6*F*G):LPRINT ""
110 A$="##########.#":LPRINT"High P/2 f (Hz)=";USING A$;H
115 LPRINT"Phase Angle=45 Degrees"
```

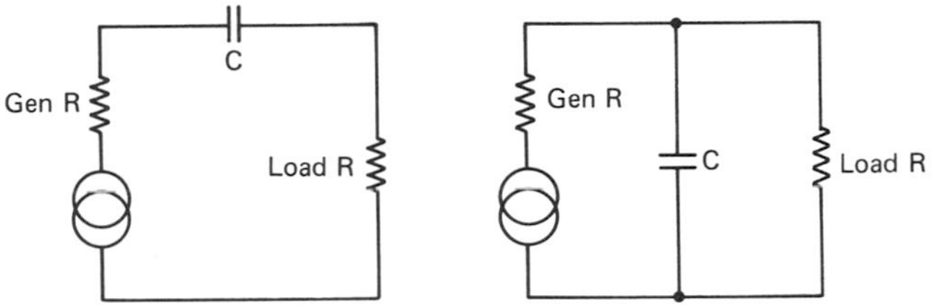

(B)

Note: The half-power frequency occurs at the point where the capacitive reactance equals the effective shunt resistance. At the half-power frequency the output voltage is −3 dB down, and the phase angle is 45°. Observe that the low P/2 frequency is determined by the coupling capacitance, and the high P/2 frequency is determined by junction capacitances, stray capacitance, and so forth.

```
HALF-POWER LOW AND HIGH CUTOFF FREQUENCIES
C in Series With Source and Sink
C in Shunt to Source and Sink
...............................

C (Mfd)= 1                              C (Mfd)= .0001
Load R (Ohms)= 10000                    Load R (Ohms)= 10000
Gen R (Ohms)= 20000                     Gen R (Ohms)= 20000

(For R and C in Series)                 High P/2 f (Hz)=    238739.5
                                        Phase Angle=45 Degrees
Low P/2 f (Hz)=         5.3
Phase Angle=45 Degrees

(For R and C in Parallel)
```

Figure 4–14(B). A half-power high- and low-cutoff frequency program.

```
5 LPRINT"HALF-POWER LOW AND HIGH CUTOFF FREQUENCIES"
10 LPRINT"C in Series With Source and Sink":LPRINT"(Survey Program)":LPRINT""
15 LPRINT"Load Incremented in 500-Ohm Steps":LPRINT ""
20 INPUT"C (Mfd)=";A
25 LPRINT"C (Mfd)=";A
30 INPUT"Gen R (Ohms)=";C
35 LPRINT"Gen R (Ohms)=";C:LPRINT "":LPRINT"...........................
......":LPRINT""
40 N=0
45 N=N+500:B=N
50 D=1/(6.283*A*10^-6*(B+C)):LPRINT ""
55 A$="######"
60 LPRINT"Load R (Ohms)=";USING A$;N
65 B$="########"
70 LPRINT"Low P/2 f (Hz)=";USING B$;D
75 IF N>10000 THEN 85
80 GOTO 45
85 END
```

C

Gen R

Load R

NOTE: Although an ac generator with internal resistance (Gen R) is shown in the equivalent circuit, the generator may be a driver transistor, and Gen R may be the output resistance of the driver transistor; in this case, Load R is the input resistance to the output stage.

This program starts with R = 500 ohms and proceeds indefinitely in increments of 500 ohms. The operator may terminate the processing at any point by pressing the BREAK key. Observe that line 35 may be written for 100-ohm or 1000-ohm increments, if desired.

```
HALF-POWER LOW AND HIGH CUTOFF FREQUENCIES
C in Series With Source and Sink
(Survey Program)

Load Incremented in 500-Ohm Steps

C (Mfd)= 1
Gen R (Ohms)= 1000

.....................................


Load R (Ohms)=    500
Low P/2 f (Hz)=       106

Load R (Ohms)=   1000
Low P/2 f (Hz)=        80

Load R (Ohms)=   1500
Low P/2 f (Hz)=        64

Load R (Ohms)=   2000
Low P/2 f (Hz)=        53

Load R (Ohms)=   2500
Low P/2 f (Hz)=        45

Load R (Ohms)=   3000
Low P/2 f (Hz)=        40

Load R (Ohms)=   3500
Low P/2 f (Hz)=        35

Load R (Ohms)=   4000
Low P/2 f (Hz)=        32

Load R (Ohms)=   4500
Low P/2 f (Hz)=        29

Load R (Ohms)=   5000
Low P/2 f (Hz)=        27

Load R (Ohms)=   5500
Low P/2 f (Hz)=        24

Load R (Ohms)=   6000
Low P/2 f (Hz)=        23
```

```
Load R (Ohms)=  6500
Low P/2 f (Hz)=        21

Load R (Ohms)=  7000
Low P/2 f (Hz)=        20

Load R (Ohms)=  7500
Low P/2 f (Hz)=        19

Load R (Ohms)=  8000
Low P/2 f (Hz)=        18

Load R (Ohms)=  8500
Low P/2 f (Hz)=        17

Load R (Ohms)=  9000
Low P/2 f (Hz)=        16

Load R (Ohms)=  9500
Low P/2 f (Hz)=        15

Load R (Ohms)= 10000
Low P/2 f (Hz)=        14

Load R (Ohms)= 10500
Low P/2 f (Hz)=        14
```

Figure 4–14(C). A half-power low-cutoff frequency-survey program.

```
5 LPRINT"HALF-POWER LOW AND HIGH CUTOFF FREQUENCIES"
10 LPRINT"C in Series With Source & Sink":LPRINT"(Heuristic Routine)":LPRINT""
15 LPRINT"........................................"
20 INPUT"C (Mfd)=";A
25 LPRINT"C (Mfd)=";A
30 INPUT"Gen R (Ohms)=";C
35 LPRINT"Gen R (Ohms)=";C
40 INPUT"Maximum Load R (Ohms)=";E
45 LPRINT"Maximum Load R (Ohms)=";E
50 INPUT"Maximum Cutoff F (Hz)=";F
55 LPRINT"Maximum Cutoff F (Hz)=";F
60 N=0
65 N=N+500:B=N
70 D=1/(6.283*A*10^-6*(B+C))
75 IF (N<E)*(D<F) THEN 90
80 IF N>E THEN 100
85 GOTO 65
90 A$="#######":LPRINT"Load R (Ohms)=";N
95 LPRINT"Low P/2 f (Hz)=";USING A$;D:END
100 LPRINT"VALUE OF C IS OUT OF RANGE; ENTER ANOTHER C VALUE":END
```

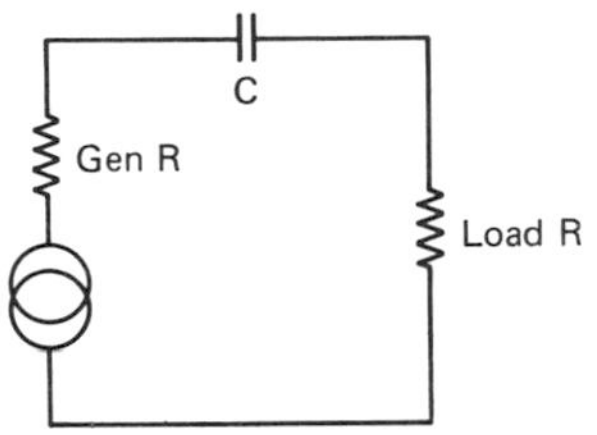

C

Gen R

Load R

Note: Observe in this example that feasible limits are assigned to Maximum Load R and Maximum Cutoff F. In turn, the solution is printed out. On the other hand, if the assigned limits do not permit a solution, the routine will end with an "out of range" message.

```
HALF-POWER LOW AND HIGH CUTOFF FREQUENCIES
C in Series With Source & Sink
(Heuristic Routine)

........................................
C (Mfd)= 1
Gen R (Ohms)= 1000
Maximum Load R (Ohms)= 4000
Maximum Cutoff F (Hz)= 40
Load R (Ohms)= 3000
Low P/2 f (Hz)=      40
```

Figure 4–14(D). A half-power low-cutoff frequency-heuristic routine.

```
5 LPRINT"HALF-POWER HIGH CUTOFF FREQUENCY"
10 LPRINT"C in Shunt to Source and Sink":LPRINT"(Survey Program)"
15 LPRINT"............................":LPRINT ""
20 INPUT"C (Mfd)=";C
25 LPRINT"C (Mfd)=";C
30 INPUT"Gen R (Ohms)=";RG
35 LPRINT"Gen R (Ohms)=";RG
40 LPRINT""
45 N=0
50 N=N+200:RL=N
55 F=(RL+RG)/(6.283*C*10^-6*RL*RG)
60 A$="############":LPRINT"Load R (Ohms)=";USING A$;N
65 LPRINT"High P/2 F (Hz)=";USING A$;F:LPRINT ""
70 IF N>2000 THEN 80
75 GOTO 50
80 END
```

```
HALF-POWER HIGH CUTOFF FREQUENCY
C in Shunt to Source and Sink
(Survey Program)
............................

C (Mfd)= .1
Gen R (Ohms)= 1000

Load R (Ohms)=          200
High P/2 F (Hz)=         9550

Load R (Ohms)=          400
High P/2 F (Hz)=         5571

Load R (Ohms)=          600
High P/2 F (Hz)=         4244

Load R (Ohms)=          800
High P/2 F (Hz)=         3581

Load R (Ohms)=         1000
High P/2 F (Hz)=         3183

Load R (Ohms)=         1200
High P/2 F (Hz)=         2918

Load R (Ohms)=         1400
High P/2 F (Hz)=         2728

Load R (Ohms)=         1600
High P/2 F (Hz)=         2586

Load R (Ohms)=         1800
High P/2 F (Hz)=         2476

Load R (Ohms)=         2000
High P/2 F (Hz)=         2387

Load R (Ohms)=         2200
High P/2 F (Hz)=         2315
```

Figure 4–14(E). A half-power high-cutoff frequency-survey program.

As noted in Chart 3–1, the designer must consider worst-case conditions, including the maximum ambient temperature rating. Note that a power-type IC is occasionally designed with automatic shutdown of the audio section to guard against excessive dissipation in the output circuit. As a practical note, accidental excessive-power dissipation could be confused with intermittent operation, inasmuch as normal operation resumes when the IC cools below the preset threshold temperature.

DESIGN OPTIMIZATION CONSIDERATIONS

A comparatively large number of options are generally available when designing an RC-coupled stage to drive a following RC-coupled stage. In many practical situations, the driven stage has high-input impedance so that the driver does not require substantial driving power or low output impedance. In turn, various driver arrangements provide viable design options, and the most economical option will usually be preferred.

The dominant consideration in design of an output stage is maximized power output with minimum distortion. A particular type of transistor has a rated power-dissipation capability that in turn establishes the obtainable output power. A particular type of transistor will also have rated characteristics that determine the minimum obtainable distortion. For example, distortion is a function of the h_{fe} degradation at high collector currents (alpha crowding). Distortion is also a function of the input-nonlinearity characteristics. Therefore, the designer should review the available types of transistors and determine their compatibilities with circuit parameters.

As previously noted, distortion in the output stage can be controlled by means of sufficient negative feedback to the driver input (provided that clipping distortion does not occur in the output transistors). Negative feedback over several stages entails incidental phase shifts in both the low-frequency cutoff region and the high-frequency cutoff region. Improper design of a negative-feedback loop over several stages may result in cumulative phase shifts that change negative feedback into positive feedback at a critical frequency. In turn, the amplifier system "takes off" and oscillates.

Unless the output transistors are operated in class AB, more or less crossover distortion will occur due to barrier potential. It was previously noted that h_{fe} suddenly falls to zero during the crossover-distortion interval. This is just another way of saying that the system gain drops to zero during this interval, with the practical result that crossover distortion cannot be reduced by negative-feedback action. The bottom line is that the designer should choose class-AB operation whenever low distortion is required. Class-B operation, with its advantage of a bit higher efficiency, is feasible in design of utility amplifiers.

Note that when substantial negative feedback is employed over several stages, the cutoff frequencies of the transistors contribute to the cumulative phase shift. As a practical rule of thumb observed by most designers, a negative-feedback loop should have reasonably small phase shift beyond the rated amplifier-frequency range. Thus, for each 10 dB of active negative feedback, comparatively little phase shift should occur for one octave beyond the amplifier-frequency range. Conservative design practice stipulates that the loop characteristic be optimized for an additional octave or two.

Optimization of output-stage design requires that temperature compensation and supplementary factors will ensure that the maximum current rating of the transistors cannot be exceeded under any condition of operation. Otherwise a "vicious circle" condition called thermal runaway may develop in which the maximum rated operating temperature of the transistors is exceeded. Thermal runaway results in either catastrophic destruction or in permanent changes in h_{fe} and leakage current values. Fuses provide supplementary insurance against the possibility of thermal runaway.

Unsuspected positive feedback due to stray coupling is a pitfall awaiting the unwary designer of high-gain systems. For this reason, the finalized prototype model should include the precise component and device placement and the precise pattern of conductor arrangement contemplated for the production run. It is evident that the preamplifier stage should be located at the maximum feasible distance from the power-output stage. In the case of miniaturized equipment, a judiciously placed shield plate can occasionally "save the day."

It is good practice—even essential—to thoroughly check the worst-case prototype models using the same power supply that will be utilized in production. In other words, a bench power supply can have considerably lower internal impedance than a consumer-product power supply. In turn, unsuspected positive feedback may occur when the high-gain system is operated from the production power supply. Tolerances on R and C values in the decoupling circuits of the high-gain system demand attention, inasmuch as no decoupling circuit is ideal, and unsuspected low-frequency phase shifts may interact to produce objectionable positive feedback at certain combinations of tolerance values.

5

TUNED AMPLIFIER DESIGN PROCEDURES

GENERAL CONSIDERATIONS

Narrow-band-tuned amplifiers are discussed in this chapter. Stated otherwise, the bandwidth of the tuned coupling circuits is a small percentage of the center frequency. Amplifier selectivity is obtained by means of parallel tuned interstage coupling networks. Series-resonant circuits are rarely employed because they are not as well adapted to impedance transformation as parallel-resonant circuits. From the viewpoint of practical design procedures, the essential properties of parallel-resonant circuits are:

1. The resonant frequency is stipulated to be

$$f_r = 1/(2\,\pi\sqrt{LC})$$

2. With reference to Figure 5–1, for a given resonant frequency, the LC product is a related constant.

3. It is stipulated that the resonant frequency is independent of the circuit resistance.

4. The Q value of a parallel-resonant circuit is the ratio of current in the tank (I_L or I_C) to the line current. This is just another way of saying that current magnification occurs at resonance, and that the circulating current in the LC loop is Q times the line current.

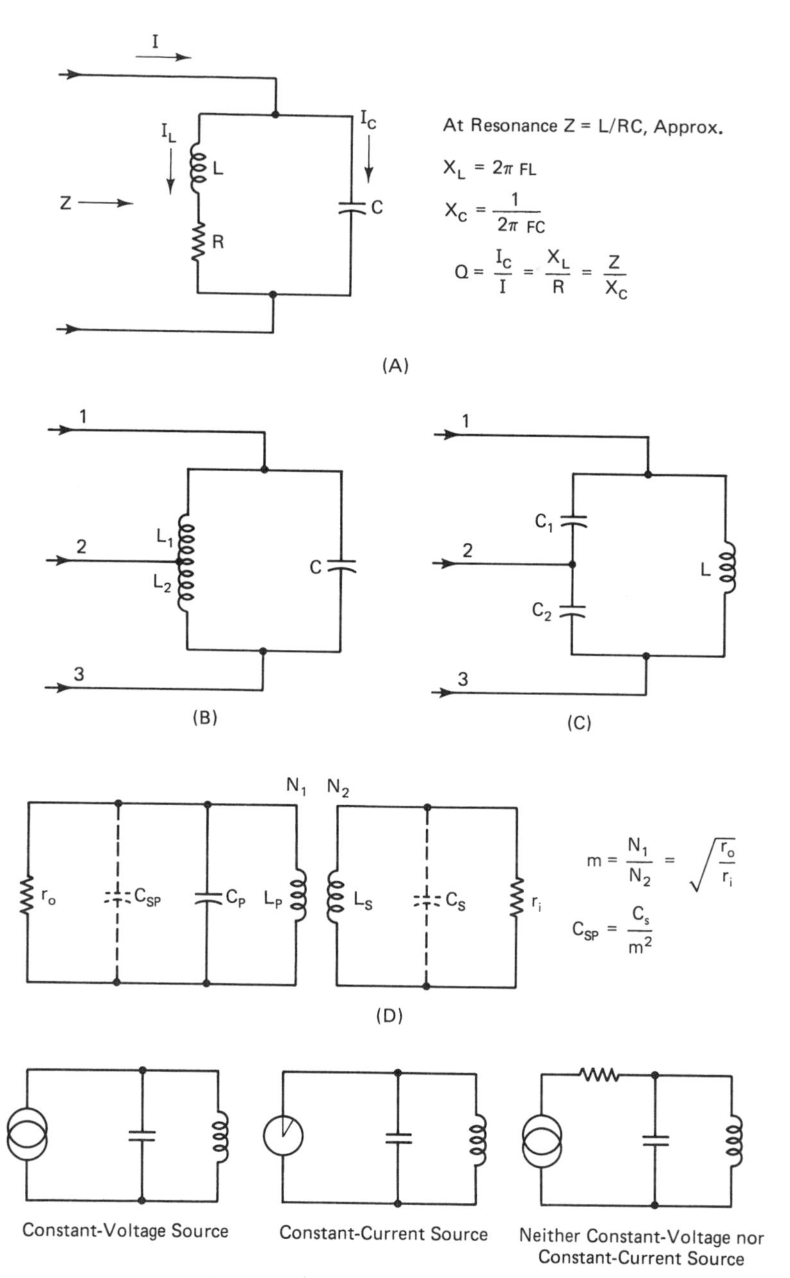

Figure 5–1. Basic parallel-tuned circuit arrangements.

NOTE: Observe in (B) that if the inductor is center-tapped, the input impedance from 2 to 3 is half of the input impedance from 1 to 3. The mutual inductance of L1 and L2 is ¼ of the total inductance.

5. The Q value is approximately related to the circuit bandwidth, as seen in Figure 5–2. In other words, the bandwidth of a parallel-resonant circuit is approximately equal to the resonant frequency divided by the Q value.

6. Observe in Figure 5–1 that the resonant frequency and the unloaded Q of a parallel-resonant circuit remain the same whether the signal is inputted across terminals 1 and 3, or across terminals 2 and 3. (However, the input impedance between 2 and 3 is much smaller than the input impedance between 1 and 3.)

7. Input impedance depends on the square of the turns ratio between the signal-input terminals. For example, if point 2 is the center tap on the inductor, the input impedance between points 1 and 3 is 4 times the impedance between points 2 and 3.

8. With reference to Figure 5–1(C), the resonant frequency and the unloaded Q of the parallel-resonant circuit remain the same whether the signal is inputted across 1 and 3 or 2 and 3. (However, the 2-to-3 input impedance is much smaller than the 1-to-3 input impedance).

9. With reference to Figure 5–1(D), in many applications we will find a tuned resonant circuit (capacitor C_p and inductor L_p) in the primary of a transformer D, coupled to a nonresonant secondary winding. In this arrangement, if N_1 denotes the number of turns on the primary, and N_2 denotes the number of turns on the secondary, the turns ratio (m) of primary to secondary under matched conditions is as formulated in the diagram.

10. If there is capacitance (C_s) in the secondary circuit, it will be reflected (referred) into the primary circuit by transformer action as a capacitance (C_{sp}) with the value formulated in the diagram.

A tuned circuit does not "stand alone" in an amplifier network. Practical sources have finite internal impedance, and practical sinks have finite impedance values. As depicted in Figure 5–1, an ideal parallel-resonant circuit may be analyzed from the viewpoint of a constant-voltage source, of a constant-current source, or of a source that is intermediate between constant-voltage and constant-current characteristics. Note the following points:

1. If driven by a constant-voltage source, the voltage across the parallel LC circuit is constant, regardless of frequency. However, the current demand of C and of L is a function of frequency; the capacitor current approaches infinity as the frequency approaches infinity; the inductor current approaches infinity as the frequency approaches zero.

2. If driven by a constant-current source, the line current is constant, regardless of frequency. However, the voltage across the LC circuit is maximum at resonance; this voltage approaches zero as the frequency approaches infinity, or as the frequency approaches zero.

3. If driven by a voltage source that has internal resistance (the usual practical design situation), the voltage across the LC circuit rises to a peak at resonance, and the circuit bandwidth decreases as the internal resistance of the source increases.

TRANSISTOR AND COUPLING NETWORK IMPEDANCES

With reference to Figure 5–2, the output impedance of a transistor can be considered as a resistance (r_o) in parallel with a capacitance (C_o). Similarly, the input impedance of a transistor can be considered as a resistance (r_i) in parallel with a capacitance (C_i). In routine design procedures, output capacitance C_o and input capacitance C_i are accounted for by considering them as components of the coupling network (Figure 5–2(B)).

As a practical example, assume that the required capacitance between terminals 1 and 2 of the coupling network is determined to be 500 pF. Assume also that capacitance C_0 is 10 pF. In turn, a capacitor with a value of 490 pF would be connected between terminals 1 and 2 so that the total capacitance would be 500 pF. This same principle is employed to compensate for capacitance C_i between terminals 3 and 4.

Maximum power transfer requirements were previously noted. To transfer maximum power from Q1 to Q2, the input impedance between terminals 1 and 2 to the coupling network must be equal to r_o, and the output impedance of the coupling network (looking into terminals 3 and 4) must be equal to r_i.

TRANSFORMER COUPLING WITH TUNED PRIMARY

A simple tuned coupling-transformer arrangement is shown in Figure 5–3. Note the following points:

1. Capacitance C_T includes the output capacitance of transistor Q1 and the input capacitance of Q2 (referred to the primary of T1). The output impedance of Q1 is matched to the input impedance of Q2 by means of an appropriate turns ratio for primary and secondary of T1.

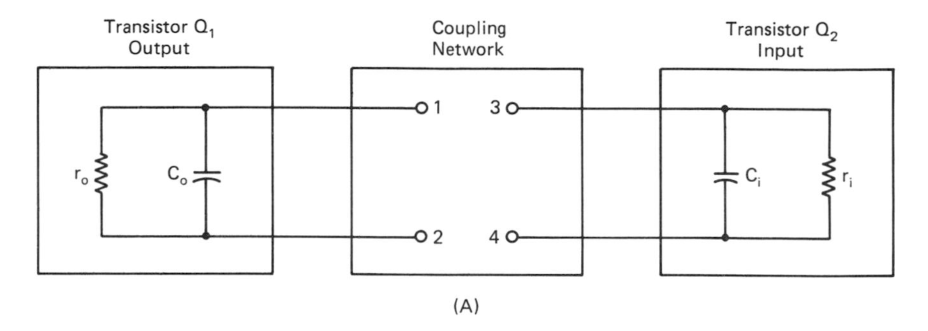

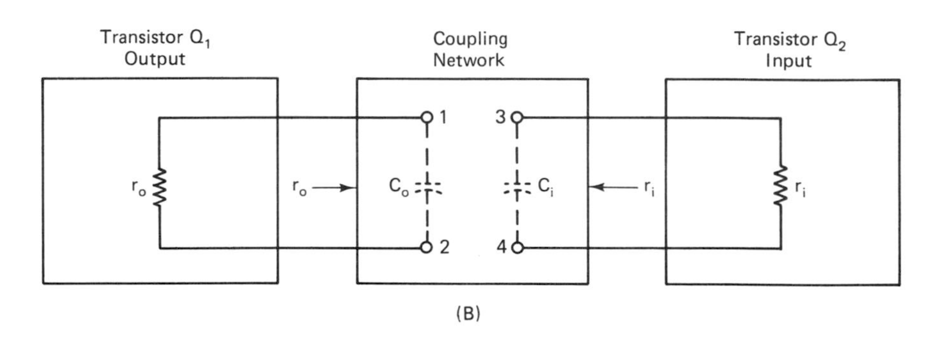

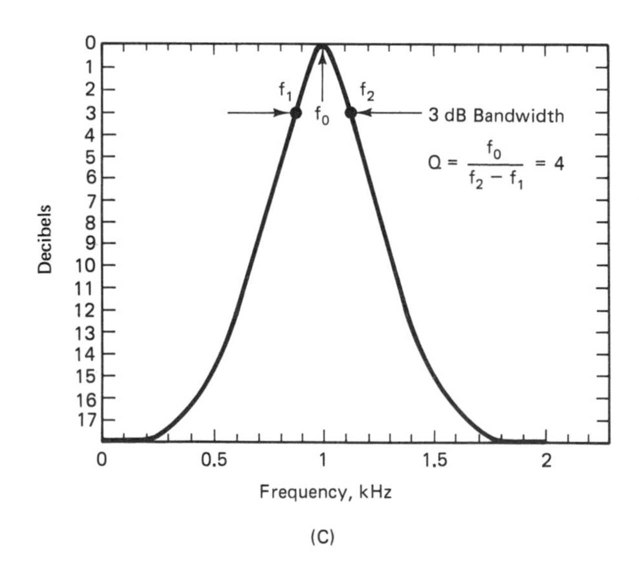

Figure 5–2. Equivalent output and input circuits for transistors with a coupling network. (See also Chart 5–1.)

NOTE: The Q of a resonant circuit is approximately equal to its resonant frequency divided by the number of Hz between the half-power frequencies (3 dB down) on its frequency response curve.

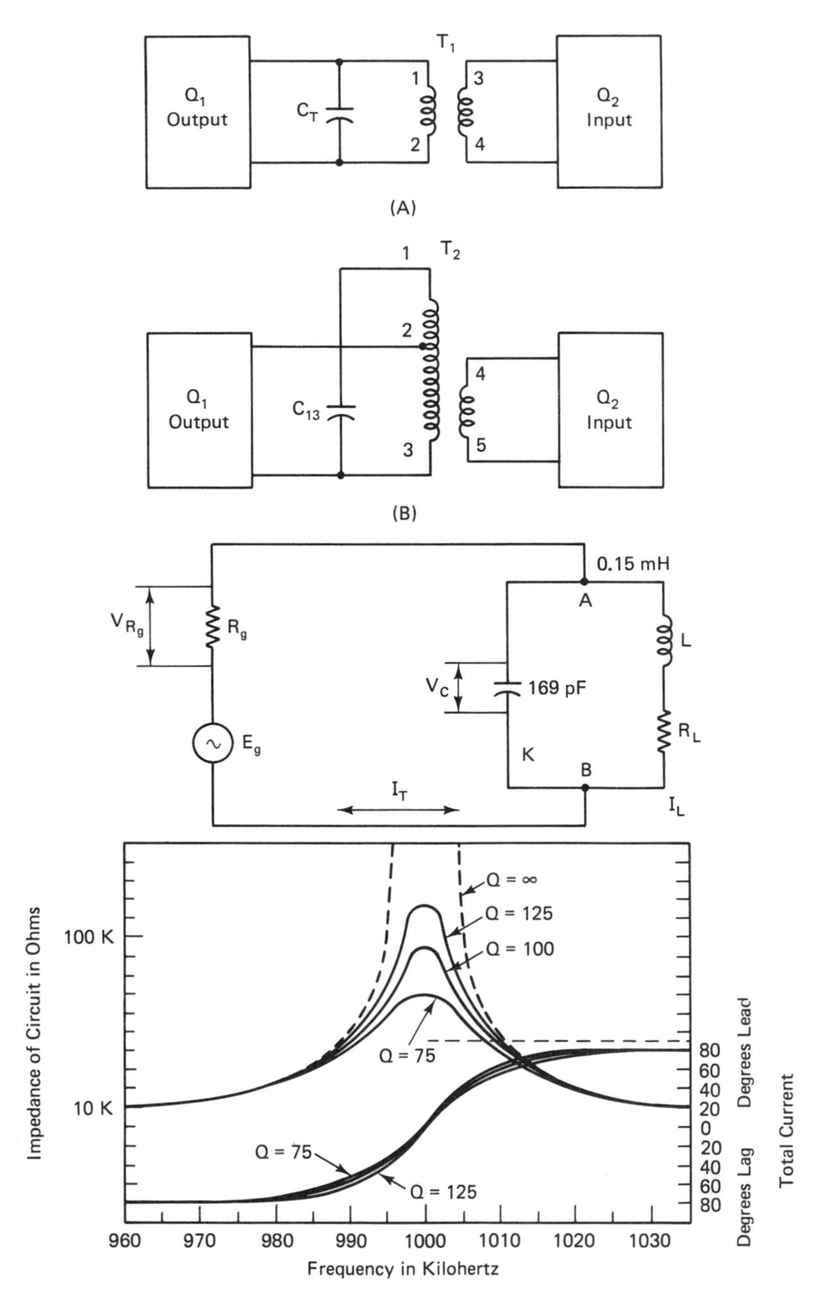

Figure 5–3. Interstage coupling transformer with tuned primary. (From *Electronics Handbook*, Goodyear Publishing Company)

NOTE: The typical insertion loss of the coupling transformer will be approximately 3 dB.

2. If L_P denotes the inductive reactance between terminals 1 and 2 of T1, then the resonant frequency (f_r) for the circuit is equal to:*

$$f_r = 1/(2\ \pi\ \Lambda \sqrt{L_p C_T})$$

Application

As a typical design application, assume that it is specified to use the coupling network of Figure 5–3(A) in an intermediate frequency amplifier employing two CE configurations. The available data is:

Resonant frequency f_r = 500 kHz
Frequency bandwidth Δf = 10 kHz
Q1 output resistance r_o = 12,000 ohms
Q1 output capacitance C_o = 10 pF
Q2 input resistance r_i = 700 ohms
Q2 input capacitance C_i = 170 pF
Coupling network loss = 3 dB
Power response at extremes of bandwidth ($f_r \pm \frac{1}{2}\Delta f$) = ¼ of response at f_r.

1. With the listed data and requirements, it can be shown that the unloaded Q of the tuned circuit before connecting the amplifier must be 288. An appropriate primary inductance (L_P) is 7.7μH. In turn, the secondary inductance must be 0.45μH. Correspondingly, C_T must have a value of 13,100 pF (this includes C_o = 10 pF, and the effective C_i referred to the primary, which is 10 pF). In practice, it is very difficult to construct a transformer with such high Q (288) and a very low inductance (7.7μH).

2. Accordingly, the design problem is relaxed by using a tapped primary winding, as seen in Figure 5–3(B). Thus, the inductance of the primary may be many times the value of inductance here calculated. As an illustration, inductance ($L_{1\text{-}3}$) between terminals 1 and 3 can be made 100 times the calculated inductance (L_p) for terminals 1 and 2 (Figure 5–3(A)). Thus:

$$\begin{aligned} L_{1\text{-}3} &= 100\ L_p \\ &= 100 \times 7.7 \\ &= 770\mu H \end{aligned}$$

*The Hewlett-Packard "Vector Impedance Calculator" provides convenience in solving resonant circuit problems.

3. To maintain the same resonant frequency, the capacitance between terminals 1 and 3 ($C_{1\text{-}3}$) must be 100 times less than the calculated capacitance (C_R):

$$\begin{aligned} C_1 - {}_3 &= C_T/100 \\ &= 13{,}100/100 \\ &= 131 \text{ pF} \end{aligned}$$

4. To maintain impedance matching for maximum energy transfer, the inductance between terminals 2 and 3 of T2 must equal the calculated inductance (L_p) between terminals 1 and 2 of T1, which in this case is 7.7μH.

AUTOTRANSFORMER COUPLING WITH TUNED PRIMARY

Inductive coupling from the output of one transistor to the input of another transistor may be provided by an autotransformer, as shown in Figure 5–4. The analysis of autotransformer action is the same as that for a transformer with separate primary and secondary windings. Capacitance C_T includes the output capacitance of Q1 and the input capacitance of Q2 referred to the primary. If $L_{1\text{-}3}$ denotes the inductance from terminal 1 to terminal 3 of T1, then the resonant frequency (f_r) of the arrangement is:

$$f_r = 1/(2\pi\sqrt{L_{1\text{-}3}C_r})$$

Note that the tap at terminal 2 is located on the winding at a point that provides impedance matching between Q1 and Q2. If, in a particular application, the inductance between terminals 1 and 3 is too small to provide the specified frequency selectivity (Q_o), the total primary inductance can be increased many times by means of the arrangement shown in Figure 5–4(B).

To maintain the same resonant frequency in this case, $C_{1\text{-}4}$ must be reduced by the same factor, so that the product of $L_{1\text{-}3}$ and C_T in (A) equals the product of $L_{1\text{-}4}$ and $C_{1\text{-}4}$ in (B). Impedance matching is maintained provided that inductances $L_{2\text{-}4}$ and $L_{3\text{-}4}$ of T2 equal inductances $L_{1\text{-}3}$ and $L_{2\text{-}3}$, respectively, of T1.

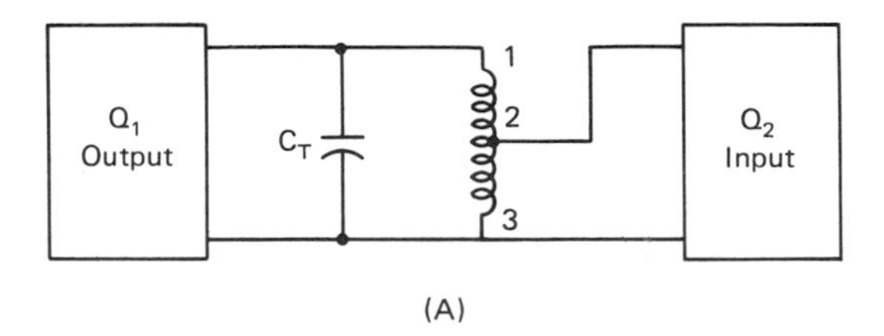

(A)

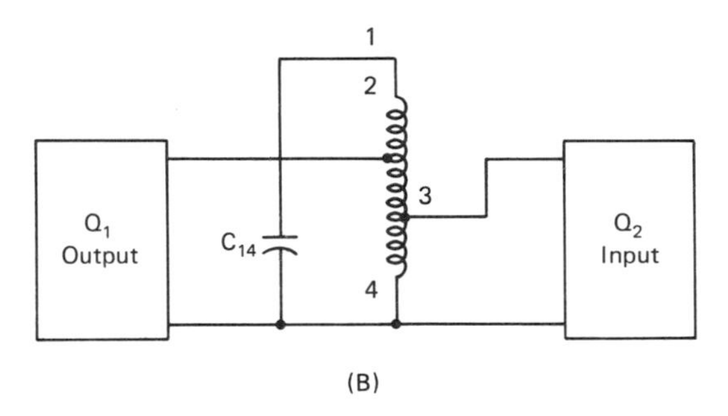

(B)

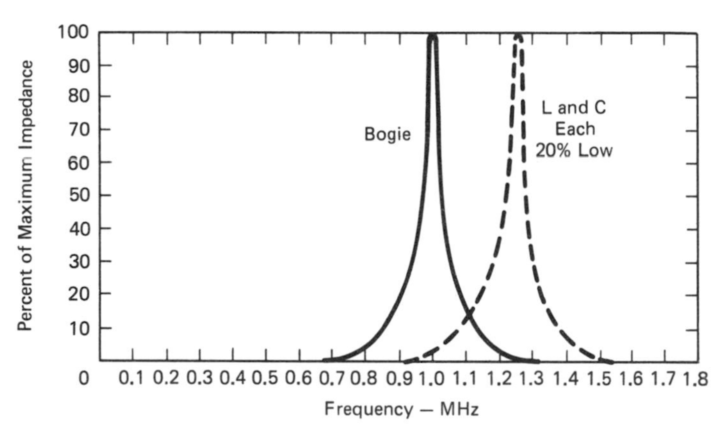

Example of Error in Resonant Frequency Due to
20 Percent Tolerance on L and C Values

Figure 5–4. Interstage coupling autotransformer with tuned primary.

NOTE: Impedance matching is obtained by tapping the Q2 input down from the top of the autotransformer. A higher Q value can be obtained by tapping the Q1 output down from the top of the winding, and adjusting C14 as required.

Computer programs for calculating the parameters of tapped tank circuits are provided in Chapter 10.

CAPACITANCE COUPLING

In many design situations, transformer coupling is not desirable due to the small number of turns required in the secondary winding. Unity coupling is sought, but can be difficult to obtain. (This problem is particularly severe in CB VHF amplifiers; such amplifiers have an input impedance of less than 75 ohms.) In such situations, capacitance coupling may be employed, as depicted in Figure 5–5(A).

1. Impedance matching of the output resistance of Q1 to the input resistance of Q2 is obtained by selecting the appropriate ratio of C1 to C2. Capacitance C2 is usually much larger than capacitance C1. Their reactance values have the opposite relationship to inductive reactance values. The total capacitance (C_T) across L1 is formulated:

$$C_T = C1C2/(C1 + C2)$$

The resonant frequency is:

$$f_r = 1/(2\pi\sqrt{L_1C_T})$$

In many applications (particularly in CB configurations), C2 is omitted inasmuch as the low input resistance effectively shunts it out of the circuit. In this case, the resonant frequency is formulated:

$$f_r = 1/(2\pi\sqrt{L_1C_1})$$

2. If inductance L1 is too small to provide the specified selectivity (Q_o), the circuit shown in Figure 5–5(B) can be utilized. The inductance can be increased by any factor and the total capacitance (C_T) reduced by the same factor to maintain the same resonant frequency. A matched condition is maintained by making the inductance between terminals 2 and 3 equal to L1, and by making the ratio of C3 to C4 equal to the ratio of C1 to C2.

3. Another arrangemennt using capacitance coupling is shown in Figure 5–5(C). The total capacitance in the circuit is:

$$C_T = C1 + C2Ci/(C2 + Ci)$$

In VHF CB amplifiers, Ci is effectively shunted by the low input resistance, and the total capacitance C_T is:

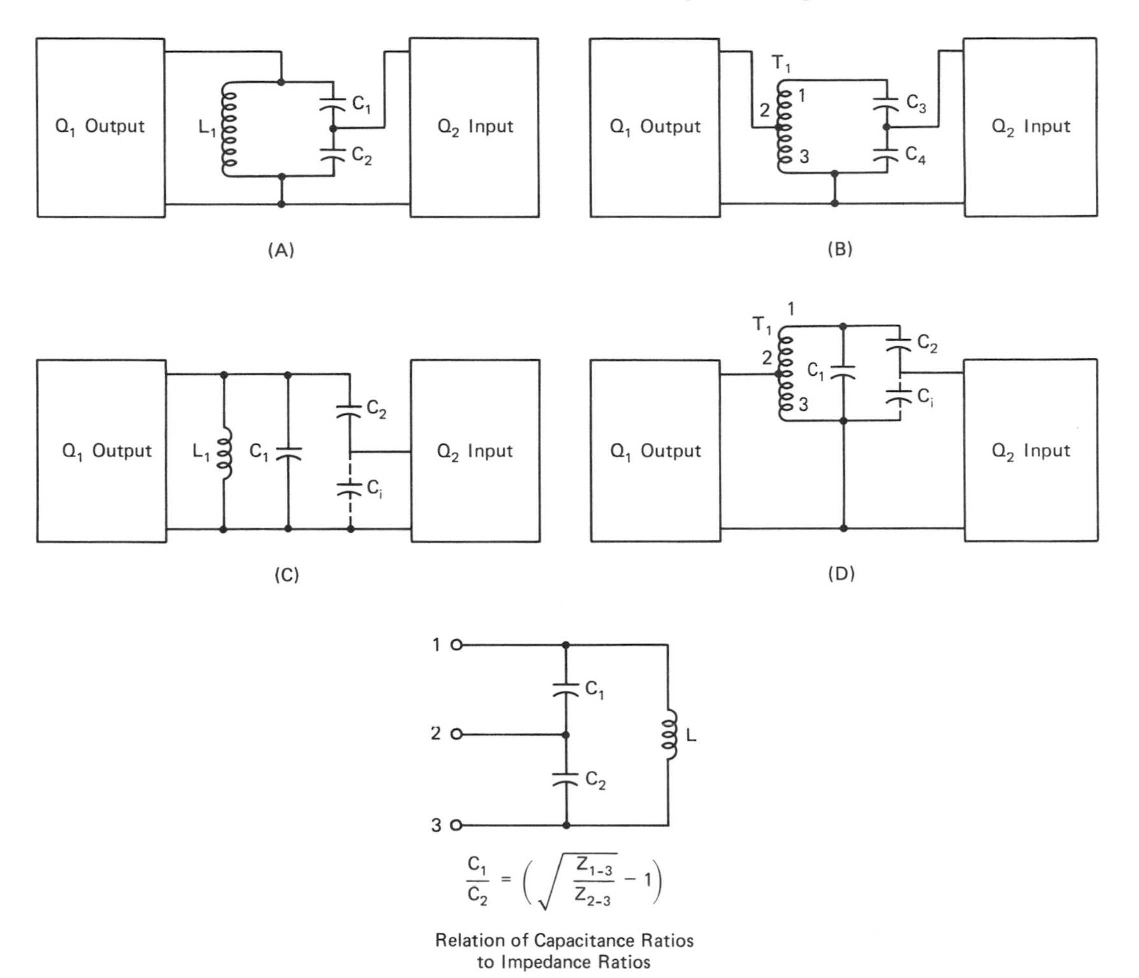

Figure 5–5. Split-capacitor coupling configuration.

NOTE: The input resistance of Q2 (particularly in the CB mode) may swamp out C4. In such a case, the designer should omit C4 and adjust the values of C1 and C2 as required.

$$C_T = C1 + C2$$

In each case the resonant frequency is:

$$f_r = 1/(2\pi\sqrt{L_1 C_T})$$

4. To obtain a high Q_o for the purpose of selectivity, the total inductance can be increased (C_T decreased accordingly), so that the arrangement appears as depicted in Figure 5–5(D).

INTERSTAGE COUPLING WITH DOUBLE-TUNED NETWORKS

Although double-tuned interstage coupling networks are comparatively costly, a better pass-band characteristic is obtained:

1. The frequency response is flatter within the pass band. (Refer to Figures 5–6, 5–7, and 5–8.)
2. The drop in output voltage versus frequency is more rapid prior to and past the ends of the pass band.
3. Better rejection of adjacent-channel frequencies is provided.

The interstage coupling networks shown in Figures 5–6(A) and (B) employ two inductively coupled circuits. In Figure 5–6(A), capacitor C1 and the primary winding inductance L_p comprise a tuned circuit. Capacitor C2 and the secondary winding inductance L_s also comprise a tuned circuit. Each circuit is (usually) tuned to the same resonant frequency (f_r) so that:

$$f_r = 1/(2\pi\sqrt{L_pC1}) = 1/(2\pi\sqrt{L_sC2})$$

Impedance matching is obtained by proper selection of the turns ratio of the primary and secondary windings. The circuit depicted in Figure 5–6(B) functions in the same manner. Tapped primary and secondary windings are employed to obtain the specified selectivity (Q_0).

The interstage coupling networks shown in Figure 5–6(C) and (D) employ two capacitance-coupled tuned circuits. Capacitor C1 and coil L1 form a resonant circuit. Capacitor C3 and coil L2 also form a resonant circuit. Each coil is tuned to the same frequency, so that:

$$f_r = 1/(2\pi\sqrt{L1C2}) = 1/(2\pi\sqrt{L2C3})$$

Impedance matching is obtained by proper selection of the ratio of the reactance of capacitor C2 to the impedance of the input parallel circuit (capacitor C3 and coil L2). The circuit shown in Figure 5–6(D) functions in the same manner as previously explained. Tapped transformers (autotransformers) are utilized to obtain the specified selectivity (Q_0).

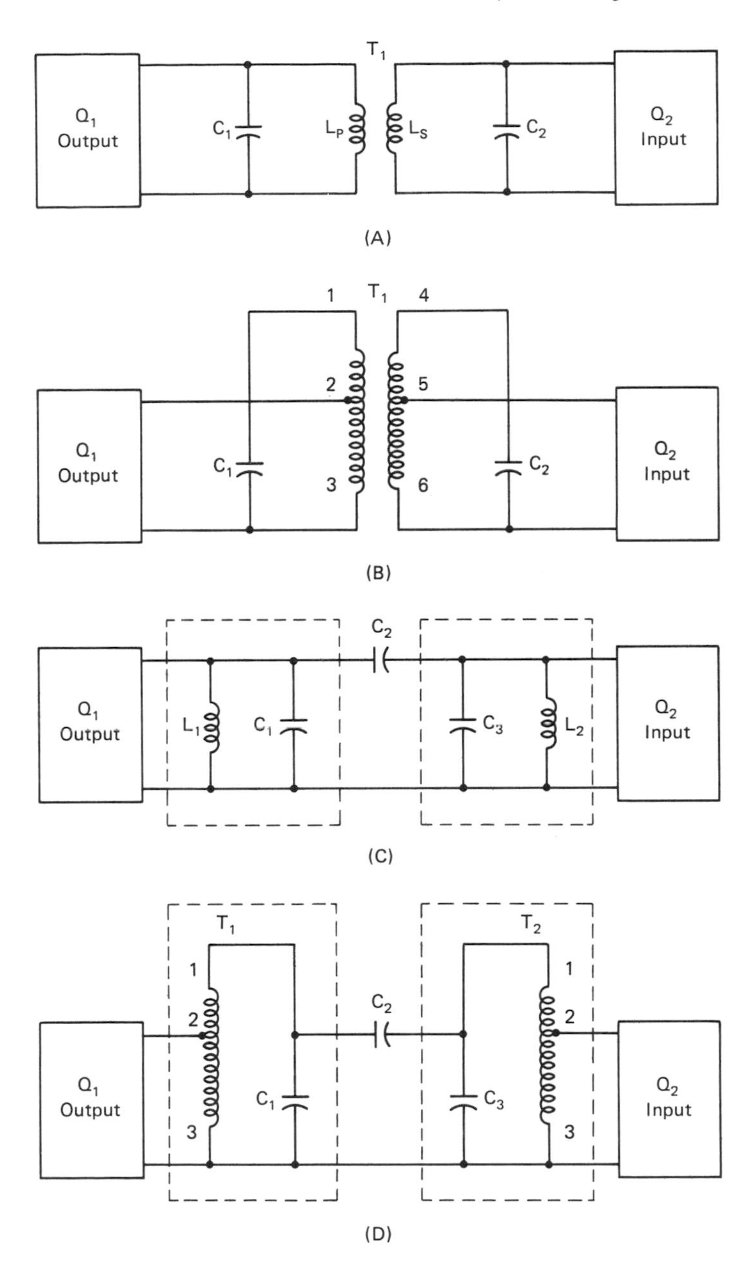

Figure 5–6. Double-tuned interstage coupling arrangements with inductive coupling or capacitive coupling.

NOTE: The primary and secondary windings may be either inductively or capacitively coupled. If capacitive coupling is used, no inductive coupling is provided. The same frequency response can be obtained with either inductive coupling or capacitive coupling.

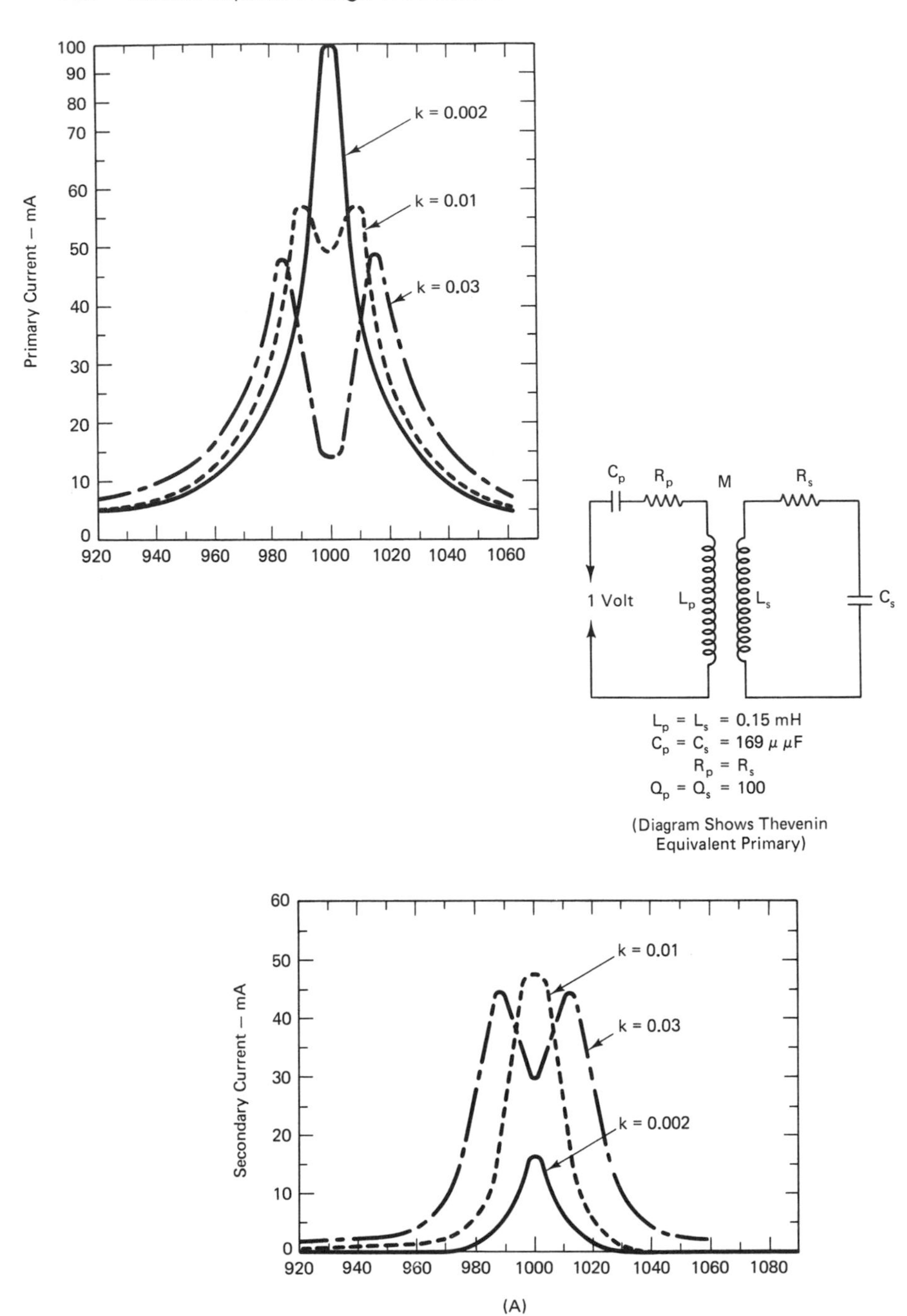

Figure 5–7(A). Current flow in coupling transformer with tuned primary and secondary.

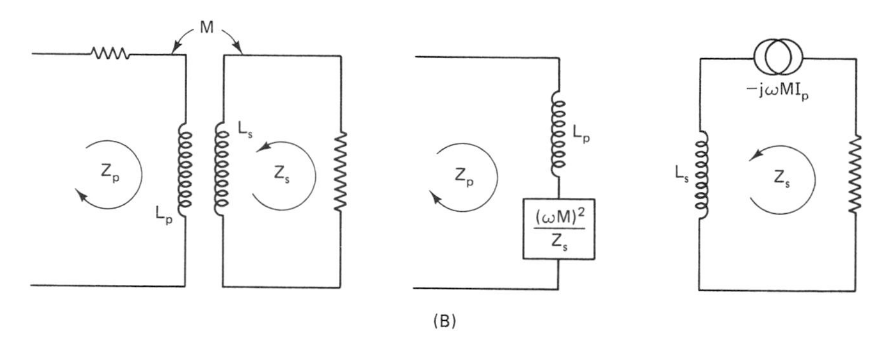

Figure 5–7(B). Parameters employed in graphical analysis of inductively coupled circuits.

NOTE: The term $(\omega M)^2$ is equal to a negative resistance; its phase angle is $-180°$. The negative resistance results from the circumstance that M causes a current demand on the primary. In turn, the positive resistance in the primary circuit is reduced by the negative resistance produced by M.

When these basic relations are observed, the designer can easily include coupled tuned circuits in chained graphical analyses of various networks.

Insofar as graphical analysis of inductively coupled circuits is concerned, the primary circuit responds to the presence of the secondary circuits as if an impedance $(\omega M)^2/Z_s$ had been connected in series with the primary where M is the mutual inductance and Z_s is the impedance of the secondary circuit by itself. The voltage induced into the secondary circuit by a primary current I_p is equal to ωMI_p and lags I_p by 90°. The secondary current that flows is the same as if $-j\omega MI_p$ were applied to the secondary without the presence of the primary.

Gain Equalization

Many design projects for tuned amplifiers require a variable range of the center frequency; a 2-to-1 or 3-to-1 range is typical. Variable tuning of a tuned amplifier is accomplished by varying the capacitance or the inductance of the coupling networks. One of practical design problems is the maintenance of constant gain over a wide frequency range. This is a comparatively difficult problem because the transistor h_{fe} drops in value with increasing frequency. To compensate, an equalizer network such as shown in Figure 5–9 may be utilized. R and C are connected in parallel and placed in series with the input of one of the stages. An equalizer attenuates the low frequencies more than the high frequencies; the reactance of C decreases as the frequency increases.

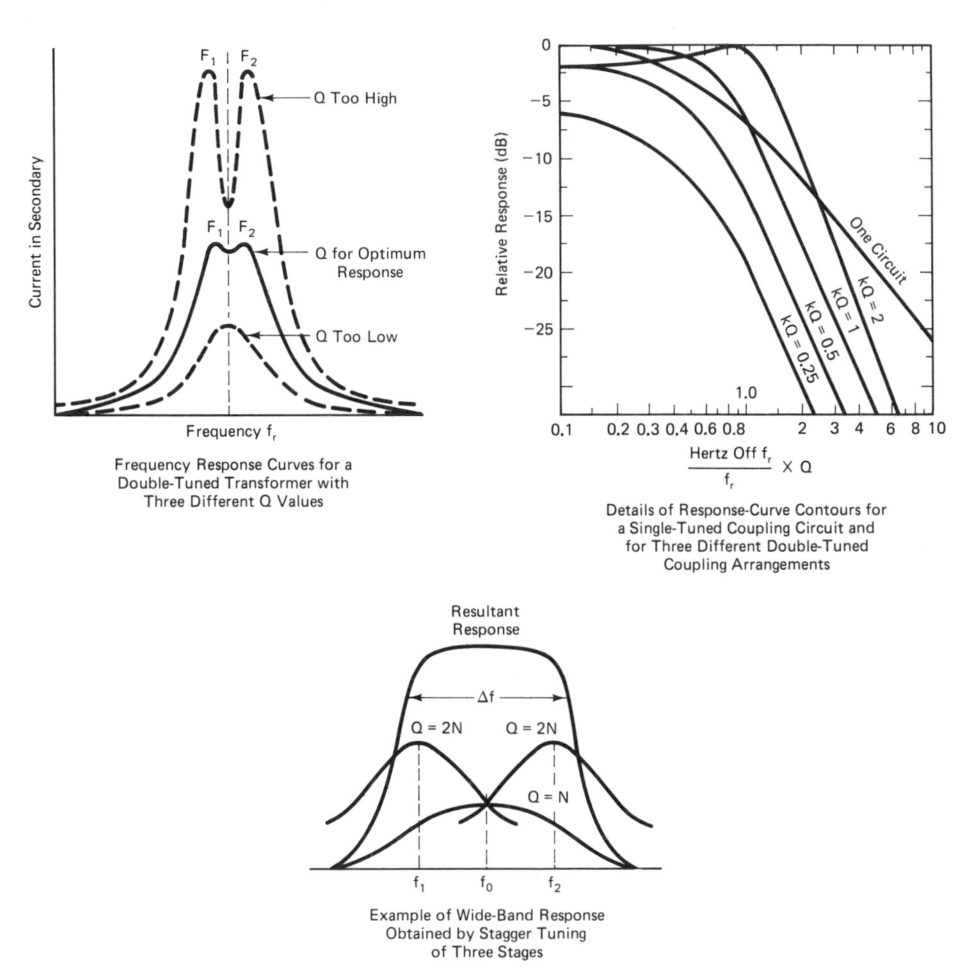

Figure 5–8. Basic factors affecting frequency response curve contours and bandwidths.

NEUTRALIZATION AND UNILATERALIZATION

A unilateral device passes energy in only one direction. A transistor is not a unilateral device (it has an h_{re} parameter). This feedback of energy from output to input is positive; it contributes to instability (regeneration) or to uncontrolled oscillation. This problem becomes of increasing concern at higher frequencies of amplifier operation. An external feedback circuit may be devised to cancel the h_{re} parameter in part, or in whole. The practical benefit of neutralization or unilateralization is prevention of self-oscillation.

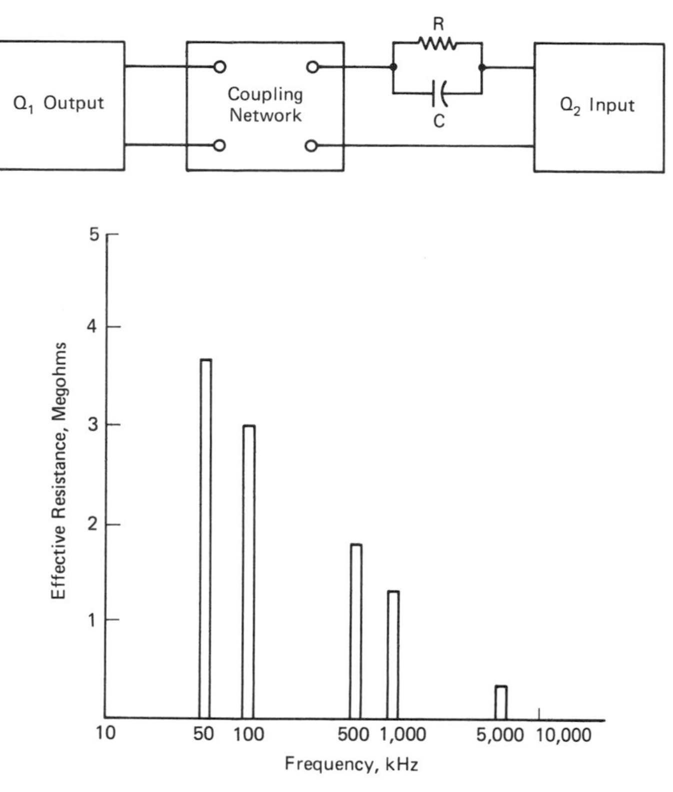

Figure 5–9. Amplifiers coupled by RC equalizer network. (Reproduced by special permission of Reston Publishing Company and Campbell Loudoun from *Handbook for Elcctronic Circuit Design.*)

NOTE: Composition resistors with high resistance values have substantial changes in effective impedance with changes in frequency. For example, a typical composition resistor with an impedance of 3.6 megohms at 50 kHz has an impedance of only 0.33 megohm at 5 MHz.

Unilateralized Common-Base Amplifier

With reference to Figure 5–10, a tuned common-base amplifier configuration is exemplified (dc biasing circuits are not shown). Transformer T1 couples the input signal into the transistor. C1 and the secondary of T1 form a parallel resonant circuit. T2 couples the amplifier output to the following stage. C2 and the primary of T2 form a parallel resonant circuit. Note that:

1. The internal elements of the transistor that may cause sufficient feedback to initiate oscillation are shown by dashed lines in Figure 5–10.

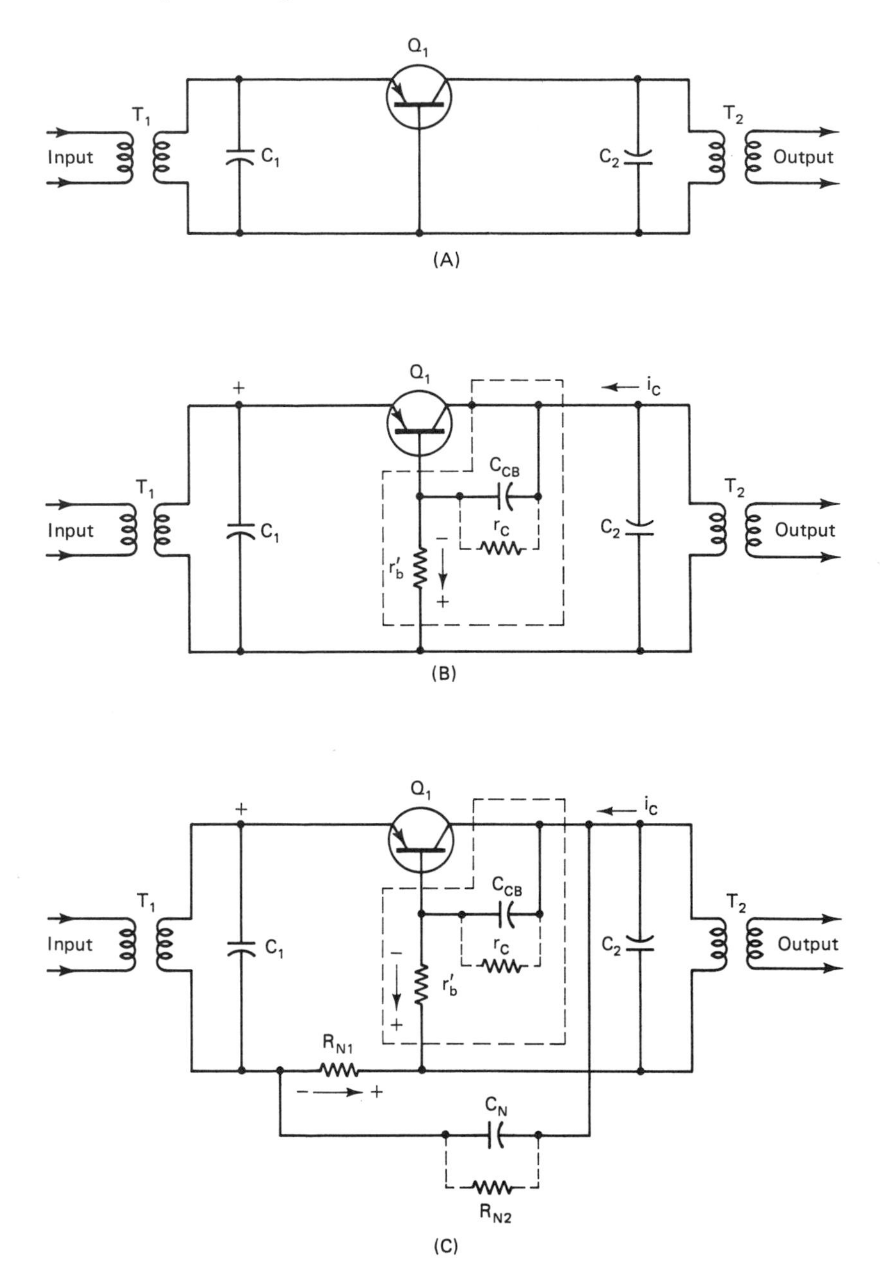

Figure 5–10. Internal feedback elements in a common-base amplifier, with external unilaterizing circuit.

Note: Neutralization and unilateralization are methods of partially or completely cancelling h_{re}. If reactance changes in the output circuit are cancelled out of the input circuit, the stage is neutralized. If both resistive and reactive changes in the output circuit are cancelled out of the input circuit, the stage is unilateralized.

2. Resistor r'_b represents the resistance of the bulk material of the base and is called the base-spreading resistance. Capacitor C_{CB} represents the capacitance of the base-collector junction. Resistor r_c represents the resistance of the base-collector junction. This is a very high resistance because the collector-base junction is reverse-biased. At very high frequencies, C_B effectively shunts r_c.

3. Assume that the incoming signal aids the forward bias and causes the emitter to go more positive with respect to the base. Collector current increases in the direction indicated. A portion of the collector current passes via capacitor C_{CB} through r'_b in the direction shown, and produces a voltage with the indicated polarity. The voltage across r'_b is in phase with the incoming signal; it constitutes positive feedback that may cause self-oscillation.

4. To overcome this possibility of oscillation, the external circuit (comprising R_{N1}, R_{N2}, and C_N) can be added to the circuit as shown in Figure 5–10(C). Resistors R_{N1} and R_{N2} and capacitor C_N correspond to resistors r'_b and r_c and capacitor C_{CB} respectively. The importance of R_{N2} depends on the signal frequency; the higher the frequency, the less is the need for R_{N2}.

5. When the incoming signal aids the forward bias, the collector current increases. A portion of the collector current passes via C_{CB} through r'_b in the direction indicated, producing a voltage with the polarity shown. A portion of the collector current also passes via C_N through R_{N1}, producing a voltage with the indicated polarity. The voltages across r'_b and R_{N1} are opposing voltages. If these voltages are equal, there is neither positive feedback nor negative feedback from output to input, and the stage is said to be unilateralized.

COMMON-EMITTER AMPLIFIER WITH PARTIAL EMITTER DEGENERATION

A CE amplifier configuration with partial emitter degeneration to unilateralize the stage is depicted in Figure 5–11. Capacitor C_N and resistors R_{N1} and R_{N2} form a unilateralizing network. Transformer T1 couples the input signal to the transistor. Resistor R1 forward-biases the base-emitter circuit. Capacitor C1 prevents short-circuiting of base-bias voltage by the T1 secondary. T2 couples the output signal to the following stage. C2 and the T2 primary form a parallel-resonant circuit. C3 blocks dc voltage from R_{N1} and couples a portion of the collector current (i_{c2}) to the emitter. L1 is an rf choke that prevents ac short-circuiting to ground of R_{N1} through C3.

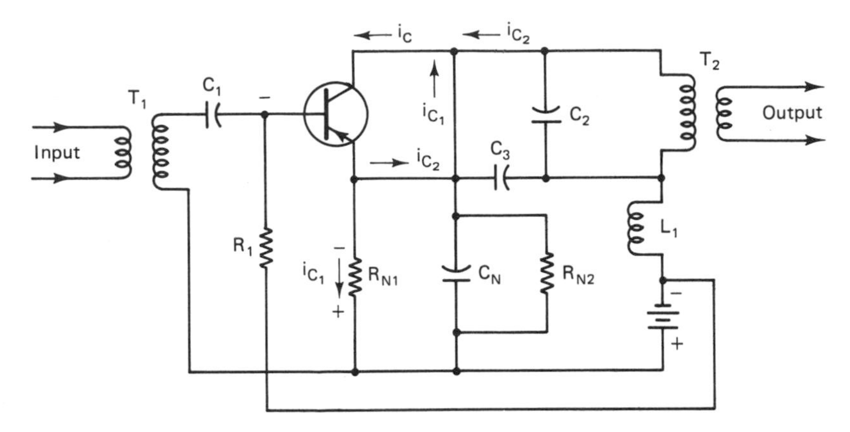

Figure 5–11. Partial emitter degeneration in a common-emitter amplifier.

NOTE: The designer may avoid the necessity for neutralizing a 455-kHz amplifier by selecting a type of transistor that has low feedback capacitance. As an illustration, a typical transistor in this category has a rated feedback capacitance of approximately 2 pF. In turn, impedance matching need not be compromised in a 455-kHz stage in order to avoid objectionable regeneration or self-oscillation.

When the input signal aids the forward bias, collector current i_c increases in the direction indicated. Internally, a portion of the collector current is coupled to the base-spreading resistance via collector-base junction capacitance. The voltage drop across the base-spreading resistance is in phase with the incoming signal and constitutes positive feedback. To offset this positive feedback, the designer may employ the circuitry shown in Figure 5–11; a portion of the collector current (i_{c1}) passes through R_{N1} and the parallel combination of C_N and R_{N2}.

The voltage drop across R_{N1} is degenerative; it is equal and opposite to the voltage drop across the base-spreading resistance. Thus, the net voltage feedback to the input circuit is zero, whereby the stage is unilateralized. The values of C_N, R_{N1}, and R_{N2} depend on the internal values of collector-base junction capacitance, base-spreading resistance, and collector resistance, respectively.

CE AMPLIFIER WITH BRIDGE UNILATERALIZATION

Historically, bridge neutralizing circuitry was first used to offset regenerative action in tuned amplifiers. Like all neutralizing techniques, precise cancellation of reverse feedback voltage can be realized at only one operating frequency. Thus, neutralizing methods are of less utility in tuned amplifiers designed to operate over an appreciable frequency range. In this situation, simple "losser" resistors and/or impedance mismatches may supplement or replace the more efficient neutralization techniques.

With reference to Figure 5–12, a CE configuration is unilateralized by means of T2 winding 2 to 3 and the combination of R_N and C_N. T1 couples the input signal to the transistor. Winding 1 to 2 on T2 couples the output signal to the following stage. R1 forward-biases the transistor. C1 prevents short-circuiting of the base-bias voltage by the secondary of T1. C2 bypasses the collector supply and places terminal 2 of T2 at ac ground potential. C3 tunes the primary of T2. Note the following points:

1. The unilateralizing circuitry has an equivalent bridge configuration.

2. This bridge does not involve T1, C1, C2, C3, R1, T2 secondary, nor the collector supply.

3. Ponts B, C, and E on the bridge denote the base, collector, and emitter terminals, respectively, of the transistor. Dashed lines indicate transistor internal feedback elements. Points 1, 2, and 3 correspond to T2 primary terminals. The voltage across terminals 1 and 3 of T2 is represented by a voltage generator with output voltage $v_{1\text{-}3}$.

4. When the bridge is balanced, no portion of $v_{1\text{-}3}$ appears between B and E, and the stage is unilateralized. Balance occurs when the voltage ratio between B and C and between B and 3 equals the ratio of voltages between C and E and between E and 3. In addition, the phase shift introduced by the circuitry between B and C must equal the phase shift introduced by the circuitry between B and 3.

5. In various applications, R_N may be omitted, and capacitor C_N suffices to prevent self-oscillation. When R_N is not included, the stage is neutralized, although it is not unilateralized.

TUNED-POWER AMPLIFIERS

Power transfer is a dominant design factor in the basic tuned power amplifier, and impedance matching is maintained in both the input and the output sections. Note that waveform distortion is of minor concern because harmonics can be largely rejected by means of

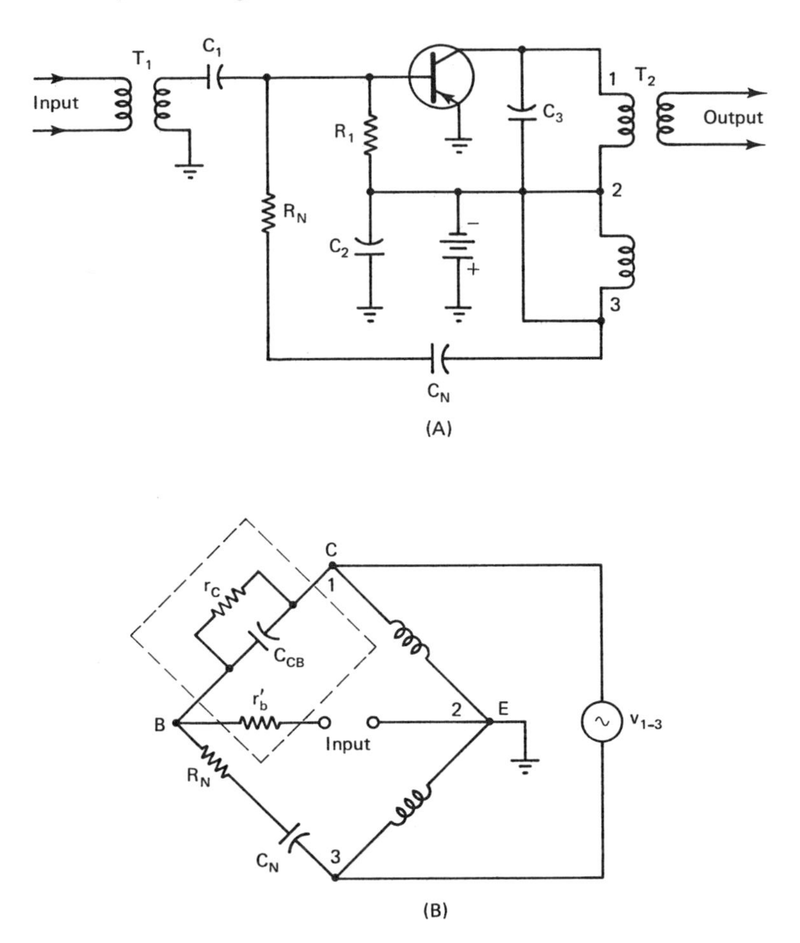

Figure 5–12. Bridge-neutralizing arrangement for common-emitter amplifier.

NOTE: As a practical consideration, unilateralization is precise only at a single operating frequency and becomes increasingly ineffective when T1 and T2 are tuned over an appreciable range of frequencies. Regeneration can be controlled by a trade-off in gain provided, for example, by a resistor in series with the base or by sufficient mismatch of transformer and transistor impedances.

tuned circuitry. An rf power amplifier may be operated in either class B or in class C. A class B amplifier can be designed to provide an rf output signal that is proportional in amplitude to the input signal level; this design is called a linear rf amplifier.

Most rf power amplifiers are operated in class C to realize maximum efficiency. A typical class C rf power amplifier has an efficiency of 65 to 70 percent (ratio of rf power output to dc power input). As an illustration, if the bogie rf power output value is 50 W, the dc power input will be approximately 70 W. In turn, if a 28-V collector supply is employed, the collector current will be 2.7 mA. (This example is simplified to the extent that the rf driving power to the transistor is ignored.)

The collector circuit may be tuned to a multiple of the driver-input frequency, and the stage then operates as a frequency multiplier. A frequency doubler has a typical efficiency of 40 percent; a frequency tripler operates at approximately 28 percent efficiency; a frequency quadrupler has a typical efficiency of 18 percent.

A basic tuned-power amplifier arrangement is shown in Figure 5–13. A relatively high-level driver signal is applied at the input of the power-amplifier circuit. This is a typical series-fed configuration. Use of shunt feed is likely to impose a design problem of parasitic oscillations that are difficult to eliminate. Parasitic oscillation is sometimes encountered also in the series-fed arrangement. In such a case, a selected rf choke may be employed and/or a ferrite bead placed on the pigtail lead of the choke.

The emitter swamping resistor has a typical value of 1 ohm. Signal leads must be kept as short as is feasible—even a straight length of wire has appreciable inductance at higher radio frequencies. The configuration shown in Figure 5–13 is essentially a class-B stage; however, it will operate in class C when sufficient drive signal is applied. Reverse bias is dropped across the emitter swamping resistor and charges the emitter bypass capacitor. The bypass capacitor may have a value of 0.1μF.

Observe that parasitic oscillations can occasionally be controlled only if a neutralizing capacitor is employed, as seen in Figure 5–13. Resistor R_B assists in avoiding parasitic oscillation involving the base rf choke. A ferrite bead on the pigtail lead is also occasionally helpful. The designer often needs to select a suitable type of choke.

In a few applications, maximum input/output linearity may be required, and the stage can then be operated in class A. Power gain is comparatively high in class A, although the operating efficiency is only 25 percent. In single-sideband (SSB) transmitters, class-B push-pull operation is commonly utilized for good linearity and reasonable efficiency.

Note that in both class-A and class-B operation, the designer must give close attention to the hazard of thermal runaway. This hazard is considerably relaxed in class-C operation. Observe also that the inherent inductance of the emitter-bypass capacitor can

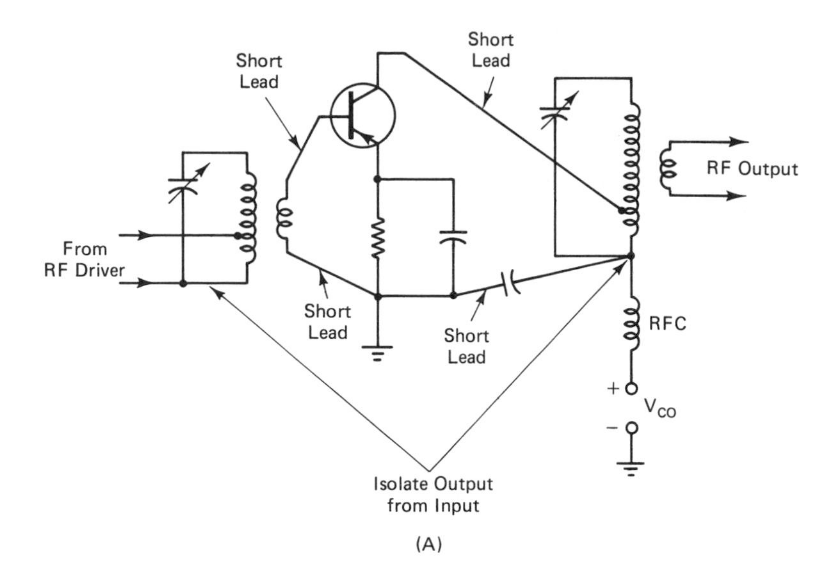

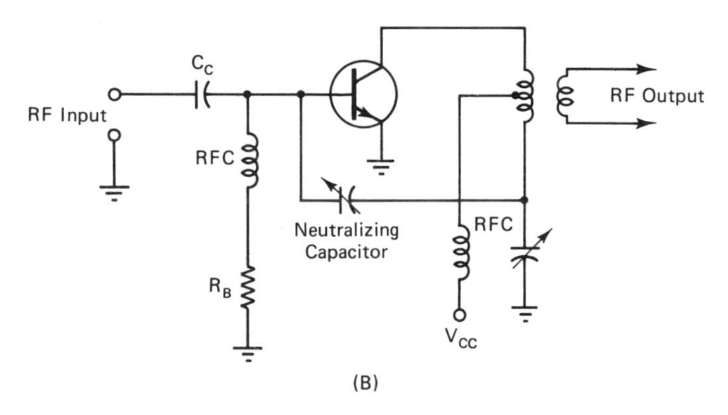

Figure 5–13. Basic tuned-power amplifier arrangement. (A) Skeleton circuitry; (B) neutralizing capacitor configuration.

NOTE: The basic configuration may operate in either class B or class C, depending upon the input drive level. The neutralizing capacitor incidentally assists in control of parasitic oscillations.

impose a design problem. Typically, several capacitors may be connected in parallel to minimize lead inductance. Alternatively the inherent inductance of the bypass capacitor with its leads can be cancelled by critical adjustment for series resonance with the capacitor at the operating frequency.

Good design practices include adequate bypassing of power supply leads, preferably with a pair of paralleled capacitors. One of these capacitors should have a high value to present low reactance at low (potentially parasitic) frequencies; the transistor has relatively high gain at lower frequencies. The other capacitor should have a comparatively small value with a low rf reactance. (A large capacitor will have excessive series impedance at high radio frequencies.) Sintered electrode tantalum capacitors are suitable for rf bypassing at frequencies up to approximately 75 MHz. However, high-Q ceramic capacitors are preferred for UHF bypassing.

Components can be checked for self-resonances or parasitic impedances on an rf impedance bridge. When the designer observes these potential trouble sources, their possible effects on adverse operation of the prototype model should be carefully evaluated. In turn, the circuit designer can make a judicious selection of available inductors and capacitors.

Feedthrough capacitors and mica "postage-stamp" capacitors have superior characteristics in VHF operation. Resistors that carry rf current should have low-series inductance and low-shunt capacitance. Small composition resistors are usually satisfactory. However, as previously noted, stray shunt capacitance can have a drastic action on the effective value of a composition resistor with a high value and operating in the rf range. Note also that whenever possible, all of the rf grounds should be connected to the ground plane within a small area. (See also Chart 5–1 and Chart 5–2.)

IMPEDANCE MATCHING NETWORKS

Impedance-matching networks comprising L or pi sections with inductors and capacitors are used to obtain maximum power transfer, as from a radio transmitter to an antenna transmission line. Basic impedance-matching networks with required parameters are shown in Figure 5–14. Note that the Q values in matching networks must be selected in accordance with the pertinent equations. In Figure 5–14(A) the Q value is equal to X_L/R at the operating frequency, or to R_{in}/X_C. In other words, after values of R_{in} and R are assigned, specific values for L and C follow from the values of X_L and X_C at the operating frequency. In Figure 5–14(B), the Q value is equal to X_L/R_{in}, or to R/X_C. In Figure 5–14(C), the Q value is equal to R_1/X_{C1}. (See also Figure 5–15.)

CHART 5–1

Characteristics of Series- and Parallel-Resonant Circuits

Quantity	Series Circuit	Parallel Circuit
At resonance: Reactance ($X_L - X_C$)	Zero; because $X_L = X_C$	Zero; because nonenergy currents are equal
Resonant frequency	$\frac{1}{2\pi\sqrt{LC}}$	$\frac{1}{2\pi\sqrt{LC}}$
Impedance	Minimum; $Z = R$	Maximum; $Z = \frac{L}{CR}$, approx.
I_{line}	Maximum	Minimum value
I_L	I_{line}	$Q \times I_{line}$
I_C	I_{line}	$Q \times 1_{line}$
E_L	$Q \times E_{line}$	E_{line}
Phase angle between E_{line} and I_{line}	0°	0°
Angle between E_L and E_C	180°	0°
Angle between I_L and I_C	0°	180°
Desired value of Q	10 or more	10 or more
Desired value of R	Low	Low
Highest selectivity	High Q, low R, high $\frac{L}{C}$	High Q, low R
When f is greater than f_o: Reactance	Inductive	Capacitive
Phase angle between I_{line} and E_{line}	Lagging current	Leading current
When f is less than f_o: Reactance	Capacitive	Inductive
Phase angle between I_{line} and E_{line}	Leading current	Lagging current

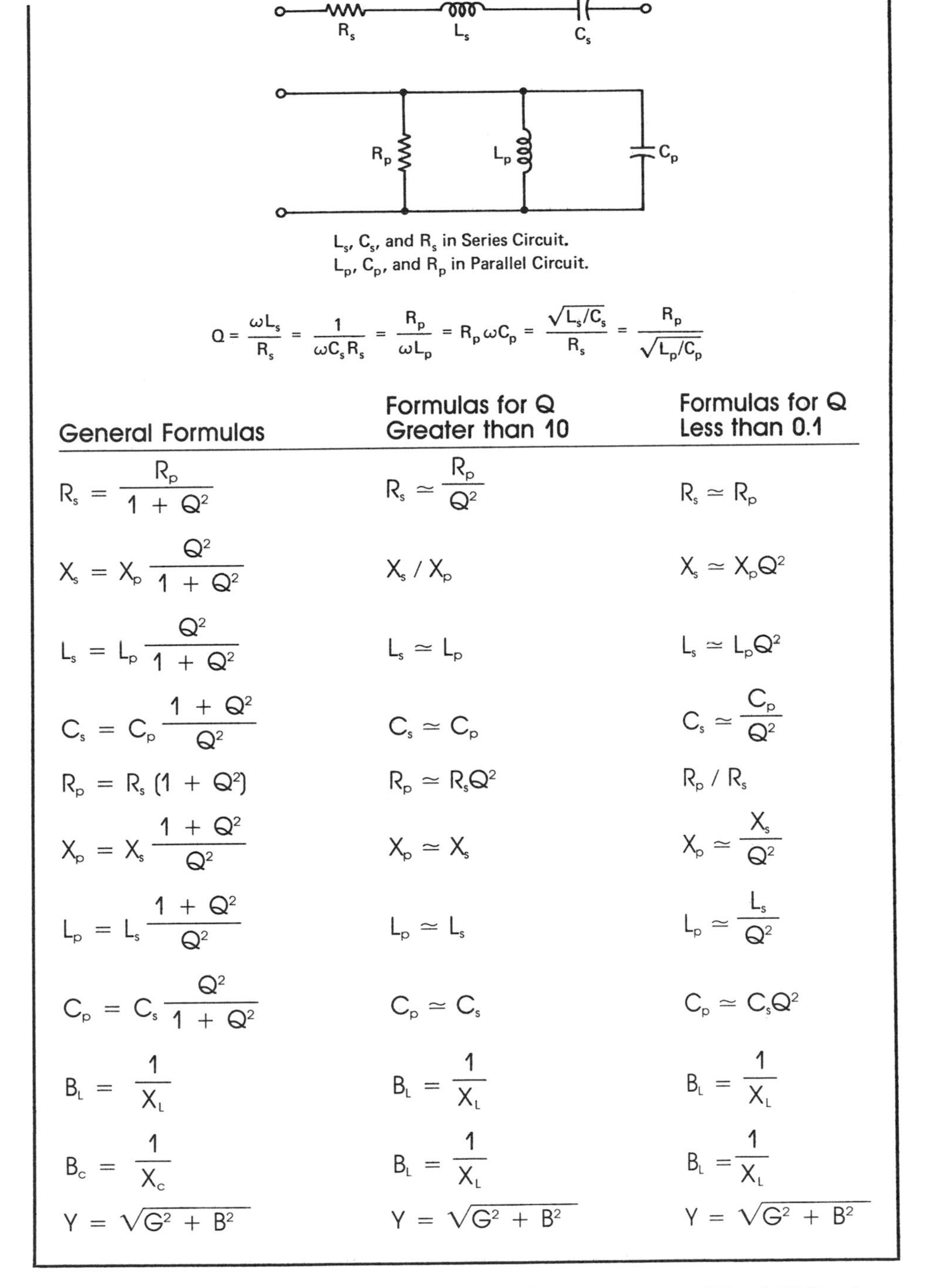

L_s, C_s, and R_s in Series Circuit.
L_p, C_p, and R_p in Parallel Circuit.

$$Q = \frac{\omega L_s}{R_s} = \frac{1}{\omega C_s R_s} = \frac{R_p}{\omega L_p} = R_p \omega C_p = \frac{\sqrt{L_s/C_s}}{R_s} = \frac{R_p}{\sqrt{L_p/C_p}}$$

General Formulas	Formulas for Q Greater than 10	Formulas for Q Less than 0.1
$R_s = \frac{R_p}{1 + Q^2}$	$R_s \simeq \frac{R_p}{Q^2}$	$R_s \simeq R_p$
$X_s = X_p \frac{Q^2}{1 + Q^2}$	X_s / X_p	$X_s \simeq X_p Q^2$
$L_s = L_p \frac{Q^2}{1 + Q^2}$	$L_s \simeq L_p$	$L_s \simeq L_p Q^2$
$C_s = C_p \frac{1 + Q^2}{Q^2}$	$C_s \simeq C_p$	$C_s \simeq \frac{C_p}{Q^2}$
$R_p = R_s (1 + Q^2)$	$R_p \simeq R_s Q^2$	R_p / R_s
$X_p = X_s \frac{1 + Q^2}{Q^2}$	$X_p \simeq X_s$	$X_p \simeq \frac{X_s}{Q^2}$
$L_p = L_s \frac{1 + Q^2}{Q^2}$	$L_p \simeq L_s$	$L_p \simeq \frac{L_s}{Q^2}$
$C_p = C_s \frac{Q^2}{1 + Q^2}$	$C_p \simeq C_s$	$C_p \simeq C_s Q^2$
$B_L = \frac{1}{X_L}$	$B_L = \frac{1}{X_L}$	$B_L = \frac{1}{X_L}$
$B_C = \frac{1}{X_C}$	$B_L = \frac{1}{X_L}$	$B_L = \frac{1}{X_L}$
$Y = \sqrt{G^2 + B^2}$	$Y = \sqrt{G^2 + B^2}$	$Y = \sqrt{G^2 + B^2}$

CHART 5–2

Tuned Amplifier Integrated Circuit Characteristics

A typical integrated circuit used in FM and TV IF networks has the following characteristics:

Voltage gain at 10.7 MHz, 61 dB

Parallel input resistance, 3 kilohms

Parallel input capacitance, 7 pF

Parallel output resistance, 31.5 kilohms

Parallel output capacitance, 4.2 pF

Noise figure at 4.5 MHz, 8.7 dB

Frequency range, 100 kHz to 20 MHz

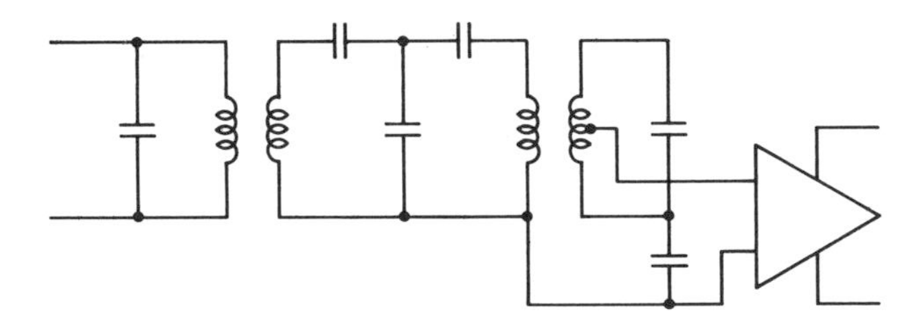

A triple-tuned coupling network between the mixer and the first integrated circuit will provide satisfactory selectivity in an FM IF section. Its voltage insertion loss will be approximately 25 dB.

This is an example of a high-insertion-loss coupling network. The resulting overall receiver noise figure is calculated as follows:

$$F = F_1 + \frac{F_{2\text{-}1}}{G_1} + \frac{F_3\text{-}1}{G_1 G_2}$$

where F_1, F_2, and F_3 are the noise figures of the rf, mixer, and third IF stages, respectively; G_1 and G_2 are the power gains of the rf and mixer stages.

If a noise figure F_3 of 26 dB is assumed for the IF section, 10 dB for the mixer, and a mixer power gain of 30 dB, the effect of IF noise on mixer noise is calculated as follows:

$$F_2' = F_2 + \frac{F_3 - 1}{G_2} = 10 + \frac{26 - 1}{1000} = 10.025 \text{ dB}$$

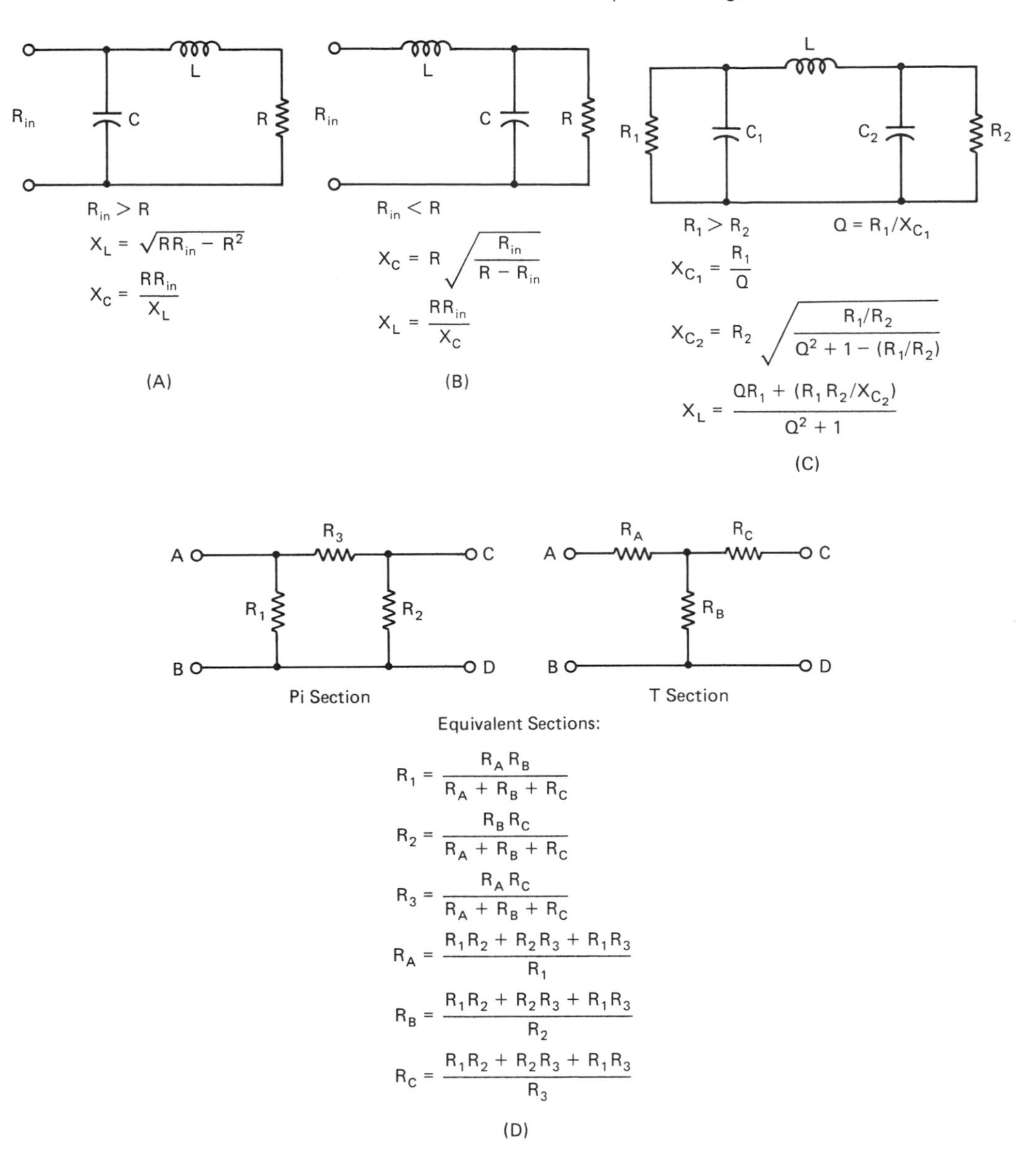

Figure 5–14. Basic impedance-matching networks. (A) L network, high Z to low Z; (B) L network, low Z to high Z; (C) pi network, high Z to low Z; (D) basic pi-T equivalent sections. (Reproduced by special permission of Reston Publishing Company and Campbell Loudoun from *Handbook for Electronic Circuit Design.* Illustration D is courtesy of Goodyear.)

NOTE: Short computer programs for solution of these equations are provided in Figure 5–15.

```
5 LPRINT"L SECTION IMPEDANCE MATCHING NETWORKS"
10 LPRINT"(High Z to Low Z, and Low Z to High Z)"
15 LPRINT"Ref: Fig. 5-14(a)"
20 LPRINT"...................................":LPRINT""
25 INPUT"Rin (Ohms)=";A
30 LPRINT "Rin (Ohms)=";A
35 INPUT"R (Ohms)=";B
40 LPRINT"R (Ohms)=";B
45 C=(A*B-B*B)^.5
50 D=A*B/C
55 E=A*(B/(A-B)^.5)
60 F=A*B/E:LPRINT""
65 A$="#####.##"
70 LPRINT"XL (Ohms)=";USING A$;C
75 LPRINT"XC (Ohms)=";USING A$;D
80 LPRINT"":LPRINT"....................................":LPRINT""
85 LPRINT "Ref: Fig. 5-14(b)"
90 LPRINT""
95 LPRINT"XC (Ohms)=";USING A$;E
100 LPRINT"XL (Ohms)=";USING A$;F
```

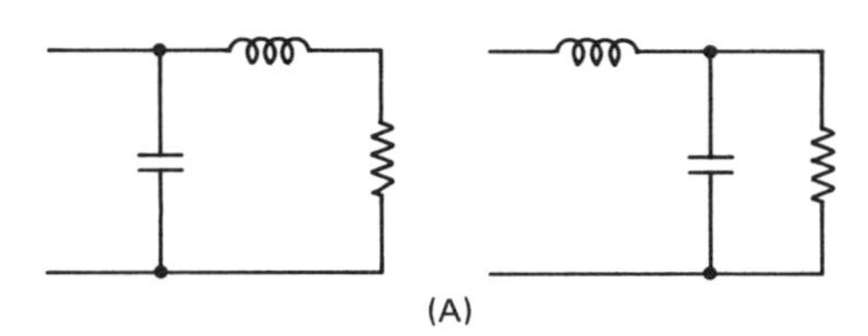

(A)

NOTE: For the purpose of simplicity, resistive pi and T sections are illustrated. However, any or all of the resistances may be replaced by reactances or by impedances, and the same equivalent equations will apply. Resistive sections can be calculated arithmetically; when reactive elements are present, the sections must be calculated by means of complex algebra.

Series inductance and shunt capacitance assist in reduction of harmonic output. The inductive reactance increases with frequency, and the capacitive reactance decreases with frequency.

A matching network is operated at its resonant frequency. In this example, L/RC = 300 ohms for the configuration in (A), and L/RC = 75 ohms for the configuration in (B).

```
L SECTION IMPEDANCE MATCHING NETWORKS
(High Z to Low Z, and Low Z to High Z)
Ref: Fig. 5-14(a)
...................................

Rin (Ohms)= 300
R (Ohms)= 75

XL (Ohms)=   129.90
XC (Ohms)=   173.21

....................................

Ref: Fig. 5-14(b)

XC (Ohms)=  1500.00
XL (Ohms)=    15.00
```

Figure 5–15(A). Program for calculation of L-section impedance-matching networks depicted in Figure 5–14.

```
5 LPRINT"PI SECTION IMPEDANCE MATCHING NETWORK"
10 LPRINT"(High Z to Low Z, and Low Z to High Z)"
15 LPRINT"Ref: Fig. 5-14(c)"
20 LPRINT"....................................":LPRINT""
25 INPUT"R1 (Ohms)=";A
30 LPRINT "R1 (Ohms)=";A
35 INPUT"R2 (Ohms)=";B
40 LPRINT"R2 (Ohms)=";B
45 INPUT"Xc1 (Ohms)=";C
50 LPRINT"Xc1 (Ohms)=:C
55 INPUT"f (Hz)=";F
60 LPRINT"f (Hz)=";F
65 Q=A/C:LPRINT"":LPRINT"....................................":LPRINT""
70 D=B*((A/B)/(Q*Q+1-(A/B))^.5)
75 E=(Q*A+(A*B)/D)/(Q*Q+1)
80 N=1/(6.283*F*C):NN=1/(6.283*F*D):L=E/(6.283*F)
85 A$="##.#############"
90 LPRINT"C1 (Farads)=";USING A$;N:LPRINT""
95 B$="#####.##"
100 LPRINT"Xc2 (Ohms)=";USING B$;D
105 C$="##.############"
110 LPRINT"C2 (Farads)=";USING C$;NN
115 D$="#####.##"
120 LPRINT"XL (Ohms)=";USING D$;E
125 E$="##.#########"
130 LPRINT"L (Hy)=";USING E$;L
```

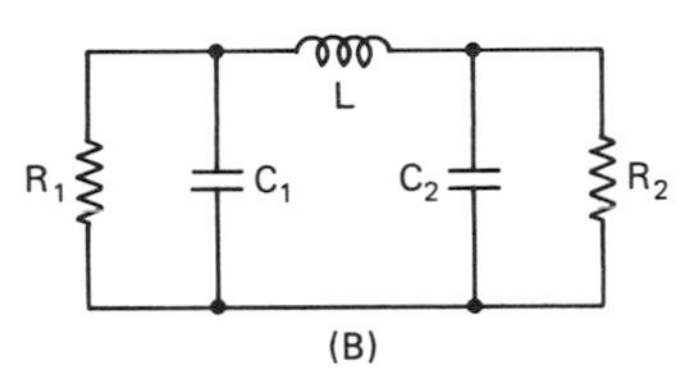

(B)

Note: The matching section is operated at its resonant frequency. Harmonics are attenuated inasmuch as XL becomes greater at higher frequencies, and Xc becomes less at higher frequencies.

```
PI SECTION IMPEDANCE MATCHING NETWORK
(High Z to Low Z, and Low Z to High Z)
Ref: Fig. 5-14(c)
....................................

R1 (Ohms)= 300
R2 (Ohms)= 75
Xc1 (Ohms)=:C
f (Hz)= 1000000

....................................

C1 (Farads)= 0.000000001592

Xc2 (Ohms)=  122.47
C2 (Farads)= 0.000000001300
XL (Ohms)=  108.37
L (Hy)= 0.000017248
```

Figure 5–15(B). Program for calculation of pi-section impedance-matching network depicted in Figure 5–14(C).

```
5 LPRINT"L SECTION IMPEDANCE MATCHING NETWORK"
10 LPRINT"(High Z to Low Z)"
15 LPRINT"Mismatch Due to Component Tolerances":LPRINT"Ref. Fig. 5-14(a)"
20 LPRINT"":LPRINT"....................................":LPRINT""
25 INPUT"Rin (Ohms)=";A
30 LPRINT "Rin (OHMS)=";A
35 INPUT"R (Ohms)=";B
40 LPRINT"R (Ohms)=";B
45 INPUT"XL (Ohms)=";C
50 LPRINT"XL (Ohms)=";C
55 INPUT"XL High Tolerance (%)=";D
60 LPRINT"XL High Tolerance (%)=";D
65 INPUT"XL Low Tolerance (%)=";DD
70 LPRINT"XL Low Tolerance (%)=";DD
75 INPUT "Xc (Ohms)=";E
80 LPRINT "Xc (Ohms)=';E
85 INPUT "Xc High Tolerance (%)=";F
90 LPRINT"Xc High Tolerance (%)=";F
95 INPUT "Xc Low Tolerance (%)=";FF
100 LPRINT"Xc Low Tolerance (%)=";FF
105 RH=(E+.01*F*E)*(C+.01*D*C)/A
110 RL=(E-.01*FF*E)*(C-.01*DD*C)/A
115 MH=(100*(RH-B))/B
120 ML=(100*(B-RL))/B
125 A$="###.#":LPRINT"........................":LPRINT""
130 LPRINT"High Tol. Mismatch (%)=";USING A$;MH
135 B$="###.##"
140 LPRINT"Low Tol. Mismatch (%)=";USING B$;ML
```

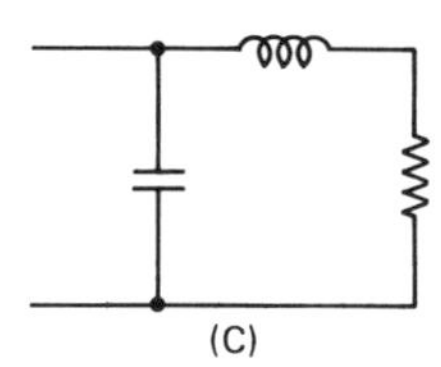

(C)

NOTE: This program computes the impedance mismatch to the load that occurs as a result of commercial tolerances on the inductors and capacitors used in an L-section impedance-matching network.

```
L SECTION IMPEDANCE MATCHING NETWORK
(High Z to Low Z)
Mismatch Due to Component Tolerances
Ref. Fig. 5-14(a)

....................................

Rin (OHMS)= 300
R (Ohms)= 75
XL (Ohms)= 129.9
XL High Tolerance (%)= 15
XL Low Tolerance (%)= 5
Xc (Ohms)=';E
Xc High Tolerance (%)= 10
Xc Low Tolerance (%)= 3
........................

High Tol. Mismatch (%)= 26.5
Low Tol. Mismatch (%)=  7.86
```

Figure 5–15(C). Program for calculating mismatch due to component tolerances on L-section impedance-matching network.

DESIGN OPTIMIZATION CONSIDERATIONS

Although it is impossible to design ideal pass-band characteristics in a tuned amplifier, double-tuned interstage coupling networks are superior to single-tuned coupling networks. Triple-tuned interstage coupling networks provide improvement over double-tuned networks. However, the cost of triple-tuned networks cannot be justified unless maximum possible performance is the dominant consideration.

Unilateralization involves somewhat increased production costs. However, this feature provides practical benefits. As an illustration, if a multistage stagger-tuned amplifier is unilateralized, its frequency response (pass-band) is unaffected by a deterioration in h_{fe} value of a transistor (only the overall gain is reduced). By the same token, the pass-band will be unaffected by transistor replacement, regardless of h_{fe} value. This observation assumes that the replacement transistor has internal elements C_{CB}, r_c, and r'_b with the same values as in the original transistor.

Thermal runaway is one of the potential troublemakers in design of tuned-power amplifiers. Accordingly, the designer should thoroughly check transistor operating temperatures under all rated operating conditions with respect to a worst-case prototype model. Calculation of worst-case parameters should include a margin of tolerance for gradual deterioration of device parameters over a reasonable period of time. Although fail-safe measures can be included, increased production costs may prohibit this approach (except for fuses).

In addition to thermal runaway, potential hazards of drive-signal loss and output-load loss may be taken into account, provided that increased production costs do not rule out the associated fail-safe circuitry. Adequate shielding of tuned-power amplifiers is mandatory. Line filters and output wave traps are often necessary. Serviceability is a marketing factor and contributes also to the company image. Serviceability is an aspect of "human engineering" in practical design procedures.

6

AUTOMATIC GAIN CONTROL AND RELATED FACTORS

GENERAL OVERVIEW

Tuned amplifiers are employed in a very wide range of electronic equipment, and it is often desired to automatically vary the amplifier gain in accordance with the incoming signal level. A tuned amplifier is ordinarily followed by a detector that can develop a dc current and voltage that is proportional to the level of the carrier. Accordingly, the detector is often operated to do double duty as an agc rectifier.

With reference to Figure 6–1, it is evident that one method of controlling the gain of a transistor amplifier is to vary the dc emitter current. The power gain rises to a maximum as the emitter current is increased and then falls with further increase of current. The diagram also shows how the power gain varies with increase of the collector voltage from a very small value to approximately 20 volts.

AUTOMATIC CONTROL OF EMITTER CURRENT OR COLLECTOR VOLTAGE

1. *DC Emitter Current Control:* The power gain of a CE amplifier, Figure 6–2(A), can be controlled by feeding agc voltage to the base of the transistor in order to vary the dc emitter current. In Figure 6–2(A) the

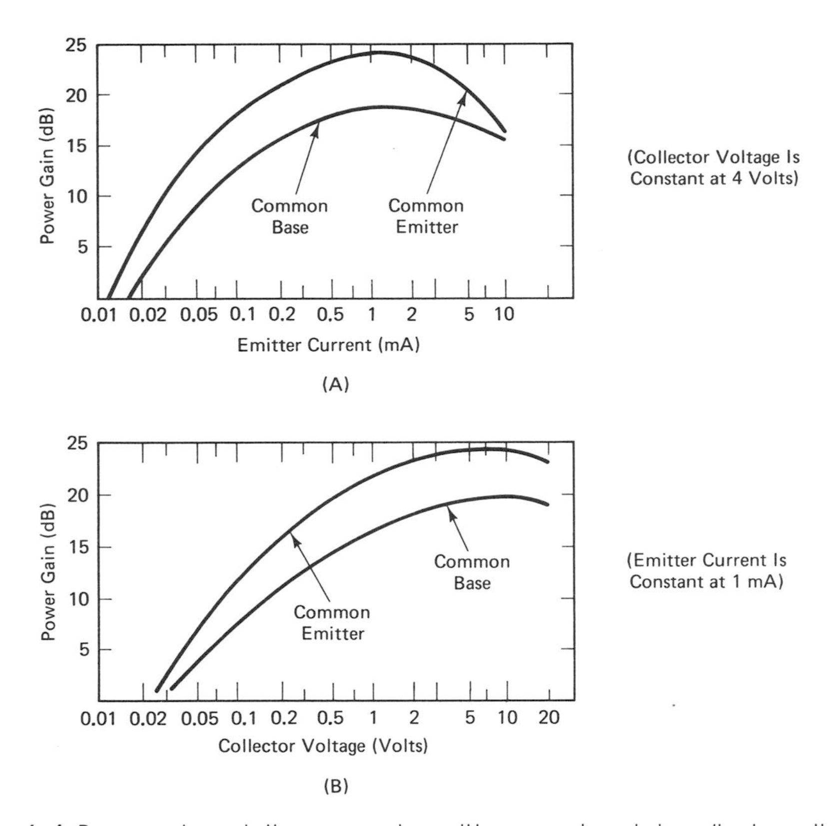

Figure 6–1. Power-gain variation versus dc emitter current and dc collector voltage.

NOTE: The gain of a bipolar transistor amplifier can be controlled either by varying the emitter current, or by varying the collector voltage, or both. In (A), the collector voltage is maintained constant at 4 volts. In (B), the emitter current is maintained constant at 1 mA. These curves are typical for transistors operating in 455-kHz tuned amplifiers.

tuned amplifier is also operating as a dc amplifier to increase the dc-current output from the second detector (amplified agc). If the detector develops sufficient control current, amplified agc need not be employed.

In the circuit illustrated in Figure 6–2(B), resistors R1 and R4 form a voltage divider to establish the no-signal negative (forward) bias on the base. (An example of reverse agc design.) The agc voltage from the second detector is positive with respect to ground; it is fed to the base via dropping resistor R2. When the dc output from the second detector increases in response to an increase in carrier level, the positive dc voltage fed to the base of Q1 via R2 reduces the net negative (forward) bias on the base and decreases the emitter current. In turn, the amplifier gain is decreased. When the dc output from the second detector decreases, opposite response occurs.

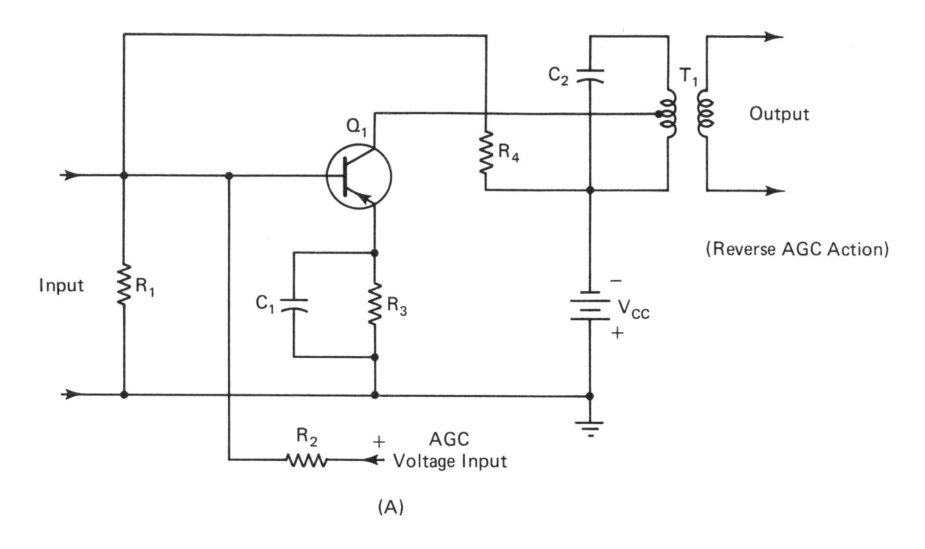

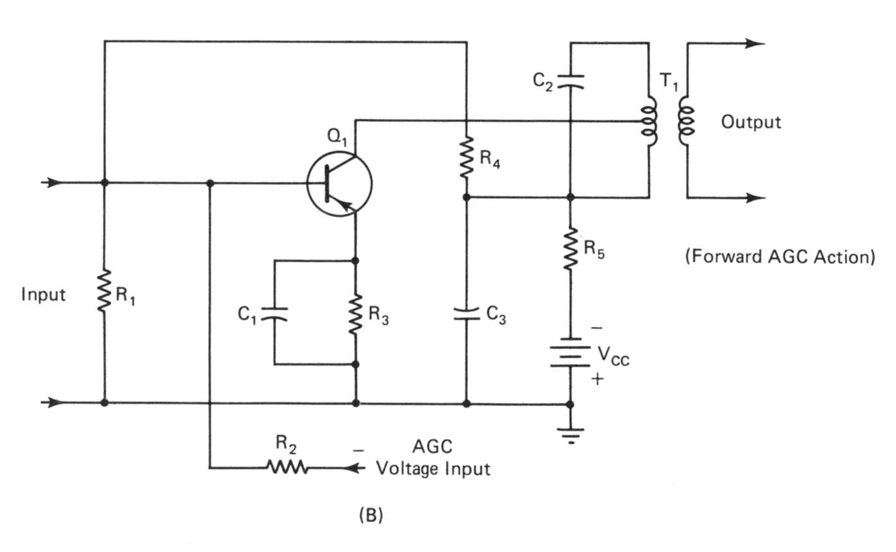

Figure 6–2. Automatic gain control of emitter current and collector voltage in a common-emitter amplifier.

NOTE: Reverse agc action biases the transistor toward cutoff as the carrier level increases. On the other hand, forward agc action biases the transistor toward saturation as the carrier level increases. The designer may utilize almost any variety of transistor for reverse-agc operation. However, it is often necessary to select a suitable variety of transistor for forward agc operation. As explained subsequently, reverse and forward agc characteristics differ with respect to cross-modulation interference.

Resistor R3 in Figure 6–2 is ac bypassed by C1; R3 is the emitter swamping resistor. Transformer T1 matches the collector output of Q1 to the input of the following stage. T1 operates at 455 kHz in this example. Specified selectivity is obtained by tapping the primary of T1.

2. *DC Collector Voltage Control:* The power gain of the CE stage in Figure 6–2(B) is controlled by feeding the agc voltage to the base of the transistor to vary the dc-emitter current, which in turn varies the dc-collector voltage. Variation of the dc-collector voltage is accomplished by passing the dc-collector current through R5 (R5 should have a value of 10 kilohms, or higher). In this example, the tuned amplifier is also operating as a dc amplifier to increase the dc voltage output from the second detector (amplified agc). In case that the second detector develops ample control voltage, amplified agc will not be required. In the exemplified circuit, R1 and R4 form a voltage divider to establish the no-signal (forward) bias on the base. The agc-control voltage from the second detector is negative with respect to ground and is fed to the base via R2. When the dc output from the second detector increases due to an increasing carrier level, the negative dc voltage fed to the base of Q1 via R2 increases the net negative (forward) bias on the base, thereby increasing the emitter current and the collector current. As a result, the drop across R5 reduces the collector voltage and reduces the stage gain. Capacitor C3 bypasses R5 to ground.

TYPICAL TUNED AMPLIFIER WITH AGC

With reference to Figure 6–3, a 455-kHz amplifier stage (Q1) and a second detector (Diode CR1) is exemplified. The input signal to the amplifier is an IF carrier with amplitude modulation at audio frequencies. The output from the diode is an audio signal. Observe that Q1 is neutralized in this example, to avoid development of regeneration or oscillation. Stage gain is controlled by reverse agc action.

Observe that two filtering actions are provided. Stated otherwise, C7 and R4 provide low-pass filter action that effectively removes the IF carrier component from the rectified signal, leaving a pulsating dc signal consisting of the audio information superimposed on a dc component that corresponds to the carrier level. In turn, C3 and R1 provide further low-pass filter action that effectively removes the audio component from the detected signal, leaving a dc-control voltage that changes value only in response to an increase or a decrease in carrier level.

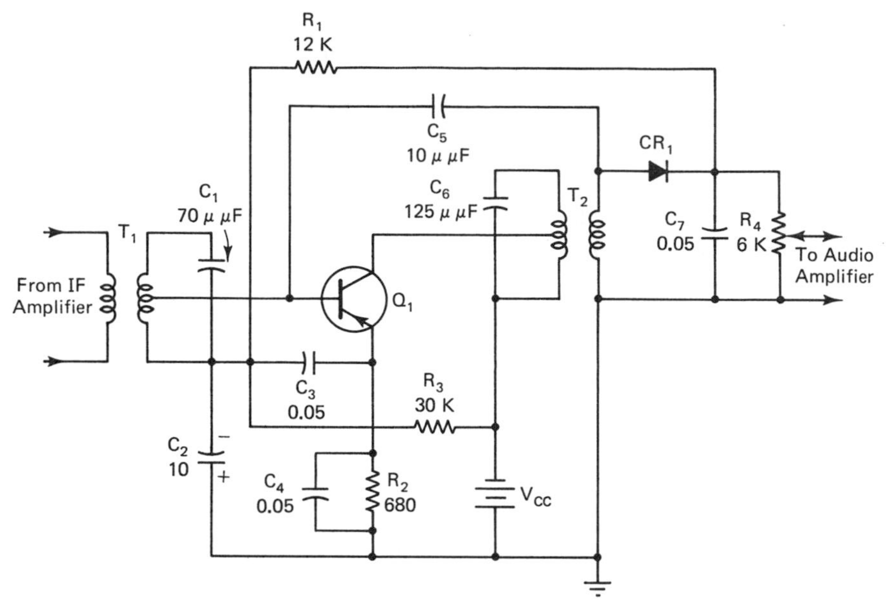

NOTE: Unless otherwise indicated, capacitances are in μF; resistances are in ohms.

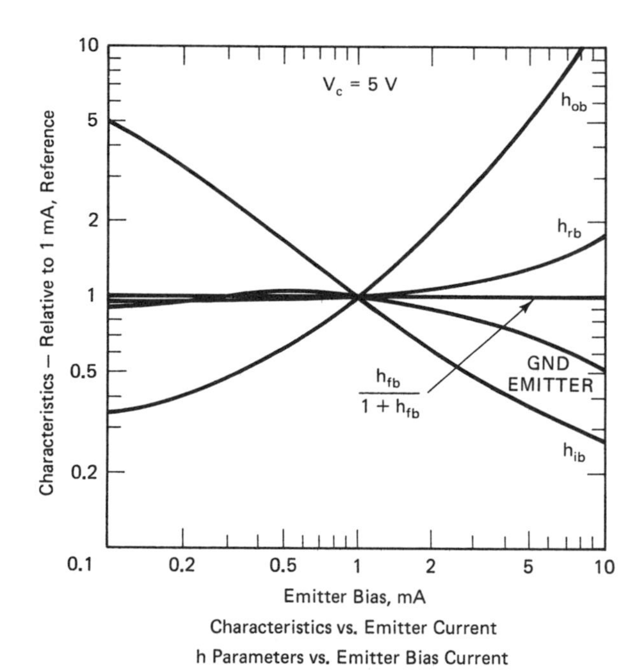

Characteristics vs. Emitter Current

h Parameters vs. Emitter Bias Current

Figure 6–3. Example of IF amplifier and second detector with agc and neutralizing circuits. (Reprinted by permission of General Electric Company.)

NOTE: An IF filter is provided by C7 and R4; an audio filter is provided by C3 and R1. The time constant of the IF filter is 300 microseconds; the time constant of the audio filter is 600 microseconds. However, these time constants are reduced to some extent by the loading action of the audio filter on the IF filter and by the shunting action of bias resistor R3.

This diagram exemplifies the changes in h_{ob}, h_{rb}, h_{ib}, and $h_{fb}/(1 + h_{fb})$ versus emitter-bias current with reference to 1 mA bias current and 5 V collector voltage. (Courtesy, GE)

Automatic gain control is more effective when the mixer or converter is preceded by an rf stage, for the following reasons:

1. In a system with converter input, agc power is taken from the detector and is applied to the first IF stage. Thus, there is only one stage of gain between the controlled stage and the detector (the second IF). However, in a system with an rf stage, there will be at least two stages of gain between the controlled stage and the detector (the converter and the IF). Accordingly, there will be more loop gain within the agc closure, with resulting better agc action. Agc voltage may also be applied to the IF section for additional control.

2. A converter is not ordinarily included in the controlled stages, inasmuch as a bias change could cause an appreciable shift in oscillator frequency. If a converter is the input stage, it must also be operated at maximum gain for optimum weak-signal performance. Under strong signal conditions, the converter will continue to operate at high gain and will tend to cause overloading of the signal channel. However, when an rf stage is employed, agc control voltage is applied to the input stage, which is the most effective point for agc action.

3. Observe that blocking of the converter on strong input signals is greatly relaxed by agc control of the preceding rf stage; the signal applied to the converter transistor may then be less than the magnitude of the incoming signal. From a design viewpoint, there are a few disadvantages (on cost, alignment, and tracking) in employing an rf stage: a three-gang tuning capacitor is required, which is a comparatively costly component. Alignment and tracking of the tuning system also requires more time and attention, which increases production costs.

EMITTER CURRENT CONTROL

As seen in Figure 6–3, a decrease in emitter current from 1 mA to 0.1 mA, for example, results in changes of h parameter values. The effect of these changes is chiefly:

1. A change in maximum available gain.

2. A change in impedance matching, since both h_{ob} and h_{ib} vary greatly.

Both of these factors result in a considerable change in power gain, as previously noted. In many cases, the collector-leakage current (I_{co}) is sufficient to prevent the transistor from being cut off completely. Typical performance data for this commonly used system is shown in Figure 6–4. Note in passing that the second detector may be slightly forward-biased as exemplified in Figure 6–5. This factor improves weak-signal detector action.

Observe that C5 functions in combination with the resistors to provide audio filtering, so that practically pure dc voltage is present on the agc line. In the event that agc voltage were taken directly from the detector, without audio filtering, the amplifier gain would be varied at an audio-frequency rate, with the result that the modulation information on the carrier would be smoothed out—a most undesirable condition.

The designer should note that employment of forward agc will almost always yield better immunity to cross-modulation interference and will ordinarily have better signal-handling capability than reverse agc. Transistors that are suitable for forward-agc operation have comparatively remote current-cutoff characteristics. On the debit side, forward agc generally involves more bandpass shift and response-curve tilt with changing signal levels than does reverse agc. If this factor is objectionable, the designer can utilize passive coupling circuits with parameters that largely compensate for shift and tilt.

It is evident that cross-modulation in rf circuits can be combatted effectively by providing improved tuned-circuit selectivity. This benefit, however, may be offset by the disadvantage of increased production costs. Mismatching of the reflected antenna impedance at the base of the rf transistor results in a slight degradation in the noise factor. However, this disadvantage of forward agc is generally disregarded in practical design procedures. Note that the gain of a conventional IF transistor can be varied over a range of approximately 20 dB, either by means of forward agc or reverse agc.

AGC SYSTEM ANALYSIS

With reference to Figure 6–6, a typical family of collector characteristics is shown for the region of collector potential below 1 V. A

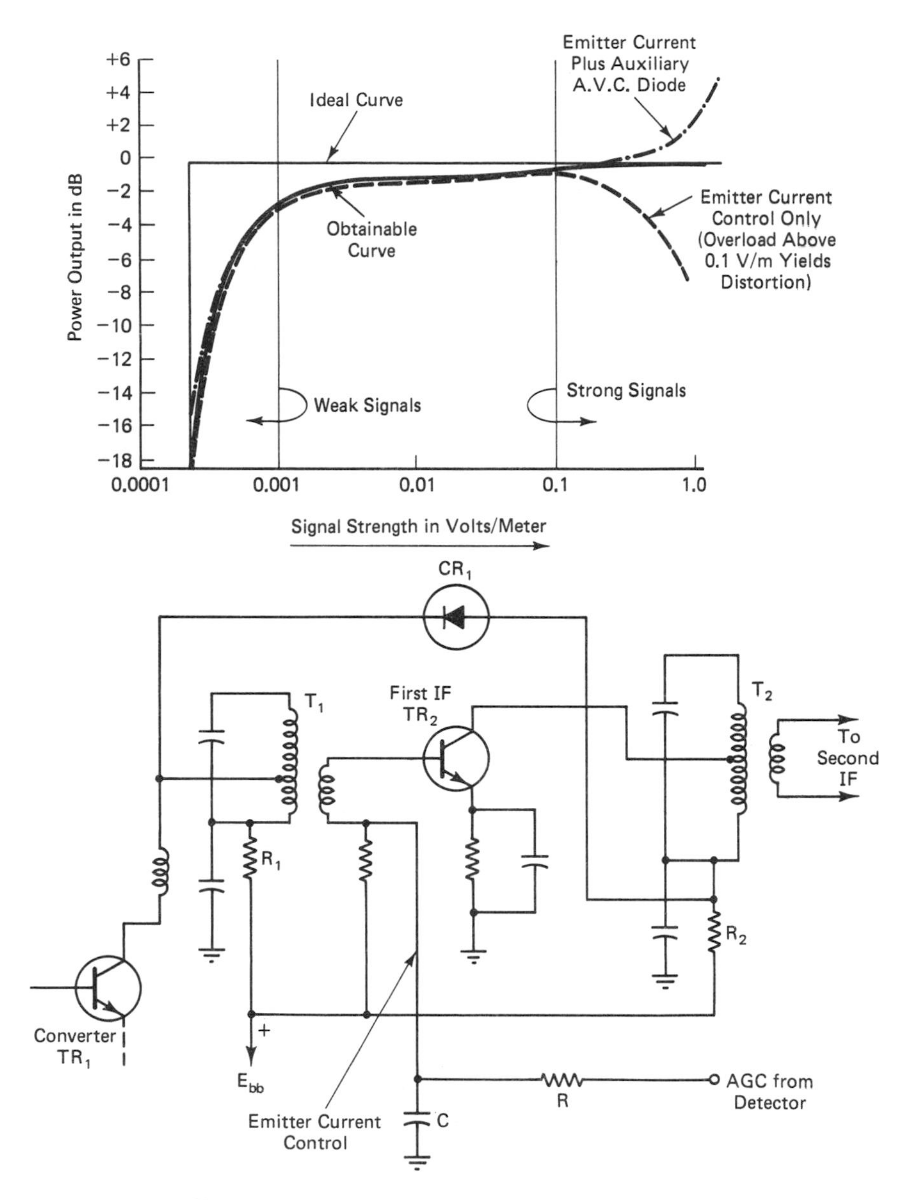

Figure 6–4. Typical agc control characteristics.

Note: Since most agc systems are somewhat limited in dynamic range, auxiliary diode agc may also be included. This arrangement shunts some of the signal to ground when a high-level signal is inputted. CR1 is back-biased by the voltage drops across R1 and R2 and presents a high impedance across T1 at low signal levels. As the signal level increases, agc action reduces the emitter current; this is accompanied by impedance mismatch and power gain is reduced. At still higher signal levels, CR1 becomes forward-biased, due to reduced voltage drop across R2. A low impedance is then shunted across T1, with accompanying increase in agc action, thereby avoiding overload distortion. (Reprinted by permission of General Electric Company.)

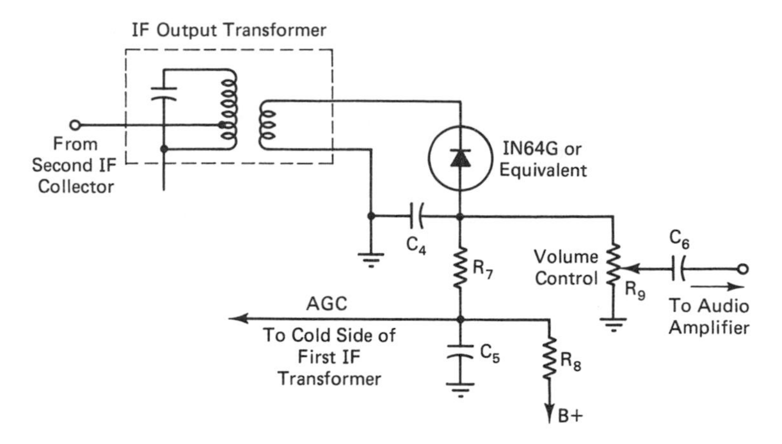

Figure 6–5. Example of detector configuration with a small forward bias. (Reprinted with permission of General Electric Company).

NOTE: A diode detector has a forward characteristic that is approximately logarithmic and that is effectively a linear function at high signal levels. On the other hand, the diode characteristic must be considered as a square-law function at low signal levels. Not only is the efficiency of rectification poorer along the square-law portion of the characteristic, but distortion also occurs due to detection nonlinearity with resulting intermodulation of the audio-component frequencies. Both disadvantages are minimized by biasing the diode out of its square-law region. The small forward bias that is required is provided by R8.

240-ohm load line (A) is drawn from $V_{ce} = 1$; the 240-ohm load line (B) is drawn from $V_{ce} = 0.6$, and the 240-ohm load line (C) is drawn from $V_{ce} = 0.2$. In turn, an I_b swing from 0.02 to 0.3 mA produces different I_C swings on the three load lines. There is a swing of 0.6 mA on (A), 0.3 mA on (B), and 0.1 mA on (C). It is evident that as the collector voltage is reduced to a small value, the transistor gain also decreases.

In Figure 6–7, the agc arrangement controls the gain of both the rf section and the IF section. Delayed agc is employed in many systems. Delayed agc denotes an operating characteristic in which the rf section operates at maximum gain on weak signals, and the gain reduction starts with application of agc voltage to the IF section. On stronger signals, the rf gain is reduced somewhat, and the IF gain continues to be reduced. The advantage of delayed agc is that noise voltages are prevented from impairing the gain of the system when tuned to weak signals. Delayed agc is obtained by electronic switching action in the rf-agc line. A reverse-biased diode is commonly used as a switch.

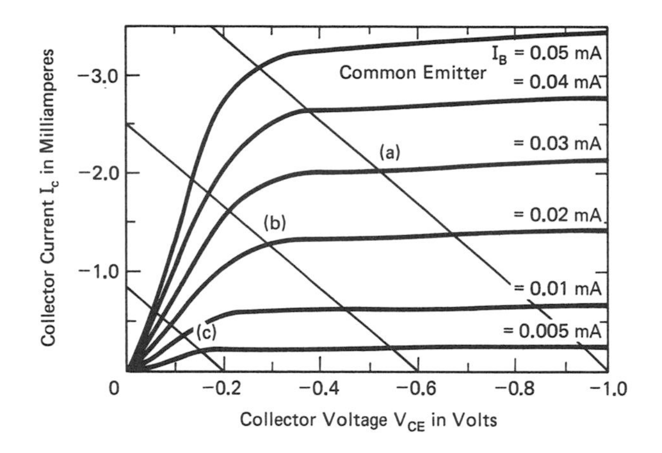

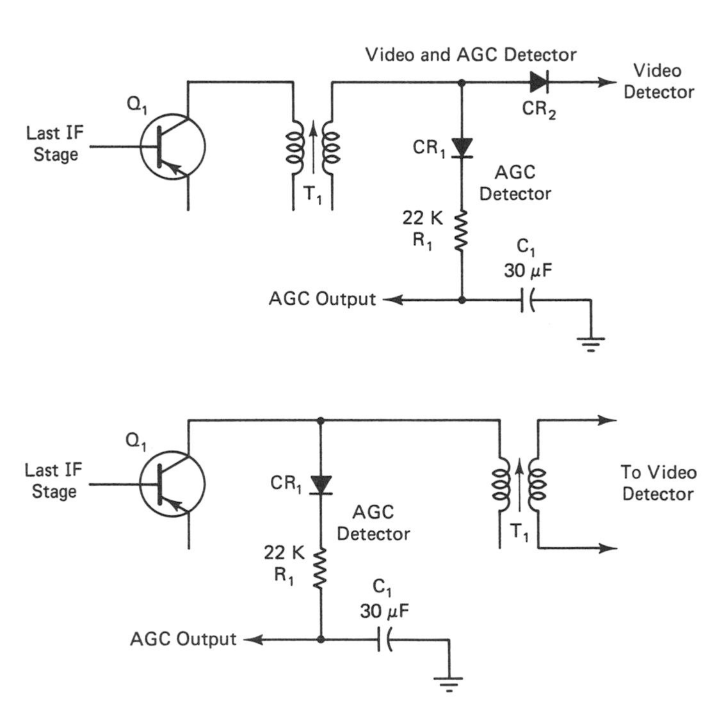

Figure 6–6. Typical collector characteristics with three load lines for analysis of agc action (Reproduced by special permission of Reston Publishing Company from *Television Theory and Servicing.*)

Note: In TV-receiver design, for example, separate agc detectors may be utilized. In the upper arrangement, the agc detector isolates the agc system from the video-amplifier channel so that agc voltage cannot back up from the RC filter into the video-detector diode. In the lower arrangement, the agc takeoff is on the primary side of the last IF transformer. Its operating features are essentially the same as described above.

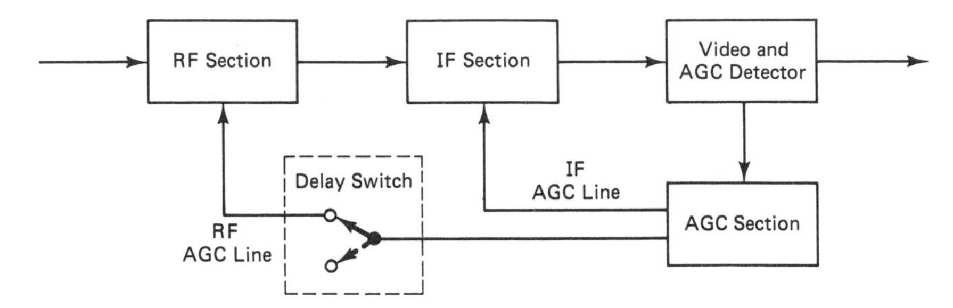

Figure 6–7. Plan of a delayed-agc arrangement.

NOTE: The delay switch may be implemented either as a reverse-biased diode or as a reverse-biased transistor. A reverse-bias potentiometer is often provided to give a maintenance facility for compensation of device and component tolerances in production and in field-servicing procedures.

PROGRAMMED FORMULAS

Designers are routinely concerned with reactance, impedance, and phase values in RC circuitry. A program for calculation of reactances, impedances, and phase angles for R and C in series and in parallel is provided in Figure 6–8. A program for calculation of R and X_C values in equivalent series or parallel circuits for a given impedance and phase angle is provided in Figure 6–9. Tolerances on R and C values are taken into account for calculation of impedances and phase angles for R and C in series and in parallel by the program provided in Figure 6–10.

DESIGN OPTIMIZATION CONSIDERATIONS

Automatic gain control is employed in a wide variety of consumer-electronic and commercial-electronic equipment. For example, automatic level control is commonly employed in tape-recorder preamplifiers. Automatic volume control is utilized in radio receivers. Automatic gain control is used in television receivers. One of the optimization considerations common to these various fields of application is the time constant of the particular control circuit. Thus, a comparatively long time constant is used in automatic-level-control (ALC) circuitry. A moderately long time constant is employed in automatic-volume-control (AVC) circuitry. A comparatively short time constant is utilized in TV automatic gain control (AGC) circuitry.

```
5 LPRINT"IMPEDANCE AND PHASE ANGLE FOR R AND C"
10 LPRINT"IN SERIES AND IN PARALLEL"
15 LPRINT"X, Z, and Phase, With Corresponding"
20 LPRINT"B, Y, and Phase Values":LPRINT""
25 INPUT"R (Ohms)=";R
30 LPRINT"R (Ohms)=";R
35 INPUT"C (Mfd)";C
40 LPRINT"C (Mfd)=";C
45 INPUT "f (Hz)=";F
50 LPRINT"f (Hz)=";F
55 X=1/(6.283*F*C*10^-6)
60 A$="######.##"
65 LPRINT"Capacitive Reactance (Ohms)=";USING A$;X
70 ZS=(R*R+X*X)^.5
75 LPRINT"Series Impedance (Ohms)=";USING A$;ZS
80 TH=-ATN(X/R)
85 LPRINT"Series Impedance Phase Angle (Degrees)=";USING A$;(TH*360/6.283)
90 GS=R/(R*R+X*X)
95 B$="##.#####"
100 LPRINT"Series Conductance (Siemens)=";USING B$;GS
105 XS=X/(R*R+X*X)
110 LPRINT"Series Susceptance (Siemens)=";USING B$;XS
115 YS=(GS*GS+XS*XS)^.5
120 LPRINT"Series Admittance (Siemens)=";USING B$;YS
125 C$="###.##"
130 LPRINT"Series Admittance Phase Angle (Degrees)=";USING C$;(-TH*360/6.283)
135 LPRINT"":LPRINT"........................................"
140 AB=R*X:BA=-6.283/4:AC=(R*R+X*X)^.5:CA=-ATN(X/R):AD=AB/AC:DA=(-6.283/4-CA)
145 LPRINT"Parallel Impedance (Ohms)=";USING C$;AD
150 LPRINT"Parallel Impedance Phase Angle (Degrees)=";USING C$;(DA*360/6.283)
155 RP=AD*COS(DA)
160 LPRINT"Parallel Impedance Resistive Component (Ohms)=";USING C$;RP
165 XP=AD*SIN(DA)
170 LPRINT"Parallel Impedance Reactive Component (Ohms)=";USING C$;-XP
175 GP=RP/(RP*RP+XP*XP)
180 D$="###.####"
185 LPRINT"Parallel Conductance (Siemens)=";USING D$;GP
190 BP=XP/(RP*RP+XP*XP)
195 LPRINT"Parallel Susceptance (Siemens)=";USING D$;-BP
200 YP=(GP*GP+BP*BP)^.5
205 LPRINT"Parallel Admittance (Siemens)=";USING D$;YP
210 TP=ATN(BP/GP)
215 E$="###.##"
220 LPRINT"Parallel Admittance Phase Angle (Degrees)=";USING E$;(-TP*360/6.283)
```

```
IMPEDANCE AND PHASE ANGLE FOR R AND C
IN SERIES AND IN PARALLEL
X, Z, and Phase, With Corresponding
B, Y, and Phase Values

R (Ohms)= 100
C (Mfd)= 1.65
f (Hz)= 1000
Capacitive Reactance (Ohms)=     96.46
Series Impedance (Ohms)=    138.94
Series Impedance Phase Angle (Degrees)=    -43.97
Series Conductance (Siemens)= 0.00518
Series Susceptance (Siemens)= 0.00500
Series Admittance (Siemens)= 0.00720
Series Admittance Phase Angle (Degrees)= 43.97
```

Figure 6–8. (Continues on next page.)

```
..........................................
Parallel Impedance (Ohms)= 69.43
Parallel Impedance Phase Angle (Degrees)=-46.03
Parallel Impedance Resistive Component (Ohms)= 48.20
Parallel Impedance Reactive Component (Ohms)= 49.97
Parallel Conductance (Siemens)=  0.0100
Parallel Susceptance (Siemens)=  0.0104
Parallel Admittance (Siemens)=  0.0144
Parallel Admittance Phase Angle (Degrees)= 46.03
```

Figure 6–8. Program for calculation of impedances and phase angles for R and C in series and in parallel.

```
5 LPRINT"R and XC Values in Equivalent Series or Parallel"
10 LPRINT"Circuits for a Specified Impedance"
15 LPRINT"And Phase Angle":LPRINT""
20 LPRINT"..........................................":LPRINT""
25 INPUT"Z (Ohms)=";Z
30 LPRINT"Z (Ohms)=";Z
35 INPUT"Phase Angle (Degrees)=";TH
40 LPRINT"Phase Angle (Degrees)=";TH
45 RS=Z*COS(TH*6.283/360)
50 XS=Z*SIN(TH*6.283/360)
55 RP=Z/COS(TH*6.283/360)
60 XP=Z/SIN(TH*6.283/360):LPRINT""
65 A$="######.##"
70 LPRINT"Equiv. Series R (Ohms)=";USING A$;RS
75 LPRINT"Equiv. Series X (Ohms)=";USING A$;XS
80 LPRINT"Equiv. Parallel R (Ohms)=";USING A$;RP
85 LPRINT"Equiv. Parallel X (Ohms)=";USING A$;XP
```

θ

Z

```
R and XC Values in Equivalent Series or Parallel
Circuits for a Specified Impedance
And Phase Angle
..........................................
Z (Ohms)= 100
Phase Angle (Degrees)= 65

Equiv. Series R (Ohms)=     42.26
Equiv. Series X (Ohms)=     90.63
Equiv. Parallel R (Ohms)=    236.60
Equiv. Parallel X (Ohms)=    110.34
```

Figure 6–9. Program for calculation of R and X_C values in equivalent series or parallel circuits for a given impedance and phase angle.

```
5 LPRINT"Z and Phase for R and C in Series and in Parallel"
10 LPRINT"As a Function of Tolerances":LPRINT""
15 LPRINT"..............................................":LPRINT""
20 INPUT "R (Ohms)=";R
25 LPRINT "R (Ohms)=";R
30 INPUT"C (Mfd)=";C
35 LPRINT "C (Mfd)=";C
40 INPUT "f (Hz)=";F
45 LPRINT "f (Hz)=";F
50 INPUT"R High Tolerance (%)=";RH
55 LPRINT"R High Tolerance (%)=";RH
60 INPUT"R Low Tolerance (%)=";RL
65 LPRINT"R Low Tolerance (%)=";RL
70 INPUT"C High Tolerance (%)=";CH
75 LPRINT"C High Tolerance (%)=";CH
80 INPUT"C Low Tolerance (%)=";CL
85 LPRINT"C Low Tolerance (%)=";CL
90 X=1/(6.283*F*C*10^-6)
95 A$="######.##":LPRINT""
100 LPRINT"Bogie Capacitive Reactance (Ohms)=";USING A$;X
105 Z=(R*R+X*X)^.5
110 LPRINT"Bogie Series Impedance (Ohms)=";USING A$;Z
115 T=ATN(X/R)
120 LPRINT"Bogie Series Phase Angle (Degrees)=";USING A$;-T*360/6.283
125 ZP=R*SIN(T)
130 LPRINT"Bogie Parallel Impedance (Ohms)=";USING A$;ZP
135 PH=6.283/4-T
140 LPRINT"Bogie Parallel Phase Angle (Degrees)=";USING A$;-PH*360/6.283
145 XX=1/(6.283*F*(C*10^-6+.01*CH*C*10^-6))
150 LPRINT"High Tol. Capacitive Reactance (Ohms)=";USING A$;XX
155 XL=1/(6.283*F*(C*10^-6-.01*CL*C*10^-6))
160 LPRINT"Low Tol. Capacitive Reactance (Ohms)=";USING A$;XL
165 ZZ=((R+.01*RH*R)^2+XX*XX)^.5
170 LPRINT"High Tol. Series Impedance (Ohms)=";USING A$;ZZ
175 ZL=((R-.01*RL*R)^2+XL*XL)^.5
180 LPRINT"Low Tol. Series Impedance (Ohms)=";USING A$;ZL
185 PQ=(R+.01*RH*R):TT=ATN(XX/PQ)
190 LPRINT "High Tol. Series Phase Angle (Degrees)=";USING A$;-TT*360/6.283
195 QP=(R-.01*RL*R):TL=ATN(XL/QP)
200 LPRINT"Low Tol. Series Phase Angle (Degrees)-";USING A$;-TL*360/6.283
205 PZ=(R+.01*RH*R)*SIN(TT)
210 LPRINT"High Tol. Parallel Impedance (Ohms)=";USING A$;PZ
215 HP=6.283/4-TT
220 LPRINT"High Tol. Parallel Phase Angle (Degrees)=";USING A$;-HP*360/6.283
225 K=(R-.01*RL*R)*SIN(TL)
230 LPRINT"Low Tol. Parallel Impedance (Ohms)=";USING A$;K
235 PP=6.283/4-TL
240 LPRINT"Low Tol. Parallel Phase Angle (Degrees)=";USING A$;-PP*360/6.283
```

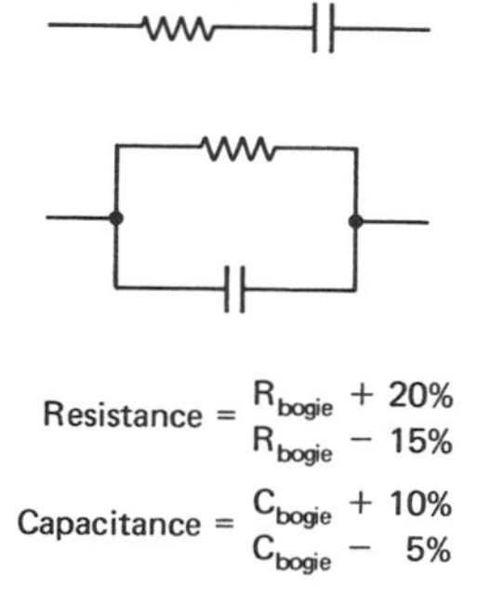

Figure 6–10. (Continues on next page.)

```
Z and Phase for R and C in Series and in Parallel
As a Function of Tolerances

.................................................

R (Ohms)= 100
C (Mfd)= 1.65
f (Hz)= 1000
R High Tolerance (%)= 20
R Low Tolerance (%)= 15
C High Tolerance (%)= 10
C Low Tolerance (%)= 5

Bogie Capacitive Reactance (Ohms)=     96.46
Bogie Series Impedance (Ohms)=    138.94
Bogie Series Phase Angle (Degrees)=    -43.97
Bogie Parallel Impedance (Ohms)=     69.43
Bogie Parallel Phase Angle (Degrees)=    -46.03
High Tol. Capacitive Reactance (Ohms)=     87.69
Low Tol. Capacitive Reactance (Ohms)=    101.54
High Tol. Series Impedance (Ohms)=    148.63
Low Tol. Series Impedance (Ohms)=    132.42
High Tol. Series Phase Angle (Degrees)=    -36.16
Low Tol. Series Phase Angle (Degrees)-    -50.07
High Tol. Parallel Impedance (Ohms)=    70.80
High Tol. Parallel Phase Angle (Degrees)=    -53.84
Low Tol. Parallel Impedance (Ohms)=    65.18
Low Tol. Parallel Phase Angle (Degrees)=    -39.93
```

Figure 6–10. Program for calculation of impedances and phase angles for R and C in series and in parallel, taking tolerances on R and C values into account.

The time constant is still shorter in TV keyed-agc circuitry. This agc system does not process the dc component of the composite video signal. Instead, agc voltage is obtained by sampling the tips of the horizontal sync pulses. In turn, the output from the sampler consists of narrow pulses with a 15,750-Hz repetition rate. These are dc pulses with an amplitude corresponding to the prevailing level of the horizontal sync pulses in the composite video signal.

Sampled pulses in a keyed-agc circuit are passed through an RC filter and smoothed into nearly pure dc, which in turn functions as the agc voltage. Note that the chief advantage of keyed agc is that 60-Hz vertical-sync pulses do not enter the sampled pulse train, so that the agc filter can have a shorter time constant than otherwise. Accordingly, the receiver gain responds with maximum rate of change to rapidly varying signal levels; for example, airplane flutter in a TV signal is much less annoying to the viewer when keyed agc is employed.

ALC is commonly provided in tape-recorder amplifiers in order to obtain optimum recording characteristics without the necessity for "riding gain" by the operator. Inasmuch as ALC varies the gain of an audio-frequency amplifier, a comparatively long time constant is required. The optimum time constant provides a "natural" tracking of changing sound levels without objectionable "pumping" of background sounds between high-level signal intervals.

Time-constant optimization in single-sideband (SSB) receivers involves different attack and decay rates. In other words, the signal characteristics are most compatible with agc transient response involving rapid attack on increasing signal levels, but slow decay on decreasing signal levels. Because of the semicritical nature of the transient response, designers often include adjustable control of attack and decay rates.

Note that in SSB reception, the carrier component is absent from the incoming signal. In other words, the agc detector derives its drive solely from the modulation envelope of the incoming signal. Rapid attack is required to avoid transient overload at the start of a word; otherwise, each syllable in the audio signal will start with an obtrusive "agc thump." Then, the agc voltage must remain essentially constant at a value corresponding to the average value of syllabic variations in the signal over a period of seconds. Observe that if the agc voltage responds rapidly to syllabic peaks, background noise may obtrude as a result of "agc pumping." As a rough rule of thumb, an attack time in the range from 50 to 200 milliseconds, and a decay time in the range from 0.5 to 3 seconds serve to optimize this type of agc system.

7

WIDE-BAND AMPLIFIER REQUIREMENTS

GENERAL CONSIDERATIONS

The term "wide-band amplifier" denotes a frequency response that exceeds the audio range (20 MHz). Thus, video amplifiers (4 MHz) and oscilloscope amplifiers (typically 20 MHz) are familiar examples of wide-band amplifiers. In this chapter, treatment is primarily directed to video amplification. Practically all video amplifiers employ some form of modified RC coupling between stages. Design of coupling circuits is concerned with uniformity of frequency response and linearity of the phase characteristic over the rated operating range.

It was previously noted that the low-frequency limit in an RC-coupled amplifier is determined by the values of the coupling capacitors (in relation to associated circuit resistance) and by the values of bypass and decoupling capacitors (in relation to associated-circuit resistance). In this regard, emitter bypass capacitors must have very large values to avoid impairment of the low-frequency limit. This is just another way of saying that an emitter-bypass capacitor generally operates with comparatively low-shunt resistance.

The high-frequency limit in an RC-coupled amplifier is primarily determined by stray and device capacitances, shown as C_o and C_i in Figure 7–1. Since the reactances of C_o and C_i decrease as the frequency increases, gain falls off as the frequency increases. Gain falls off at a slower rate, however, if the circuit resistance is reduced; for example, if collector load resistances are halved, the high-frequency response is extended. On the other hand, halving of the collector-load resistances reduces the amplifier gain, and additional stages are then required to restore the original gain.

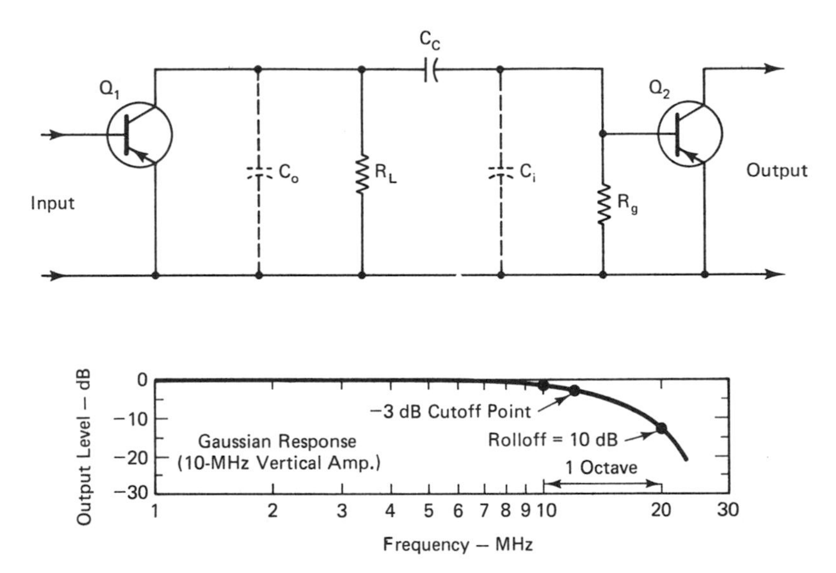

Figure 7–1. Skeleton RC-coupled amplifier showing inherent capacitances that affect high-frequency response.

NOTE: As a practical design guideline, when basic RC-coupled stages are cascaded, the rate of falloff in high-frequency response increases as the number of stages is increased. However, the rate of falloff approaches a limit as the number of stages is increased indefinitely. This limit is designated as a Gaussian response, and the limit is approached rather rapidly as stages are cascaded. The Gaussian response is of significance in design of oscilloscope amplifiers with somewhat limited bandwidth. In other words, if the input waveform contains significant harmonics beyond the amplifier passband, minimum ringing and overshoot of the output waveform will result if the amplifier has a Gaussian cutoff characteristic.

The Gaussian response has the same shape as that of the standard probability curve (bell curve).

Low-Frequency and Phase Characteristics

Wide-band amplifiers such as video amplifiers are often required to reproduce complex waveforms such as square waves with minimum distortion. This ability requires that the bandwidth be sufficient to include all of the significant harmonics in the waveform. A linear phase-versus-frequency characteristic is also required to avoid tilt distortion and related responses such as overshoot and ringing. Delay time through the amplifier is constant at any frequency, and phase shift (if linear) varies directly with frequency. Nonlinear phase shift versus frequency results in waveform distortion, regardless of frequency response.

A variation in the phase of the low-frequency response is more detrimental than a variation in gain. The gain of the low-frequency response will drop to only 99.4 percent of its midfrequency value when the phase shift varies 2° from the phase at midfrequency. In typical applications, a phase shift of 2° is the maximum tolerable value. With reference to Figure 7–2, phase shift distortion can be minimized by designing for direct variation of phase shift with frequency. In the illustration, the fundamental frequency and its second harmonic are depicted in the input. Next, the fundamental frequency and second harmonic are shown at the output, but shifted in phase (45° for the fundamental and 90° for the second harmonic).

Note that the amount of phase shift depicted in Figure 7–2 is exaggerated for clarification. A delay time of 1/800 second causes a 45° phase shift in the fundamental frequency. Since the delay time is constant and phase shift varies directly with frequency in this example, the second harmonic is also delayed 1/800 second. The resulting phase shift of the second harmonic is 90°. As the frequency doubles, the phase shift doubles, and the fundamental frequency and the second harmonic remain in the same relative phase relationship (due to the linear phase characteristic).

HIGH-FREQUENCY COMPENSATION

Shunt Compensation: With reference to Figure 7–3, inductor L_1 is included in series with load resistor R_L for shunt compensation. Its inductive reactance compensates for the shunting effect of output-impedance capacitance C_o and input-impedance capacitance C_i.

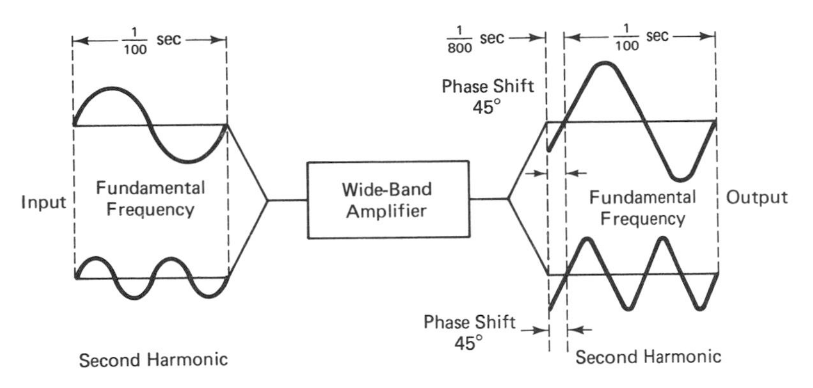

Figure 7–2. Phase relationship comparison, showing basic delay-time characteristic.

NOTE: Phase distortion in a video amplifier is much more objectionable than phase distortion in an audio amplifier. Most listeners are unable to detect the presence of phase distortion in the sound output from an audio amplifier. On the other hand, TV viewers immediately recognize that a picture has "poor quality" when there is appreciable phase distortion in the video amplifier. Similarly, oscilloscope operators immediately recognize that a displayed waveform is "distorted" when there is appreciable phase distortion in the vertical amplifier.

Low-frequency phase distortion is evidenced as tilt along the top and bottom of a displayed square wave. High-frequency phase distortion is evidenced as overshoot, often accompanied by ringing on the leading and trailing edges of a displayed square wave.

Capacitor C_C is effectively a short-circuit at high frequencies. Since inductor L_1 (in series with R_L and capacitors C_i and C_o) form a parallel-resonant circuit having a very broad response, this form of compensation is commonly called shunt peaking.

The resonant peak of the foregoing parallel combination maintains a practically uniform gain through the high-frequency range. The impedance of the parallel resonant circuit has a value that is approximately the same as that of the load resistor R_L. When the signal frequency increases, the decrease in the capacitive reactances of C_o plus C_i is precisely compensated by the increase in the inductive reactance of L_1. In turn, the amplifier frequency response is extended to a considerably higher frequency limit.

Series Compensation: With reference to Figure 7–3(B), series compensation is obtained by adding inductor L_2 in series with capacitor C_c. Considering that C_c is an effective short-circuit at high frequencies, L_2 and C_i form a series-resonant circuit at very high frequencies. As this high-frequency region is entered, the series circuit

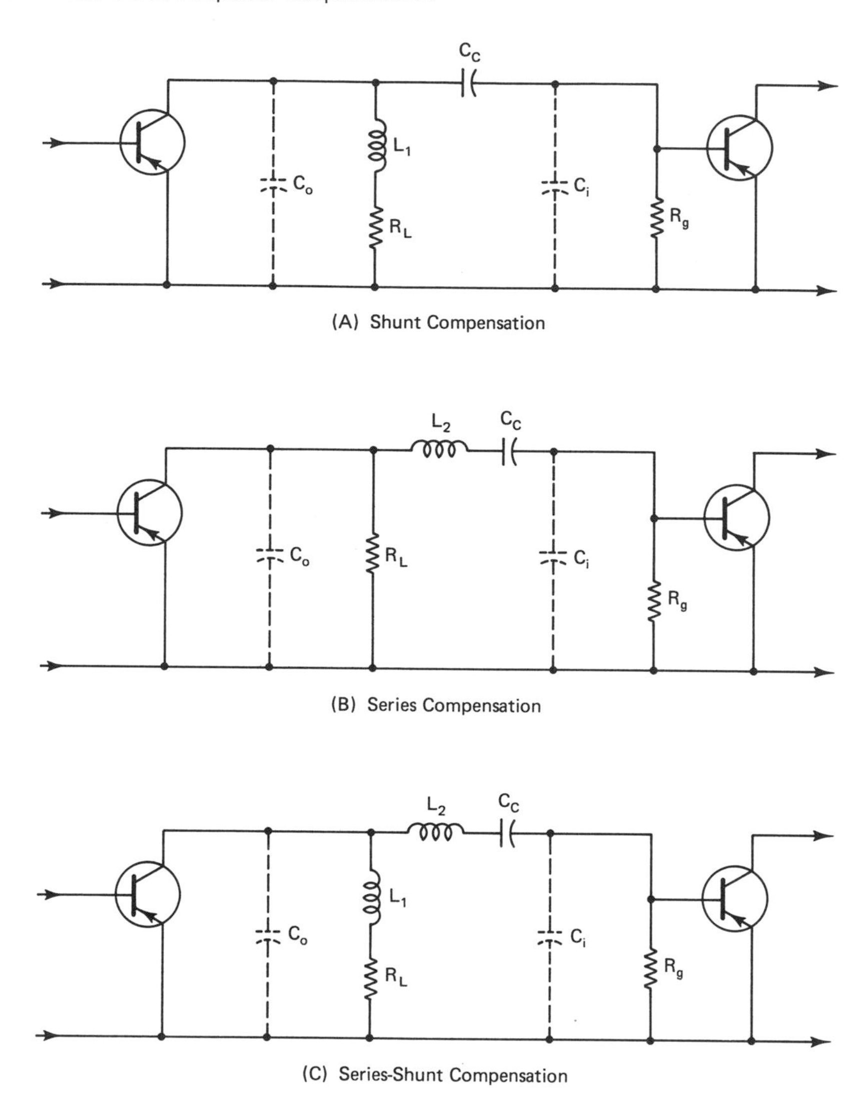

Figure 7–3. Fundamental shunt, series, and series-shunt high-frequency compensation with peaking coils in compensation coupling between stages.

NOTE: The frequency response of a wide-band amplifier at room temperature can be quite different at higher or lower temperatures. Production tolerances have a substantial effect on frequency response. The frequency response is not the same for high-signal levels as for low-signal levels and may be objectionably different. (The amplifier must be rated accordingly.) Stray capacitance variations also have a marked effect on frequency response. As a rule of thumb, a 20 percent change in value of a critical parameter can result in a change of 0.7 dB in gain for each stage over the final half-octave of the frequency response in the case of simple shunt peaking. When both series and shunt peaking are used, the relevant parameters become more critical.

approaches resonance. In turn, current flow through C_i increases as the frequency increases. The capacitive reactance of C_o decreases the voltage across R_L as the frequency increases. However, since the current flow through C_i increases with frequency, it compensates for the decrease in voltage across R_L.

When the component values are properly proportioned in a series-compensated coupling circuit, the frequency response is approximately the same as for a shunt-compensated coupling circuit. However, the high-frequency gain of a wide-band amplifier using series peaking is about 50 percent greater than for the same amplifier using shunt peaking. Designers of oscilloscope amplifiers sometimes prefer shunt peaking inasmuch as there is less tendency for the output waveform to overshoot and ring than when series peaking is used. Overshoot and ringing do not occur in any case, unless the input waveform has significant harmonics that extend beyond the amplifier passband.

LOW-FREQUENCY COMPENSATION

Over the low-frequency end of the video-amplifier frequency-response range, the input and output capacitances of the transistors have practically no effect on the frequency response. The low-frequency response is limited by C_C and R_g (Figure 7–4). The time constant (R_gC_C) must be large in order to prevent the low-frequency response from falling off and creating associated phase distortion. Loss of gain at low frequencies is minimized by adding a compensating filter in series with load resistor R_L. It comprises resistor R_F and capacitor C_F.

The function of the low-frequency filter is to increase the collector-load impedance at low frequencies and to compensate for the phase shift produced by capacitor C_C and resistor R_g. At high frequencies, C_T is practically a short-circuit. In turn, the collector load impedance is represented by R_L at high frequencies. However, as the frequency decreases, the reactance of C_F increases. At very low frequencies, C_F is practically an open circuit, and the collector load impedance consists of R_L plus R_F. This arrangement extends the frequency response to a much lower frequency limit. Both gain and phase become more uniform over the low-frequency region of amplifier response.

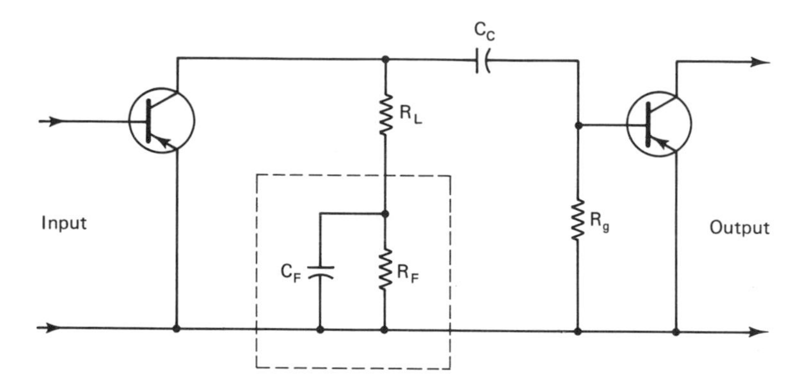

NOTE: Low-frequency and high-frequency peaking arrangements are designed to obtain maximum extension and uniformity of amplifier frequency response. These arrangements disregard the resulting phase characteristic of the amplifier, and in practice, this phase characteristic can often be neglected. On the other hand, in demanding applications, the phase characteristic may require equalization. A delay equalizer is defined as a corrective network that is designed to make the phase delay or envelope delay of a circuit or system substantially constant over a desired frequency range.

A simple bridged-T-delay equalizer is shown below. It is a constant-resistance network because the indicated reactance relations provide a resistive characteristic impedance. Its phase-versus-frequency characteristic can be varied over a wide range by choice of reactance values.

This arrangement is particularly helpful in equalizing the high-frequency phase characteristic of the amplifier. In turn, at low frequencies, the equalizer is essentially "out of the circuit."

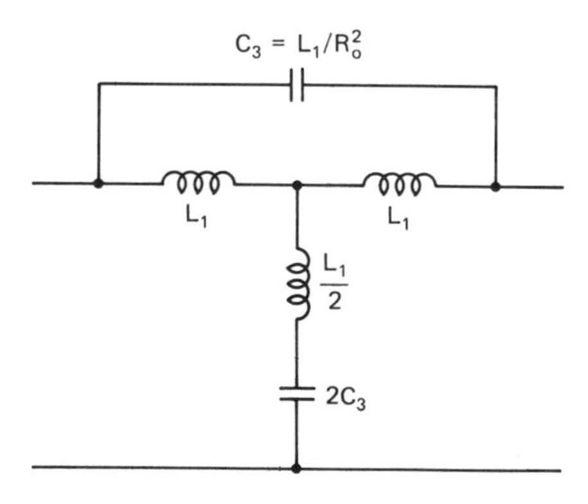

Figure 7–4. Arrangement of RC section for low-frequency compensation coupling.

SIMULTANEOUS HIGH- AND LOW-FREQUENCY COMPENSATION

A typical wide-band amplifier configuration with frequency response from 30 Hz to 4 MHz utilizing both high- and low-frequency compensation is shown in Figure 7–5. The high- and the low-frequency compensating circuits function independently and do not interfere with each other. At low frequencies, the series reactance of L_1 is very small and has no effect on the collector-load impedance. The series reactance of L_2 is also very small with no effect on the input circuit. The reactances of output-impedance capacitance C_o (not shown) and input-impedance capacitance C_i (not shown) are so large that they have no effect at low frequencies.

Stated otherwise, the combination peaking coupling circuitry has no effect on amplifier action at low frequencies. Conversely, at high frequencies, the reactance of C_F is very small and can be

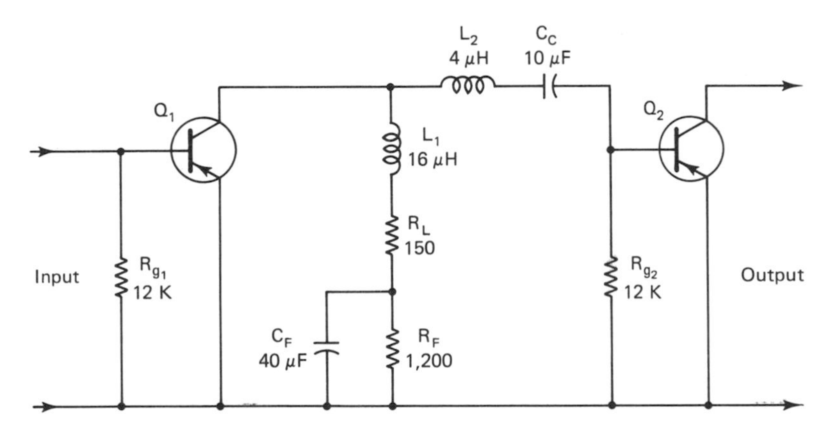

Figure 7–5. Typical low- and high-frequency-compensation coupling for a frequency response from 30 Hz to 4 MHz.

NOTE: Designers must direct close attention to production tolerances on all component values in this type of circuitry. Otherwise, both the low-frequency and the high-frequency regions of response will deviate substantially from bogie. Because close-tolerance components are costly, the designer may prefer to include maintenance adjustments. For example, peaking coils may be provided with adjustable ferrite cores, and R_F may be a 2000-ohm potentiometer. It is advisable to specify a reasonable tolerance on C_F. In turn, amplifier characteristics can be optimized at the production test station.

This is a skeleton configuration that features the frequency compensating arrangement. The supply voltage and bias circuitry are omitted. A complete configuration also requires attention to bias stabilization.

regarded as a short-circuit for ac. In turn, R_F has no effect on the collector-load impedance at high frequencies, or the compensating filter R_FC_F has no effect on high-frequency response. As noted in the diagram, tolerances are an important design factor in this type of circuitry. Since worst-case conditions involving all pertinent tolerance factors are difficult to evaluate, it is generally advisable to proceed on the basis of test measurements on worst-case prototype models over the rated temperature range.

WIDE-BAND AMPLIFIER INTEGRATED CIRCUITRY

Wide-band amplifiers can be designed around integrated circuits, such as the 3005 differential amplifier package noted in Figure 7–6. Peaking coils are not utilized; the external circuitry comprises only resistors and capacitors. From an applications viewpoint, an IC video amplifier such as exemplified cannot provide a substantial output-voltage swing. Thus, the 3005 is rated for operation from two 12-volt sources, with the result that the maximum available output-voltage swing is less than 24 volts. When IC wide-band amplifier capabilities are adequate, a production-cost advantage may be realized.

PROGRAMMED FORMULAS

Designers are routinely concerned with vector addition, subtraction, multiplication, and division. In turn, it is often helpful to have

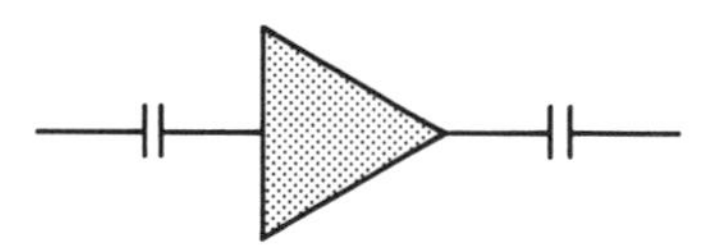

Figure 7–6. Wide-band amplifiers can be designed from integrated circuits.

NOTE: A video amplifier can be configured from a 3005 IC, for example, and will provide a transconductance of 20 millisiemens over a 10-MHz frequency range, with a midband voltage gain of 19 dB. The IC package contains a balanced differential amplifier driven from a controlled constant-current source. The noise figure is 7.8 dB.

Encyclopedia of Integrated Circuits *by Walter H. Buchsbaum, Sc.D., provides access to more than 250 different functions of integrated circuits from the user's viewpoint. Also included are numerous practical tips that can save many hours of design time.*

a computer program available for solving equations with vectors in rectangular or polar form, as provided in Figure 7–7.

This program inputs a pair of vectors in either rectangular or polar format; calculates their sum, difference, product, or quotient;

```
1 LPRINT"Addition, Subtraction, Multiplication, or Division"
2 LPRINT"Of a Pair of Vectors in Polar or Rectangular Form"
3 LPRINT"":LPRINT"............................................."
4 LPRINT""
5 INPUT"Polar 1; Rect. 0";X
6 IF X=0 THEN GOTO 8
7 IF X=1 THEN GOTO 12
8 INPUT"Z1 Resistance (Ohms)=";A
9 INPUT"Z1 Reactance (Ohms)=";B
10 INPUT"Z2 Resistance (Ohms)=";C
11 INPUT"Z2 Reactance (Ohms)=";D:GOTO 20
12 INPUT"Z1 Magnitude (Ohms)=";E
13 INPUT"Z1 Angle (Degrees)=";F
14 INPUT"Z2 Magnitude (Ohms)=";G
15 INPUT"Z2 Angle (Degrees)=";H
16 A=E*COS(F*6.283/360)
17 B=E*SIN(F*6.283/360)
18 C=G*COS(H*6.283/360)
19 D=G*SIN(H*6.283/360)
20 A$="#########"
21 LPRINT"R1 (Ohms)=";USING A$;A
22 LPRINT"X1 (Ohms)=";USING A$;B
23 LPRINT"R2 (Ohms)=";USING A$;C
24 LPRINT"X2 (Ohms)=";USING A$;D:LPRINT""
25 NJ=B/A
26 PH=ATN(NJ)
27 E=(A*A+B*B)^.5
28 LPRINT"Z1 (Ohms)=";USING A$;E
29 LPRINT"Phase Angle (Degrees)=";USING A$;PH*360/6.283
30 LPRINT"(Polar)":LPRINT""
31 G=(C*C+D*D)^.5
32 JJ=D/C
33 HP=ATN(JJ)
34 LPRINT"Z2 (Ohms)=";USING A$;G
35 LPRINT"Phase Angle (Degrees)=";USING A$;HP*360/6.283
36 LPRINT"(Polar)":LPRINT""
37 Q=A+C
38 QQ=B+D
39 A$="######.##":LPRINT"Z1+Z2 (Ohms Resistance)=";USING A$;Q
40 LPRINT"Z1+Z2 (Ohms Reactance)=";USING A$;QQ
41 LPRINT"(Rect.)":LPRINT""
42 NN=E*G
43 MM=PH+HP
44 LPRINT"Z1*Z2 (Ohms)=";USING A$;NN
45 LPRINT"Phase Angle (Degrees)=";USING A$;MM*360/6.283
46 LPRINT"(Polar)":LPRINT""
47 Y=(Q*Q+QQ*QQ)^.5
48 XY=QQ/Q
49 TH=ATN(XY)
50 LPRINT"Z1+Z2 (Ohms)=";USING A$;Y
51 LPRINT"Phase Angle (Degrees)=";USING A$;TH*360/6.283
52 LPRINT"(Polar)":LPRINT""
53 RZ=NN*COS(MM)
54 XZ=NN*SIN(MM)
55 LPRINT"Z1*Z2 (Ohms Resistance)=";USING A$;RZ
```

Figure 7–7. (Continues on next page.)

```
56 LPRINT"Z1*Z2 (Ohms Reactance)=";USING A$;XZ
57 LPRINT "(Rect)": LPRINT""
58 LY=A-C
59 YL=B-D
60 LPRINT"Z1-Z2 (Ohms Resistance)=";USING A$;LY
61 LPRINT"Z1-Z2 (Ohms Reactance)=";USING A$;YL
62 LPRINT"(Rect.)":LPRINT""
63 QL=C-A
64 LQ=D-B
65 LPRINT"Z2-Z1 (Ohms Resistance)=";USING A$;QL
66 LPRINT"Z2-Z1 (Ohms Reactance)=";USING A$;LQ
67 LPRINT"(Rect)":LPRINT""
68 XZ=E/G
69 ZX=PH-HP
70 LPRINT"Z1/Z2 (Ohms)=";USING A$;XZ
71 LPRINT"Phase Angle (Degrees)=";USING A$;ZX*360/6.283
72 LPRINT"(Polar)":LPRINT""
73 CZ=G/E
74 ZC=HP-PH
75 A$="######.##":LPRINT"Z2/Z1 (Ohms)=";USING A$;CZ
76 LPRINT"Phase Angle (Degrees)=";USING A$;ZC*360/6.284
77 LPRINT"(Polar)":LPRINT""
78 MX=(LY*LY+YL*YL)^.5
79 XX=ATN(YL/LY)
80 LPRINT"Z1-Z2 (Ohms)=";USING A$;MX
81 LPRINT"Phase Angle (Degrees)=";USING A$;XX*360/6.283
82 LPRINT"(Polar)":LPRINT""
83 JX=(QL*QL+LQ*LQ)^.5
84 XJ=LQ/QL
85 JQ=ATN(XJ)
86 LPRINT"Z2-Z1 (Ohms)=";USING A$;JX
87 LPRINT"Phase Angle (Degrees)=";USING A$;JQ*360/6.283
88 LPRINT"(Polar)":LPRINT""
89 PO=XZ*COS(ZX)
90 OP=XZ*SIN(ZX)
91 LPRINT"Z1/Z2 (Ohms Resistance)=";USING A$;PO
92 LPRINT"Z1/Z2 (Ohms Reactance)=";USING A$;OP
93 LPRINT"(Rect.)":LPRINT""
94 PR=CZ*COS(ZC)
95 RP=CZ*SIN(ZC)
96 LPRINT"Z2/Z1 (Ohms Resistance)=";USING A$;PR
97 LPRINT"Z2/Z1 (Ohms Reactance)=";USING A$;RP
98 LPRINT"(Rect.)":LPRINT""
99 LPRINT"When angles greater than 90 degrees are being processed,"
100 LPRINT"there is a possible 180 degree ambiguity in the final answer."
101 LPRINT"To check for 180 degree ambiguity, make a rough sketch of"
102 LPRINT"the vector diagram.":LPRINT""
103 LPRINT"Slight inaccuracies in computed values may occur due to"
104 LPRINT"single-precision and rounding off processing"
```

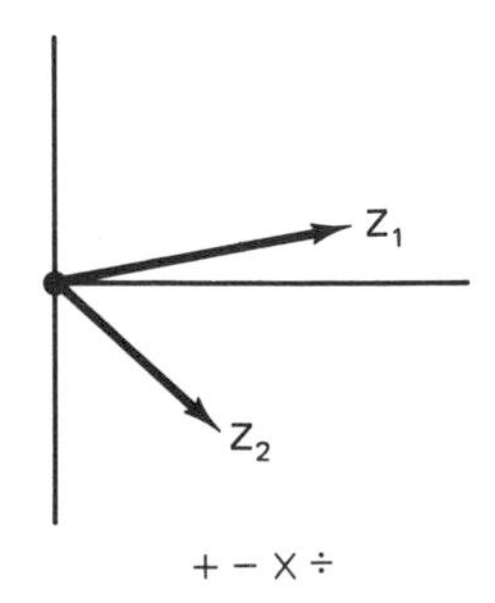

+ − X ÷

```
Addition, Subtraction, Multiplication, or Division
Of a Pair of Vectors in Polar or Rectangular Form

...............................................

R1 (Ohms)=        181
X1 (Ohms)=         85
R2 (Ohms)=        127
X2 (Ohms)=        272

Z1 (Ohms)=        200                          Z2-Z1 (Ohms Resistance)=    -54.48
Phase Angle (Degrees)=          25             Z2-Z1 (Ohms Reactance)=    187.37
(Polar)                                        (Rect)

Z2 (Ohms)=        300                          Z1/Z2 (Ohms)=        0.67
Phase Angle (Degrees)=          65             Phase Angle (Degrees)=    -40.00
(Polar)                                        (Polar)

Z1+Z2 (Ohms Resistance)=     308.04            Z2/Z1 (Ohms)=        1.50
Z1+Z2 (Ohms Reactance)=    356.41              Phase Angle (Degrees)=     40.00
(Rect.)                                        (Polar)

Z1*Z2 (Ohms)= 59998.22                         Z1-Z2 (Ohms)=     195.13
Phase Angle (Degrees)=     90.00               Phase Angle (Degrees)=    -73.79
(Polar)                                        (Polar)

Z1+Z2 (Ohms)=    471.08                        Z2-Z1 (Ohms)=     195.13
Phase Angle (Degrees)=     49.17               Phase Angle (Degrees)=    -73.79
(Polar)                                        (Polar)

Z1*Z2 (Ohms Resistance)=       -0.01           Z1/Z2 (Ohms Resistance)=      0.51
Z1*Z2 (Ohms Reactance)= 59998.22               Z1/Z2 (Ohms Reactance)=     -0.43
(Rect)                                         (Rect.)

Z1-Z2 (Ohms Resistance)=      54.48            Z2/Z1 (Ohms Resistance)=      1.15
Z1-Z2 (Ohms Reactance)=   -187.37              Z2/Z1 (Ohms Reactance)=      0.96
(Rect.)                                        (Rect.)

When angles greater than 90 degrees are being processed,
there is a possible 180 degree ambiguity in the final answer.
To check for 180 degree ambiguity, make a rough sketch of
the vector diagram.

Slight inaccuracies in computed values may occur due to
single-precision and rounding off processing
```

Figure 7–7. Program for addition, subtraction, multiplication, or division of a pair of vectors in polar or in rectangular form.

NOTE: The programmer can reduce inaccuracies in computed values by using an approximation of 3.141593 instead of 3.1416 in converting degrees into radians.

and prints out the answers in both rectangular and polar formats. The printout includes Z2 − Z1, and Z2/Z1, as well as Z1 − Z2 and Z1/Z2. Observe that in the case that a zero-resistance value may be encountered by the processor, the IBM PC would ordinarily stop and print out an error message (division by zero is a forbidden arithmetical operation). However, the program is written so that in the case that a zero resistance value is encountered, a jump will occur and the processing will continue on the basis that the computed angle would have been 90°. Since the associated reactance may be either positive or negative, an advisory (+ or −) note is included in the printout. In turn, the operator must decide whether the computed 90° angle is in fact a positive angle, or a negative angle.

Observe also that the IBM PC is basically designed to compute trigonometric angles. Thus, an angle of −45° denotes a line that lies in the second and fourth quadrants. On the other hand, a vector is confined to a single quadrant, and an angle of −45° denotes a vector that lies in the fourth quadrant; an angle of +135°, in turn, would denote a vector that lies in the second quadrant. The bottom line is that when R is negative and X is positive, for example, the printout will indicate a −45° angle (instead of a +135° angle). This is just another way of saying that the operator must decide whether an angle that is printed out with a value in the range from zero to +90° or −90° does in fact describe a vector in the first or fourth quadrants, or perhaps a vector in the second or third quadrants.

The foregoing evaluations are quite similar to the evaluations that we observe when calculating with pencil and paper. Stated otherwise, we are guided by the ground rule that stipulates that a rough sketch of the vector relations be drawn before starting calculation. Then, when the value of the angle has been calculated, we consult our sketch of the vector relations to determine whether the vector does in fact lie in the fourth quadrant, or perhaps in the second quadrant, for example.

Next, a program for calculating the resultant of two impedances connected in parallel is provided in Figure 7–8. (The printout also includes the resultant of the two impedances connected in series.) Observe that this program is also applicable to the problem of matching an impedance to the resultant of two impedances that are connected in parallel. As an illustration, suppose that a source has an internal impedance of 200 ohms at 20°, which is connected in parallel with a circuit impedance of 300 ohms at 60 degrees. In turn, this program shows that this resultant source impedance will be matched by an impedance of 127.36 ohms at 35.83°. (This answer is for a magnitude match; a conjugate match for maximum power transfer will be 127.36 ohms at −35.83°, in this example.)

```
1 LPRINT"TOTAL IMPEDANCE OF TWO VECTORS IN PARALLEL"
2 LPRINT"Vectors May Be Either in Polar or Rectangular Form"
3 LPRINT"":LPRINT"..............................................."
4 LPRINT""
5 INPUT"Polar, 1; Rect., 0";X
6 IF X=0 THEN GOTO 8
7 IF X=1 THEN GOTO 12
8 INPUT"Z1 Resistance (Ohms)=";A
9 INPUT"Z1 Reactance (Ohms)=";B
10 INPUT "Z2 Resistance (Ohms)=";C
11 INPUT"Z2 Reactance (Ohms)=";D:GOTO 20
12 INPUT"Z1 Magnitude (Ohms)=";E
13 INPUT "Z1 Phase Angle (Degrees)=";F
14 INPUT"Z2 Magnitude (Ohms)=";G
15 INPUT "Z2 Phase Angle (Degrees)=";H
16 A=E*COS(F*6.283/360)
17 B=E*SIN(F*6.283/360)
18 C=G*COS(H*6.283/360)
19 D=G*SIN(H*6.283/360)
20 A$="######.##"
21 LPRINT"R1 (Ohms)=";USING A$;A
22 LPRINT"X1 (Ohms)=";USING A$;B
23 LPRINT"R2 (Ohms)=";USING A$;C
24 LPRINT"X2 (Ohms)=";USING A$;D:LPRINT""
25 E=(A*A+B*B)^.5
26 NJ=B/A
27 PH=ATN(NJ)
28 LPRINT"Z1 (Ohms)=";USING A$;E
29 LPRINT"Phase Angle (Degrees)=";USING A$;PH*360/6.283:LPRINT""
30 G=(C*C+D*D)^.5
31 JJ=D/C
32 HP=ATN(JJ)
33 LPRINT"Z2 (Ohms)=";USING A$;G
34 LPRINT "Phase Angle (Degrees)=";USING A$;HP*360/6.283:LPRINT""
35 Q=A+C
36 QQ=B+D
37 LPRINT"Z1+Z2 (Ohms Resistance)-";USING A$;Q
38 LPRINT"Z1+Z2 (Ohms Reactance)=";USING A$;QQ:LPRINT""
39 NN=E*G
40 MM=PH+HP
41 LPRINT"Z1*Z2 (Ohms)=";USING A$;NN
42 LPRINT"Phase Angle (Degrees)=";USING A$;MM*360/6.283:LPRINT""
43 Y=(Q*Q+QQ*QQ)^.5
44 XY=QQ/Q
45 TH=ATN(XY)
46 LPRINT"Z1+Z2 (Ohms)=";USING A$;Y
47 LPRINT "Phase Angle (Degrees)=";USING A$;TH*360/6.283:LPRINT""
48 RZ=NN*COS(MM)
49 XZ=NN*SIN(MM)
50 LPRINT"Z1*Z2 (Ohms Resistance)=";USING A$;RZ
51 LPRINT"Z1*Z2 (Ohms Reactance)=";USING A$;XZ:LPRINT""
52 AQ=NN/Y
53 QA=MM-TH
54 LPRINT"Z1 and Z2 in Parallel (Ohms)=";USING A$;AQ
55 LPRINT"Phase Angle (Degrees)=";USING A$;QA*360/6.283:LPRINT""
56 LPRINT"When angles greater than 90 degrees are being processed,"
57 LPRINT"there is a possible 180 degree ambiguity in the final answer."
58 LPRINT"To check for 180 degree ambiguity, make a rough sketch of"
59 LPRINT"the vector diagram.":LPRINT""
60 LPRINT"Slight inaccuracies in computed values may occur due to"
61 LPRINT"single-precision and rounding-off programming and"
62 LPRINT"processing."
```

Figure 7–8. (Continues on next page.)

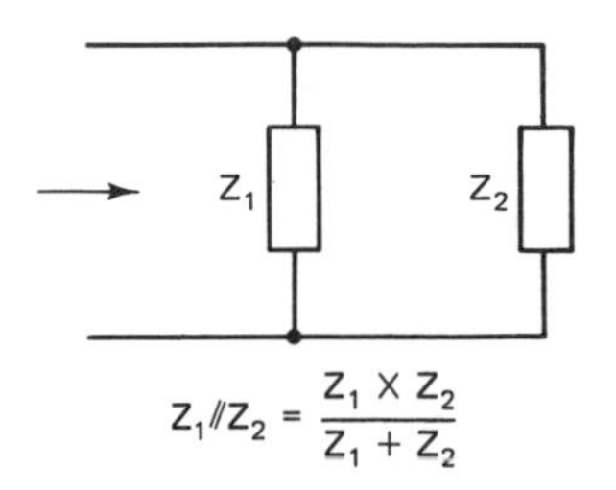

$$Z_1 /\!/ Z_2 = \frac{Z_1 \times Z_2}{Z_1 + Z_2}$$

```
TOTAL IMPEDANCE OF TWO VECTORS IN PARALLEL
Vectors May Be Either in Polar or Rectangular Form

.........................................

R1 (Ohms)=    187.94
X1 (Ohms)=     68.40
R2 (Ohms)=    150.01
X2 (Ohms)=    259.80

Z1 (Ohms)=    200.00
Phase Angle (Degrees)=     20.00

Z2 (Ohms)=    300.00
Phase Angle (Degrees)=     60.00

Z1+Z2 (Ohms Resistance)-     337.95
Z1+Z2 (Ohms Reactance)=     328.21

Z1*Z2 (Ohms)= 60000.00
Phase Angle (Degrees)=     80.00

Z1+Z2 (Ohms)=     471.09
Phase Angle (Degrees)=     44.16

Z1*Z2 (Ohms Resistance)= 10421.32
Z1*Z2 (Ohms Reactance)= 59088.04

Z1 and Z2 in Parallel (Ohms)=     127.36
Phase Angle (Degrees)=     35.84

When angles greater than 90 degrees are being processed,
there is a possible 180 degree ambiguity in the final answer.
To check for 180 degree ambiguity, make a rough sketch of
the vector diagram.

Slight inaccuracies in computed values may occur due to
single-precision and rounding-off programming and
processing.
```

Figure 7–8. Program for calculating the resultant of two impedances connected in parallel.

Parallel impedances can be solved to advantage using admittance values instead of impedance values, inasmuch as parallel admittances are additive. Accordingly, the designer frequently needs to convert impedance values into admittance values. This conversion is facilitated by the program provided in Figure 7–9.

Either impedance values or admittance values may be inputted in rectangular form, and corresponding admittance or impedance values will be printed out. The printout includes the inputted data as a reminder to the operator.

Tolerances on vector impedances are a matter for concern in practical design work. If resistive and reactive tolerances are equal to 10 percent, for example, then the tolerance on their series or parallel impedance is also 10 percent. On the other hand, if the resistive tolerance is equal to 10 percent, and the reactive tolerance is equal to 5 percent, for example, then the tolerance on their series or parallel impedance becomes a function of the impedance phase angle. This dependence is evident from the extreme cases wherein the impedance is almost a pure resistance and wherein the impedance is almost a pure reactance.

Preliminary evaluation of tolerances on vector impedances can be facilitated by employment of a survey/heuristic routine such as provided in Figure 7–10. In the exemplified run, a bogie resistive value of 175 ohms with a tolerance of 10 percent is inputted, and a bogie reactive value of 100 ohms with a tolerance of 5 percent is inputted. A Z upper limit of 225 ohms is specified, and a Z lower limit of 175 ohms is specified.

The survey printout starts from bogie and proceeds in steps of higher resistive values and higher reactive values with their corresponding impedance values. The heuristic terminates at the step that exceeds the specified upper-Z limit, notes the corresponding resistive and reactive values, the impedance phase angle, and the tolerance values that yield the specified upper Z limit. These tolerance values are worst-case tolerance limits on R and X for the specified upper-Z limit. (If the original inputted tolerance values are allowed to stand, then the specified upper-Z limit must be relaxed accordingly to determine the worst-case Z value.)

The survey printout continues in Section 2 from bogie and proceeds to the specified lower-Z limit. The heuristic terminates at the step that passes the specified lower-Z limit, and notes the corresponding resistive and reactive values, the impedance phase angle, and the tolerance values that yield the specified lower-Z limit. As before, these tolerance values are worst-case tolerance limits on R and X for the specified lower-Z limit. (If the original inputted tolerance values are allowed to stand, then the lower-specified-Z limit must be relaxed accordingly to determine the worst-case lower-Z value.) (See also Chart 7–1.)

```
1 LPRINT "IMPEDANCE TO ADMITTANCE VECTORS":LPRINT"ADMITTANCE TO IMPEDANCE VECTOR
S":LPRINT""
3 LPRINT"...........................................................................
":LPRINT""
10 INPUT"Z to Y, 1; Y to Z, 0:";K
20 IF K=1 THEN 35
30 IF K=0 THEN 70
35 INPUT"Resistance (Ohms)";R
45 INPUT"Reactance (Ohms)=";X
55 P=(R*R+X*X):G=R/P:B=X/P
60 LPRINT"Z (Ohms Resistance)=";R:LPRINT"Z (Ohms Reactance)=";X:LPRINT""
65 LPRINT"Y (Siemens Conductance)=";G:LPRINT"Y (Siemens Susceptance)=";-B:END
70 INPUT"G (Siemens Conductance)";G
80 INPUT"B (Siemens Susceptance)=";BB:B=-BB
90 Q=(G*G+B*B):R=G/Q:X=B/Q
95 LPRINT"Y (Siemens Conductance)=";G:LPRINT"Y (Siemens Susceptance)=";-B:LPRINT
""
100 LPRINT"Z (Ohms Resistance)=";R:LPRINT"Z (Ohms Reactance)=";X
```

```
IMPEDANCE TO ADMITTANCE VECTORS
ADMITTANCE TO IMPEDANCE VECTORS

....................................................

Z (Ohms Resistance)= 200
Z (Ohms Reactance)=-100

Y (Siemens Conductance)= .004
Y (Siemens Susceptance)= .002


IMPEDANCE TO ADMITTANCE VECTORS
ADMITTANCE TO IMPEDANCE VECTORS

....................................................

Y (Siemens Conductance)= .004
Y (Siemens Susceptance)= .002

Z (Ohms Resistance)= 200
Z (Ohms Reactance)=-99.99999
```

Figure 7–9. Program for converting impedance to admittance, or converting admittance to impedance.

NOTE: Admittance is the reciprocal of impedance. Thus, $Y = 1/Z$. Since $Z = R + jX$, $Y = 1/(R + jX)$. In turn, $Y = (R - jX)/(R^2 + X^2)$. Accordingly, admittance can be separated into its conductance and susceptance components, or $Y = G + jB$. In other words, $Y = R^2/(R^2 + X^2) + (-jX)/(R^2 + X^2)$. Since the numerator of the susceptance is negative, the admittance equation can be written in simpler form as $Y = R^2/(R^2 + X^2) - jX/(R^2 + X^2)$. Therefore, if we write $Y = G + jB$, it is understood that B is negative.

```
1 LPRINT "Section 1"
2 LPRINT "R AND X TOLERANCES REQUIRED WITH SPECIFICATION"
3 LPRINT"OF MAXIMUM UPPER LIMIT FOR Z"
4 LPRINT "(From Bogie to Maximum)"
5 LPRINT "Section 2"
6 LPRINT "R AND X TOLERANCES REQUIRED WITH SPECIFICATION"
7 LPRINT "OF MINIMUM LOWER LIMIT FOR Z"
8 LPRINT "(From Bogie to Minimum)"
9 LPRINT "........................................":LPRINT"(Section 1)":LP
RINT""
10 INPUT"R (Ohms)=";R
15 INPUT"Tentative R Tol.(%)=";Q
20 INPUT"X (Ohms)=";X
25 INPUT"Tentative X Tol.(%)=";QQ
30 INPUT"Z Upper Limit (Ohms)";K
35 N=0:M=0:P=0
40 R=R+N*R:X=X+M*X:P=P+1
45 A$="####":PRINT"Pass No.";USING A$;P:B$="######.##":LPRINT"R=";USING B$;R
50 LPRINT"X=";USING B$;X
55 Z=(R*R+X*X)^.5:TA=(X/R):PA=ATN(TA)
60 LPRINT"Z=";USING B$;Z:LPRINT""
65 IF Z>K THEN 75
70 N=N+.001*Q:M=M+.001*QQ:GOTO 40
75 LPRINT"Maximum R=";USING B$;R:LPRINT""
80 LPRINT"Maximum X=";USING B$;X:LPRINT""
85 LPRINT"Phase Angle (Degrees)=";USING B$;PA*360/6.283:LPRINT""
90 LPRINT"Maximum R Tol.(%)=";USING B$;N*100:LPRINT"Maximum X Tol.(%)=";USING B$
;M*100:LPRINT""
95 LPRINT"R and X Tolerances Required With Specification of Minimum Lower Limit
for Z":PRINT"":PRINT"(Section 2)":PRINT""
100 LPRINT"(From Bogie to Minimum)"
105 LPRINT"(Section 2)":LPRINT""
110 INPUT "R (Ohms)=";RR
115 INPUT"Tentative R Tol.(%)=";W
120 INPUT"X (Ohms)=";XX
125 INPUT"Tentative X Tol.(%)=";WW
130 INPUT"Z Lower Limit (Ohms)=";KK
135 NN=0:MM=0:PP=0
140 RR=RR-NN*RR:XX=XX-MM*XX:PP=PP+1
145 PRINT"Pass No.";USING A$;PP: LPRINT"R=";USING B$;RR
150 LPRINT"X=";USING B$;XX
155 ZZ=(RR*RR+XX*XX)^.5:AT=XX/RR:AP=ATN(AT)
160 LPRINT"Z=";USING B$;ZZ:LPRINT""
165 IF ZZ<KK THEN 175
170 NN=NN+.001*W:MM=MM+.001*WW:GOTO 140
175 LPRINT"Minimum R=";USING B$;RR:LPRINT""
180 LPRINT"Minimum X=";USING B$;XX:LPRINT""
185 LPRINT"Phase Angle (Degrees)=";USING B$;AP*360/6.283:LPRINT""
190 LPRINT"Minimum R Tol.(%)=";USING B$;NN*100:LPRINT"Minimum X Tol.(%)=";USING
B$;MM*100
```

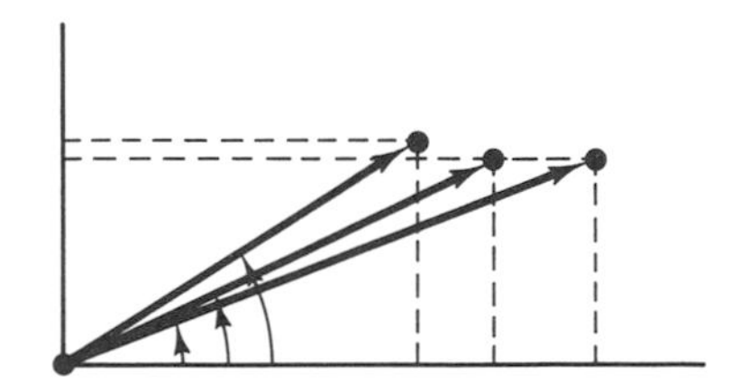

Figure 7–10. (Continues on next page.)

```
Section 1
R AND X TOLERANCES REQUIRED WITH SPECIFICATION
OF MAXIMUM UPPER LIMIT FOR Z
(From Bogie to Maximum)
Section 2
R AND X TOLERANCES REQUIRED WITH SPECIFICATION
OF MINIMUM LOWER LIMIT FOR Z
(From Bogie to Minimum)
..............................................
(Section 1)

R=   175.00          R=   180.29          R=   193.12
X=   100.00          X=   101.51          X=   105.09
Z=   201.56          Z=   206.90          Z=   219.86

R=   176.75          R=   185.69          R=   202.78
X=   100.50          X=   103.03          X=   107.72
Z=   203.32          Z=   212.36          Z=   229.61

Maximum R=   202.78

Maximum X=   107.72

Phase Angle (Degrees)=    27.98

Maximum R Tol.(%)=    5.00
Maximum X Tol.(%)=    2.50

R=   175.00          R=   164.69          R=   150.20
X=   100.00          X=    97.03          X=    92.71
Z=   201.56          Z=   191.15          Z=   176.51

R=   173.25          R=   158.10          R=   141.19
X=    99.50          X=    95.09          X=    89.93
Z=   199.79          Z=   184.49          Z=   167.39

R=   169.79
X=    98.51
Z=   196.29

Minimum R=   141.19

Minimum X=    89.93

Phase Angle (Degrees)=    32.50

Minimum R Tol.(%)=    6.00
Minimum X Tol.(%)=    3.00
```

Figure 7–10. Survey/heuristic routine for R and jX tolerance calculation with specified maximum and minimum Z limits.

DESIGN OPTIMIZATION CONSIDERATIONS

Wide-band amplifiers require closer attention to component tolerances than comparatively narrow-band low-frequency amplifiers. High-gain wide-band amplifiers present more demanding design requirements than lower-gain amplifiers. Optimization is facilitated by employment of significant negative feedback; however, additional stages are required when a large amount of negative

CHART 7–1

Wide-Band IC Amplifier Noise Figures

A typical wide-band integrated-circuit amplifier has the following characteristics:

Differential voltage gain,
(single-ended input, differential output), 35 times

−3 dB bandwidth, 45 MHz

Output voltage swing, 0.5 V_{rms}

Noise figure, 10 dB

Parallel input resistance, 150 kilohms

Parallel input capacitance, 2.2 pF

Output resistance, 125 ohms

The equivalent low-frequency noise for the wide-band IC amplifier may be expressed in nanovolts per square root of the frequency. A typical IC exhibits a limiting low-frequency noise of 125 nanovolts per $Hz^{0.5}$ and falls to 25 nanovolts per $Hz^{0.5}$ at 100 Hz. Above 200 Hz the noise component is dominated by shot noise.

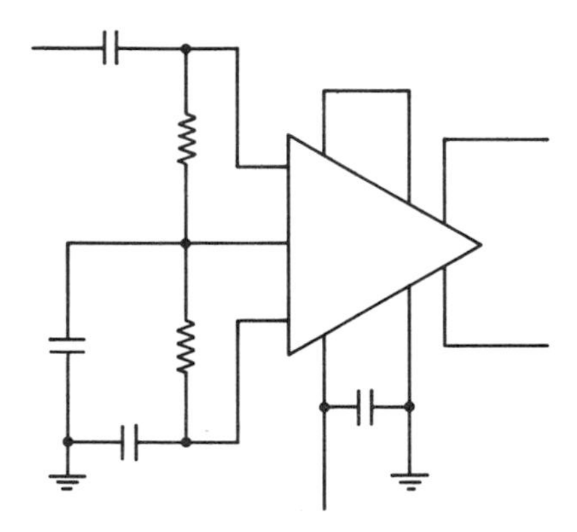

The noise figure of the wide-band IC amplifier is related to its total noise figure as follows:

$$NF = 20 \log_{10} \frac{E_{NT}}{(4KTR_S)^{0.5}}$$

$$NF = 20 \log_{10} \frac{E_{NT}}{126.6 \times 10^{-12}(Rs)^{0.5}}$$

where E_{NT} is the noise voltage is volts per $Hz^{0.5}$, K is Boltzmann's constant (1.38 × 10^{-23} joule/°K, T is the temperature in degrees Kelvin, and Rs is the external source resistance in ohms.

feedback is utilized, and the resulting production cost may dictate a trade-off in this regard.

Design optimization requires particular attention to the low-frequency reaction to component tolerances, inasmuch as the high-frequency reaction tends to dominate preliminary attention. It is important to thoroughly check the low-frequency performance and stability of a widc-band amplifier when operated from the same power supply as will be used in production. This is just another way of saying that a bench power supply usually has lower internal impedance than a power supply in a unit of consumer-electronics equipment.

Power-supply internal impedance is often a matter for concern inasmuch as the decoupling circuits in a wide-band high-gain amplifier are not ideal, and the power-supply internal impedance represents a source of low-frequency coupling between input and output stages of a high-gain amplifier. In some cases, power-supply tolerances represent a pitfall awaiting the unwary designer of high-gain wide-band amplifiers. Although extensive and conservative decoupling provisions throughout a wide-band amplifier will relax power-supply internal-impedance requirements, the resulting production cost may dictate judicious trade-offs in this regard.

It is evident that stray coupling of output to input stages can be a source of instability at high frequencies. Accordingly, the input stage should be located as far from the output stage as may be feasible. In the case of miniaturized high-gain wide-band amplifiers, input/output stray coupling may be reduced to a sufficiently low level by employment of a shield plate. A low-impedance common-ground conductor is essential, but should also be supplemented by conservative high-frequency decoupling circuits.

8

BASIC MOSFET AMPLIFIER DESIGN

COMMON-SOURCE AMPLIFIER

MOSFET (Metal Oxide Semiconductor Field Effect Transistor) devices are also called IGFET transistors and are the major category of unipolar transistors. They are characterized by very high input impedance and very wide dynamic range of input-signal capability. However, MOSFETs in the consumer-electronics category are not noted for low noise. MOSFET circuits can be configured to provide a parabolic (square-law) transfer characteristic that features exceptionally low cross-modulation interference.

With reference to Figure 8–1, MOSFETs are available in both depletion and enhancement types; an intermediate category between these types (sometimes called DE MOSFETs) features a moderate drain-current flow at zero bias. This chapter is primarily concerned with the depletion type of MOSFET. Three basic MOSFET amplifier configurations are utilized, as diagrammed in Figure 8–2. Since the common-source arrangement is in most extensive use, it is considered in this introductory topic.

A common-source amplifier is characterized by high-input impedance and high-output impedance. Unlike the bipolar transistor, the unipolar MOSFET is regarded as a voltage-operated device. In turn, amplifier performance is not calculated in terms of

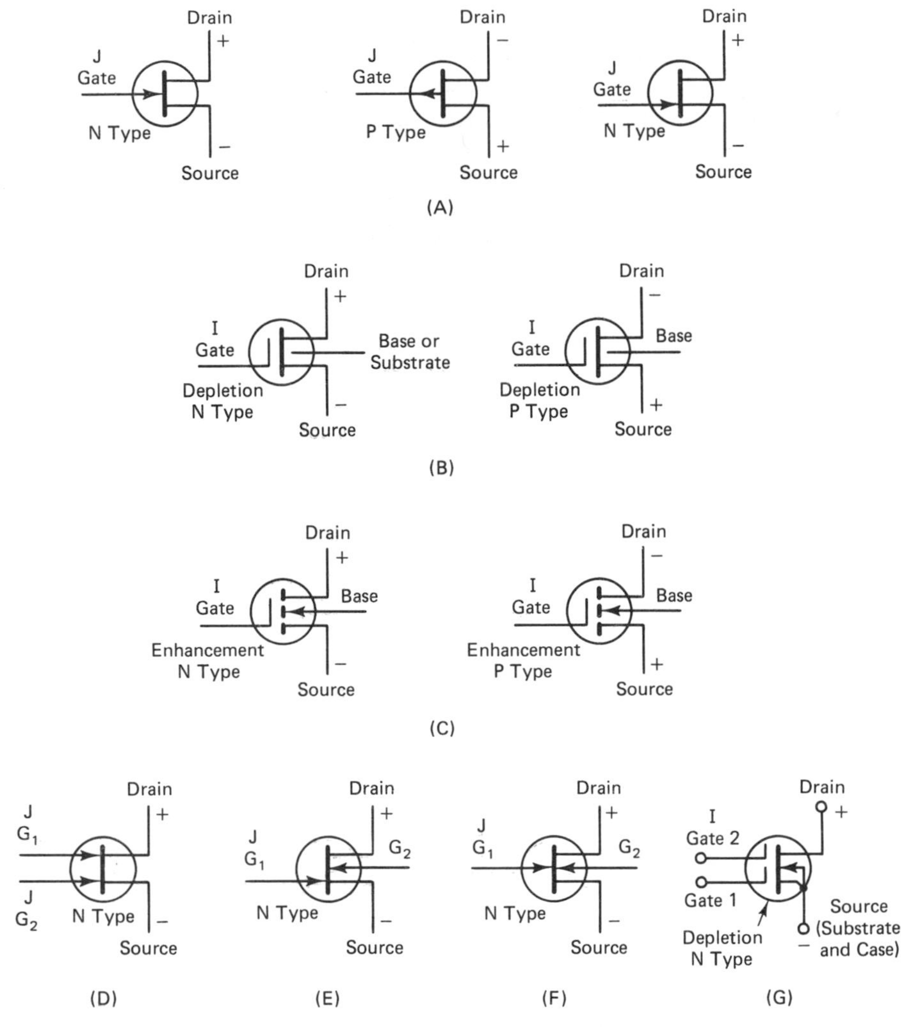

Figure 8–1. Field-effect transistors. (A) N and P types of JFETs—arrow points to N-type substance, away from P-type substance; (B) depletion-type MOSFETs—arrow points to N-type substrate, away from P-type substrate; (C) enhancement-type MOSFETs—arrow points to N-type substrate, away from P-type substrate; (D) dual-gate N-channel FET; (E) nonsymmetrical N-channel FET; (F) alternate symbol for (E); (G) dual-gate depletion N-type MOSFET. (Reproduced by special permission of Reston Publishing Company and Michael Thomason from *Handbook of Solid State Devices*.)

NOTE: Field-effect (unipolar) transistors are employed chiefly as RF amplifiers and oscillators. They are used occasionally as general-purpose amplifiers and choppers.

The unijunction transistor (UJT or JFET) is now used to a very limited extent. However, the insulated-gate FET (IGFET or MOSFET) is utilized to an appreciable extent.

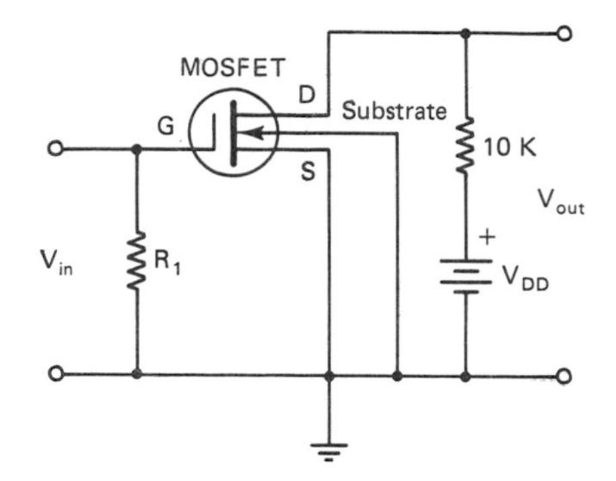

Common Source

Voltage Gain: 33 Times
Transconductance: 5,000 μ Siemens
Power Gain: 15.5 dB (33 Times)
Input Resistance: Very High
Output Resistance: 20 K

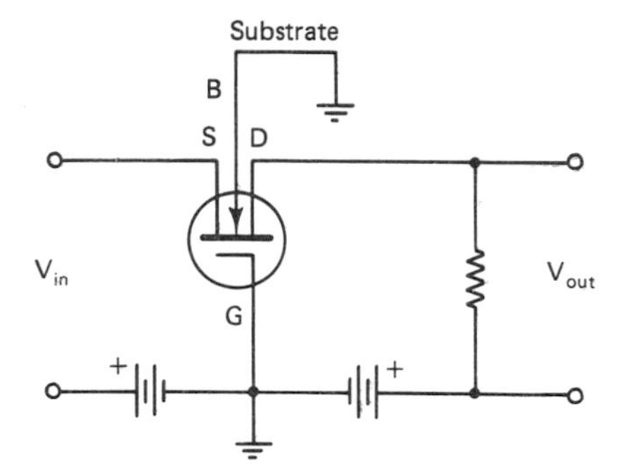

Common Gate

Voltage Gain: 2.8 Times
Input Resistance: 950 Ohms
Output Resistance: High

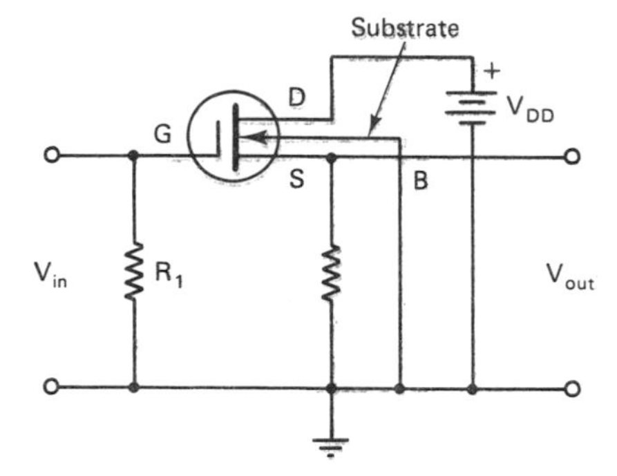

Common Drain

Voltage Gain: 0.81
Input Resistance: 2 Meg
Output Resistance: 952 Ohms

Figure 8–2. Overview of basic MOSFET amplifier configurations and typical circuit parameters.

Note: A short computer program for calculating the amplifier stage parameters in the three basic configurations is provided at the end of the chapter.

The values listed in the diagrams are typical and correspond to numerical calculations provided in the text. Each of the parameters has an appreciable range of possible values, depending upon the particular MOSFET that is used and the resistive values that are chosen by the designer.

Typically, transconductance decreases approximately 25 percent with a temperature increase of 80° from ambient. Similarly, drain current decreases approximately 25 percent with a temperature increase of 80° from ambient.

hybrid parameters, but in terms of transconductance. As indicated in Figure 8–2, a typical MOSFET has a transconductance of 5000 μSiemens. No negative feedback is present in this common-source arrangement, and the voltage gain for the stage is formulated:

$$A = \frac{g_{fs} r_{os} R_L}{r_{os} + R_L}$$

where g_{fs} denotes the gate-to-drain forward transconductance,
r_{os} denotes the common-source output resistance, and
R_L denotes the load resistance

EXAMPLE:

g_{fs} = 5000 μSiemens r_{os} = 20 kilohms R_L = 10 kilohms

A = 33.3

Current Feedback

Suppose next that an unbypassed resistor is connected in series with the source lead in the common-source configuration of Figure 8–2. This resistor will provide negative-current feedback, and the resulting voltage gain is formulated:

$$A' = \frac{g_{fs} r_{os} R_L}{r_{os} + (g_{fs} r_{os} + 1)R_s + R_L}$$

where R_s denotes the unbypassed source resistance.

EXAMPLE:

g_{fs} = 5000 μSiemens r_{os} = 20 kilohms R_L = 10 kilohms

R_s = 1000 ohms   A′ = 7.63

As would be anticipated, the output impedance in the common-source configuration is increased by current feedback. (The input impedance is very high—typically 10 megohms—even in the absence of current feedback.) The output impedance (resistance) with current feedback is formulated:

$$Z_o = r_{os} + (g_{fs}r_{os} + 1)R_S$$

EXAMPLE:

$r_{os} = 20$ kilohms $\quad g_{fs} = 5000$ μSiemens

$R_S = 1000$ ohms $\quad Z_o = 121$ kilohms

COMMON-DRAIN AMPLIFIER

The common-drain configuration (also called the source-follower arrangement) is diagrammed in Figure 8–2. It is characterized by a low-output resistance and low-voltage gain. Because of its inherent current feedback, it has low distortion. Observe that the common-drain configuration has an output that is in phase with its input, whereas the common-source arrangement has an output that is reversed in phase with respect to its input. The common-drain configuration has high-input resistance and is useful for matching a high-impedance source to a low-impedance load. It also has a greater input signal-handling capability than the common-source configuration. The voltage gain of a common-drain stage is formulated:

$$A' = \frac{R_s}{\frac{\mu + 1}{\mu} R_S + \frac{1}{g_{fs}}}$$

EXAMPLE:

$R_s = 1000$ ohms $\quad \mu = 33 \quad g_{fs} = 5000$ μSiemens

$A' = 0.81$

Inasmuch as the amplification factor (μ) of a MOSFET transistor is ordinarily much greater than 1, the formula for voltage gain of a common-drain stage can be approximately formulated:

$$A' = \frac{g_{fs}R_s}{1 + g_{fs}R_s}$$

EXAMPLE:

g_{fs} = 10,000 μSiemens R_s = 500 ohms A' = 0.82

Voltage Feedback

In the example of the common-drain configuration diagrammed in Figure 8–2, the input resistor R1 is returned to ground. However, if the designer returns R1 to the source terminal, voltage feedback occurs and the input resistance of the stage is then formulated:

$$Ri' = \frac{R1}{1 - A'}$$

where A' denotes voltage gain as in the previous example.

EXAMPLE:

R1 = 1 megohm A' = 0.82 R_i' = 55.6 megohms

The input capacitance of a high-frequency amplifier can be of greater concern than its input resistance. When the common-drain configuration works into a resistive load, the input capacitance of the MOSFET is reduced by the inherent feedback. In turn, the effective input capacitance to the common-drain arrangement is formulated:

$$C_i' = C_{gd} + (1 - A')C_{gs}$$

where C_{gd} denotes the rated gate-to-drain capacitance,
C_{gs} denotes the rated gate-to-source capacitance, and
A' denotes the voltage gain of the stage.

> EXAMPLE:
>
> $C_{gd} = 0.3$ pF $\quad C_{gs} = 5$ pF $\quad A' = 0.75 \quad C_i' = 1.5$ pF

The effective output resistance of a common-drain arrangement is formulated:

$$R_o' = \frac{r_{os}R_S}{(g_{fs}r_{os} + 1)R_s + r_{os}}$$

where r_{os} denotes the MOSFET output resistance (ohms).

> EXAMPLE:
>
> $g_{fs} = 5000$ μSiemens $\quad r_{os} = 20$ kilohms $\quad R_s = 1000$ ohms
>
> $r_o' = 952$ ohms

The output capacitance of a high-frequency amplifier can be of concern. The output capacitance of a source-follower stage is formulated:

$$C_o' = C_{ds} + C_{gs}\left(\frac{1 - A'}{A'}\right)$$

where C_{ds} denotes the rated drain-to-source capacitance,
C_{gs} denotes the rated gate-to-source capacitance, and
A' denotes the voltage gain of the stage.

> EXAMPLE:
>
> $A' = 0.75 \quad C_{gs} = 5$ pF $\quad C_{ds} = 0.25$ pF $\quad C_o' = 1.9$ pF

COMMON-GATE AMPLIFIER

With reference to Figure 8–2, the common-gate configuration finds application in transforming a low-input impedance into a high-output impedance. Its voltage gain is formulated:

$$A = \frac{(g_{fs}r_{os} + 1)R_L}{(g_{fs}r_{os} + 1)R_G + r_{os} + R_L}$$

where R_G denotes the signal-source resistance.

EXAMPLE:

$g_{fs} = 5000\ \mu$Siemens $r_{os} = 20$ kilohms $R_L = 2$ kilohms

$R_G = 500$ ohms $A = 2.8$

Observe that the input resistance of the common-gate configuration is approximately the same as the output resistance of the common-drain configuration. Note that the voltage gain of the common-gate arrangement is a function of the signal-source resistance and that the voltage gain decreases as the source resistance increases. This relation is the result of the low-input resistance to the stage.

MOSFET BIASING CIRCUITRY

The basic MOSFET bias arrangements include self-bias, fixed bias, and a combination of both methods. With reference to Figure 8–3, self-bias is derived from a resistive voltage drop across a source resistor resulting from drain-current flow. Fixed bias, as its name indicates, is derived from a dc voltage source and a resistive voltage divider. Many designers prefer a combination of self-bias and fixed bias. One of the practical design factors is the wide tolerance on the rated drain-current value for most MOSFETs. (See Figure 8–4.)

It is seen in Figure 8–3 that the exemplified depletion-type MOSFET would draw an objectionably large drain current (I_D) at zero bias. A reasonable choice of operating point for class-A operation would be at $I_D = 5$ mA and $V_{GS} = -1.1$ V. As an illustration of MOSFET drain-current tolerances, a production lot of devices could exhibit zero-bias drain-current values ranging from 5 mA to 25 mA. If a source resistor of approximately 220 ohms is utilized, the range of zero-bias drain-current values is lessened from about 4 mA to 20 mA.

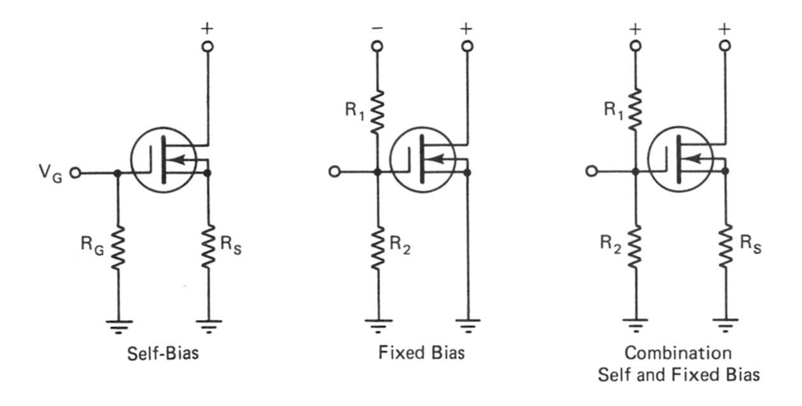

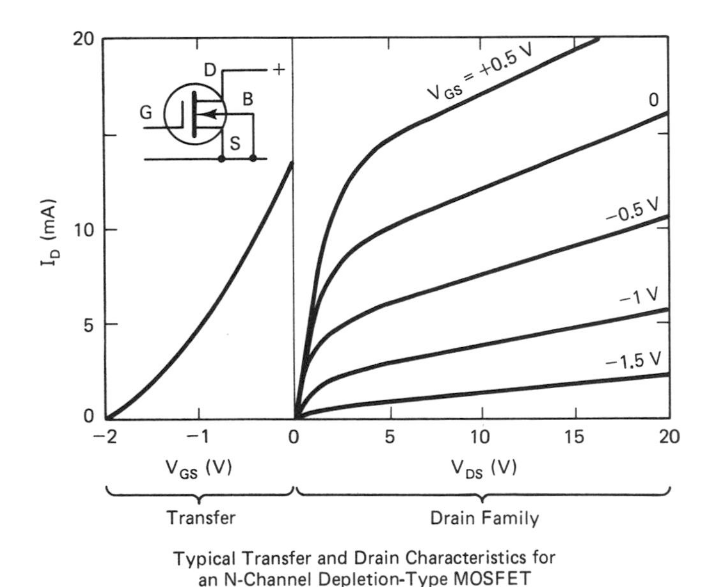

Figure 8–3. Basic MOSFET bias arrangements.

Fixed bias per se is not a practical approach to MOSFET bias-circuit design in view of the wide manufacturing tolerance on bogie drain-current value. If fixed bias is employed, a bias-adjustment control must be provided. This control is set as required in production, and the resulting production costs cannot be justified. Device

Parameter	Bipolar Transistor	JFET Transistor	MOSFET Transistor
Input Impedance	Low	High	Very High
Noise	Low	Low	Unpredictable
Aging	Not Noticeable	Not Noticeable	Noticeable
Bias Voltage Temperature Coefficient	Low and Predictable	Low and Predictable	Unpredictable
Control Electrode Current	High	0.1 nA	10 pA
Overload Capability	Reasonably Good	Reasonably Good	Good

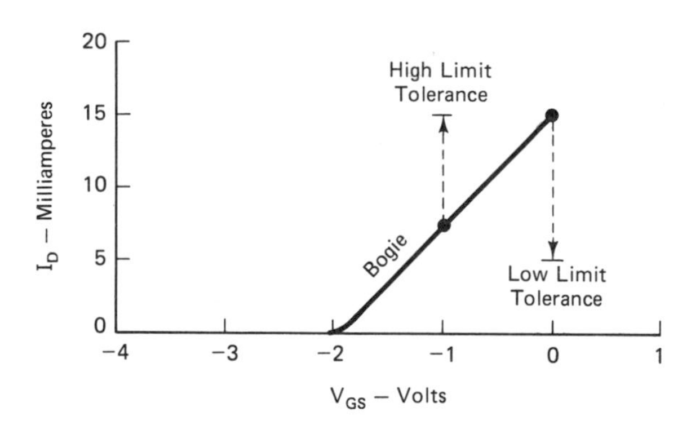

Figure 8–4. Comparative MOSFET, JFET, and bipolar transistor characteristics. (Reproduced by special permission of Reston Publishing Company and Michael Thomason from *Handbook of Solid State Devices*.)

NOTE: This example of a MOSFET I_D/V_{GS} bogie characteristic with high- and low-tolerance limits shows the essential problem confronted by the bias-circuit designer. The I_D/V_{GS} data for a MOSFET is utilized with the associated g_{fs}/I_D characteristic to facilitate preliminary bias-circuit calculations.

manufacturers have established that this tolerance problem is met to best advantage by a combination of self-bias and fixed bias:

1. A large amount of self-bias is used, so that the resulting current feedback serves to effectively reduce the device-tolerance limits.
2. The associated excessive-source voltage produced by a large amount of self-bias is then corrected by provision of an appropriate value of fixed bias, thereby establishing the desired operating point.

A source resistance of 1000 ohms provides a large amount of self-bias; a source resistance of 100 ohms provides a comparatively small amount of self-bias. In turn, although it might be supposed that a 1000-ohm source resistance would be chosen, the incidental "throwaway" of supply voltage may be considered objectionable. Thus, a 1000-ohm source resistance drops 5 volts at 5 mA, whereas a 100-ohm source resistance drops only 0.5 volt. In other words, with all other things being equal, a large amount of self-bias increases production costs by necessitating a higher supply voltage. (A practical example follows.)

Fixed bias (Figure 8–3) involves a voltage divider, and this divider introduces input circuit loading. Input loading can present a problem when very high input resistance is desired, unless unusually high values of resistors are employed in the divider. In the case of high-frequency circuitry, the divider can be bypassed, as shown in Figure 8–5.

At this point, it is helpful to consider a numerical example of bias-circuit design. Suppose that the drain-to-source voltage V_{DS} is 15 volts and that a small-signal transconductance g_{fs} of 7000 microsiemens is chosen. In turn, the corresponding drain current will be 5.1 mA, as exemplified in Figure 8–5. With this bogie value of drain current determined, the designer proceeds to find the gate-to-source voltage that is required. With reference to the bogie curve exemplified in Figure 8–4, this value will be −1.1 volts.

In turn, the source voltage V_s is calculated:

$$V_S = V_G - V_{GS} = 1.1 \text{ volts}$$

or the voltage drop across the source resistor is equal to the voltage from gate to ground minus the potential difference from gate to source.

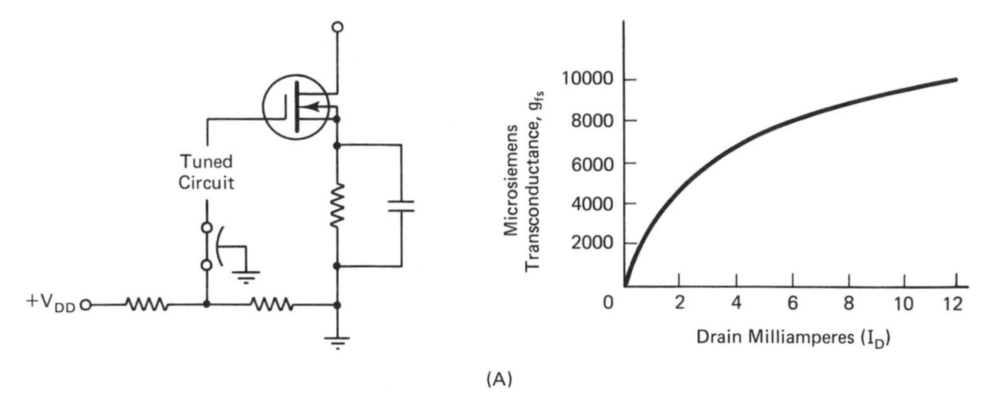

Figure 8–5(A). Fixed-bias voltage divider can be bypassed in high-frequency amplifiers to avoid input circuit loading.

NOTE: The drain current I_D that is required for a specified value of g_{fs} can be determined from a data sheet for the particular MOSFET. The G_{fs}/I_D graph is applicable at a stipulated V_{DS} such as +15V and at a stipulated temperature such as 25°C.

Next, the required value of source resistance is calculated:

$$R_S = V_s/I_D = 1.1/0.0051 = 216 \text{ ohms}$$

or the required value of source resistance is equal to the ratio of source voltage to drain current (which is also the source current).

Then, the necessary value of supply voltage is calculated:

$$V_{DD} = V_{DS} + V_s = 15 + 1.1 = 16.1 \text{ volts}$$

or the supply voltage value is equal to the sum of the drain-to-source voltage and the source voltage.

This is a basic example of a self-bias circuit. It is evident from the low and high tolerance limits exemplified in Figure 8–4 that the foregoing bogie circuit calculation is associated with an "impossible" worst-case situation. As previously noted, the worst-case problem can be relaxed by the use of a combination of self-bias and fixed bias.

As an illustration, consider the bias-circuit parameters required for the same transistor characteristics as just shown and

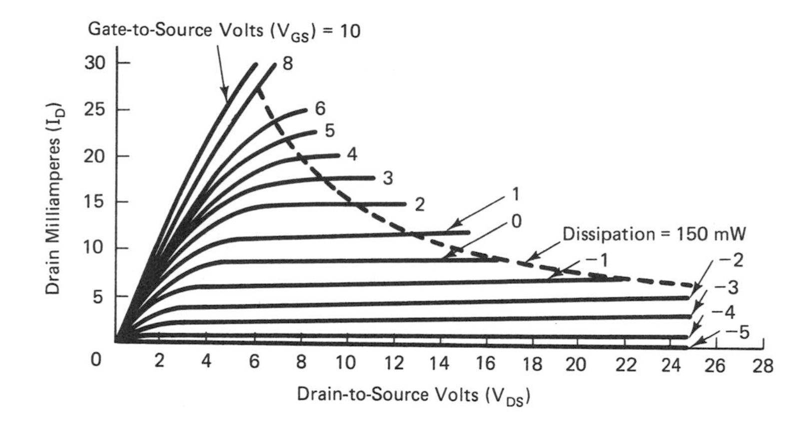

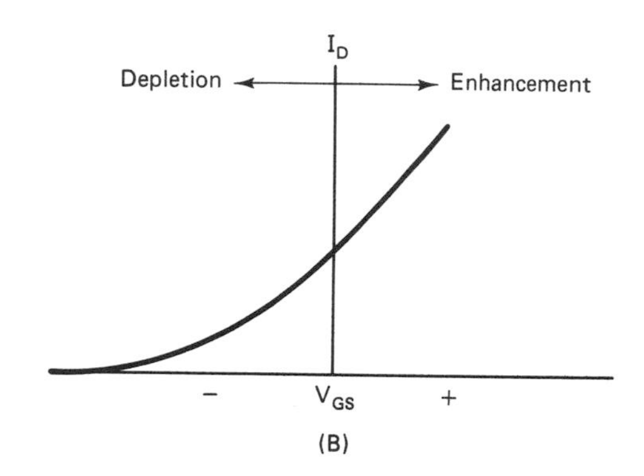

Figure 8–5(B). Family of drain characteristics for a depletion/enhancement (DE) MOSFET and associated voltage-current characteristic. (Reproduced by special permission of Reston Publishing Company and Michael Thomason from *Handbook of Solid-State Devices.*)

NOTE: The maximum rated power-dissipation locus has a value of 150 mW in this example. The locus (dotted curve) is a parabola. Observe that a depletion/enhancement MOSFET could be operated at zero bias in a class-A stage.

with a source resistance of 1000 ohms (instead of 216 ohms), supplemented by an optimum value of fixed bias. In turn, the source voltage is calculated:

$$V_s = I_D R_s = 1000 \times 0.0051 = 5.1 \text{ volts}$$

or this comparatively high value of source resistance results in a substantially higher value of source voltage than before: 5.1 volts instead of 1.1 volts.

Next, the value of fixed-bias voltage that will be required to offset this higher value of source voltage is calculated:

$$V_G = V_{GS} + V_s = 5.1 - 1.1 = 4 \text{ volts}$$

or with reference to Figure 8–3, a fixed-bias potential of 4 volts must be applied to the gate to compensate for the 5.1-volt drop across the 1000-ohm source resistor.

Then, the necessary dc-supply voltage is calculated:

$$V_{DD} = V_{DS} + V_s = 15 + 5.1 = 20.1 \text{ volts}$$

or a dc-supply voltage of 20.1 volts will be required instead of 16.1 volts with a 216-ohm source resistance.

Next, the ratio of dc-supply voltage to fixed-bias gate voltage is calculated:

$$V_{DD}/V_G = (R1 + R2)/R2 = 20.1/4 = 5.03$$

or the voltage divider ratio required to obtain 4 volts of fixed bias (Figure 8–3) is 5.03, which rounds off to 5.

The designer has now determined the ratio (R1 + R2/R2) and proceeds to establish reasonable values for R1 and R2. These values must be sufficiently high to avoid objectionable input circuit loading. It is evident from Figure 8–3 that R1 and R2 are in parallel, insofar as input circuit loading is concerned. As an illustration, if the input circuit is stipulated to have an input resistance of 75,000 ohms, the corresponding values for R1 and R2 are calculated:

$$R1R2/(R1 + R2) = 75{,}000$$

$$(R1 + R2)/R2 = 5.03$$

$$R1 = 377{,}250 \text{ ohms}$$

$$R2 = 93{,}610 \text{ ohms}$$

It is evident that if an input resistance of several megohms were desired, instead of 75,000 ohms, unusually large values of resistance would be required for R1 and R2.

DUAL-GATE MOSFETS

Dual-gate MOSFETS such as depicted in Figure 8–6 provide design flexibility in AGC circuitry, for example. Thus, gate 1 may be utilized for signal input and gate 2 may function as an AGC input. Note in passing that a dual-gate MOSFET operates satisfactorily as a single-gate MOSFET if the gate-1 and gate-2 terminals are tied together. The MOSFET arrangement shown in Figure 8–6 exemplifies the back-to-back diode gate-protection feature, which facilitates production by removing the hazard of static-electricity damage to the device.

An example of biasing circuitry for a dual-gate MOSFET in rf-amplifier application with reverse AGC is also shown in Figure 8–6. Note that neutralization is not required. A drain-to-source voltage of 15 volts is stipulated, with a transconductance of 10,000 microsiemens. In this example, the stipulated transconductance can be realized with a gate-1-to-source voltage (V_{G1S}) of -0.45 volt along with a gate-2-to-source voltage of $+4$ volts. The associated drain current is approximately 10 mA.

In rf application, the shunt resistance for the signal-input gate is typically 25 kilohms. The AGC gate is operated at rf ground potential via bypassing and is fixed-biased in combination with its AGC input. Device manufacturers have established that a source resistance of approximately 270 ohms is suitable for rf application. In turn, the designer proceeds to calculate the source voltage, as follows:

$$V_S = I_D R_s = 0.010 \times 270 = 2.7 \text{ volts}$$

or the bogie-voltage drop across the source resistor is 2.7 volts.

Next, the fixed-bias voltage required for the signal-input gate is calculated:

$$V_{G1} = V_{G1S} + V_s = 2.25 - 0.45 = 2.25 \text{ volts}$$

or the voltage divider will supply 2.25 volts of fixed bias to the signal-input gate.

Then, the fixed-bias voltage required for the AGC gate is calculated:

$$V_{G2} = V_{G2S} + V_s = 4.0 + 2.7 = 6.7 \text{ volts}$$

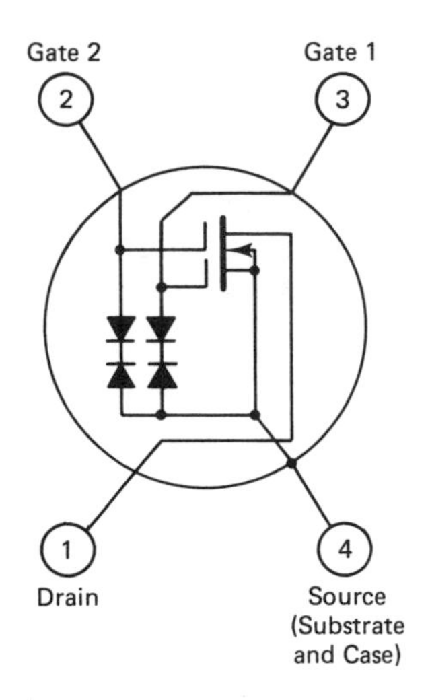

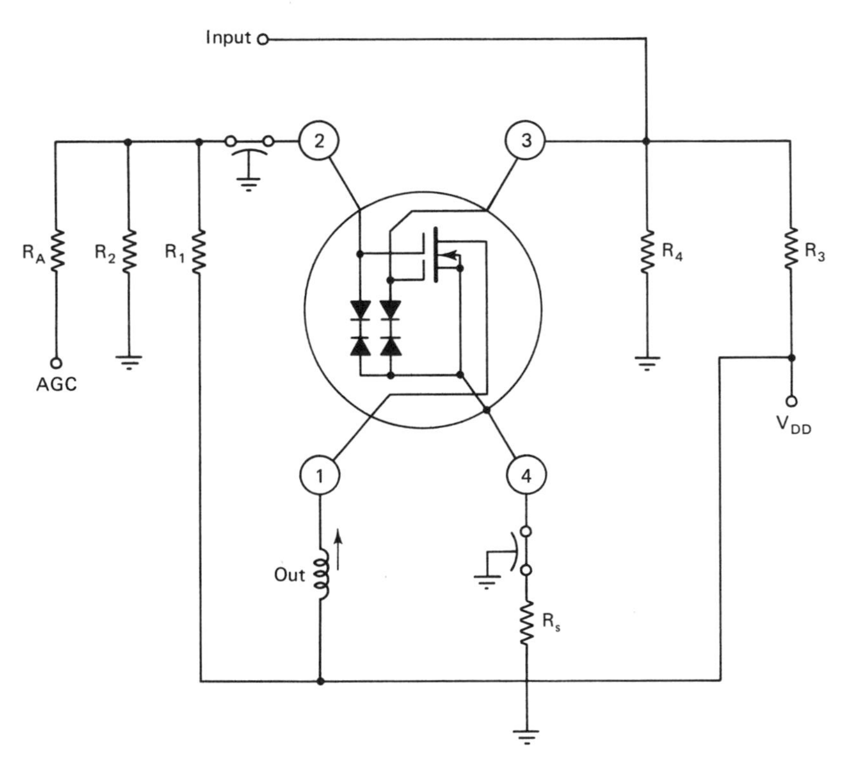

Figure 8–6. Dual-gate MOSFET with back-to-back diode-gate protection.

NOTE: This arrangement exemplifies biasing circuitry for a dual-gate MOS transistor in rf-amplifier application. Reverse AGC voltage is applied at terminal 2.

In mixer application, the oscillator injection voltage is applied at terminal 2, and AGC biasing is not used.

or in the absence of reverse AGC-bias voltage, the AGC gate should rest at 6.7 volts.

In turn, the required value of dc supply voltage is calculated:

$$V_{DD} = V_{DS} + V_s = 15 + 2.7 = 17.7 \text{ volts}$$

or the dc-supply voltage must be equal to the sum of the drain-to-source voltage and the source-resistor voltage drop.

Appropriate values for the voltage-divider resistances are calculated as exemplified previously. Thus, R4 will have a value of 28.6 kilohms, and R3 will have a value of 197 kilohms. The ratio for R1/R2 will be 11.67, and the parallel-resistance value of R1 and R2 will be assigned with respect to the internal resistance of the AGC source and the extent of AGC action that is desired.

PROGRAMMED FORMULAS

When a common-source MOSFET stage is to be evaluated for voltage gain, either with or without an unbypassed source resistor, time and effort can often be saved by employing a computer program such as provided in Figure 8–7. This program also calculates the output resistance of a stage with an unbypassed source resistor. Thus, it gives a helpful comparison of stage performance with and without current feedback.

In preliminary design procedures, it is frequently helpful to run a survey program for obtaining an overview of the voltage-gain range of a common-source MOSFET stage, either with or without an unbypassed source resistor, as provided in Figure 8–8. This survey program also calculates the range of output resistance for a stage with an unbypassed source resistor. In turn, the designer can determine tradeoffs that may be required.

In practice, the output load may be a capacitive impedance instead of resistive. In turn, the output voltage will not be in phase with the output current, and the phase shift will be a function of frequency. The phase characteristic of the amplifier is nonlinear when the output load is a capacitive impedance. (See Figure 8–9.)

Note in passing that a linear-phase characteristic (straight-line-phase-vs.-frequency plot) may have any slope from zero to infinity. Thus, zero slope corresponds to a purely resistive load; infinite slope corresponds to a purely reactive load. Simple ampli-

```
1 LPRINT "COMMON-SOURCE AMPLIFIER"
2 LPRINT"Voltage Gain With or Without an Unbypassed Source Resistor"
3 LPRINT"                               ---":LPRINT"Output Resistance With an Unbypas
sed Source Resistor":LPRINT""
5 LPRINT"..........................................................":LPRINT""
6 LPRINT"Section 1: Gain With Bypassed Source Resistor":LPRINT"Section 2: Gain W
ith Unbypassed Source Resistor":LPRINT"Section 3: Output Resistance With Unbypas
sed Source Resistor":LPRINT""
7 PRINT"Section 1: Gain With Bypassed Source Resistor":PRINT"Section 2: Gain Wit
h Unbypassed Source Resistor":PRINT"Section 3: Output Resistance With Unbypassed
 Source Resistor":PRINT""
8 LPRINT"..........................................................":LPRINT"":LP
RINT"(Section 1)":LPRINT"":PRINT"(Section 1)":PRINT""
9 LPRINT"Voltage Gain With Bypassed Source Resistor"
10 INPUT"Gfs (Siemens)=";A
15 LPRINT"Gfs (Siemens)=";A
20 INPUT"Ros (Ohms)=";B
25 LPRINT"Ros (Ohms)=";B
30 INPUT"RL (Ohms)=";C
35 LPRINT"RL (Ohms)=";C
40 D=A*B*C/(B+C):A$="#####.##"
45 LPRINT"Voltage Gain=";USING A$;D:LPRINT"":PRINT"Voltage Gain=";USING A$;D:PRI
NT""
50 LPRINT"Voltage Gain of Common-Source Amplifier With Unbypassed Source Resisto
r":LPRINT"(Section 2)":LPRINT""
55 PRINT"(Section 2)":PRINT""
60 INPUT"Gfs (Siemens)=";E
65 LPRINT"Gfs (Siemens)=";E
70 INPUT"Ros (Ohms)=";F
75 LPRINT"Ros (Ohms)=";F
80 INPUT"RL (Ohms)=";G
85 LPRINT"RL (Ohms)=";G
90 INPUT"Rs (Ohms)=";H
95 LPRINT"Rs (Ohms)=";H
100 AA=E*F*G/(F+(E*F-1)*H+G)
105 LPRINT"Voltage Gain=";USING A$;AA:LPRINT"":PRINT"Voltage Gain=";USING A$;AA:
PRINT""
110 LPRINT""
115 LPRINT"Output Resistance With Unbypassed Source Resistor":LPRINT"(Section 3)
":LPRINT"":PRINT"(Section 3)":PRINT""
120 INPUT"Ros (Ohms)=";I
125 LPRINT"Ros (Ohms)=";I
130 INPUT"Gfs (Siemens)=";J
135 LPRINT"Gfs (Siemens)=";J
140 INPUT"Rs (Ohms)=";K
145 LPRINT"Rs (Ohms)=";K
150 N=J*I+1
155 RR=I+N*K
160 LPRINT"Output Resistance (Ohms)=";RR:PRINT"Output Resistance (Ohms)=";RR
```

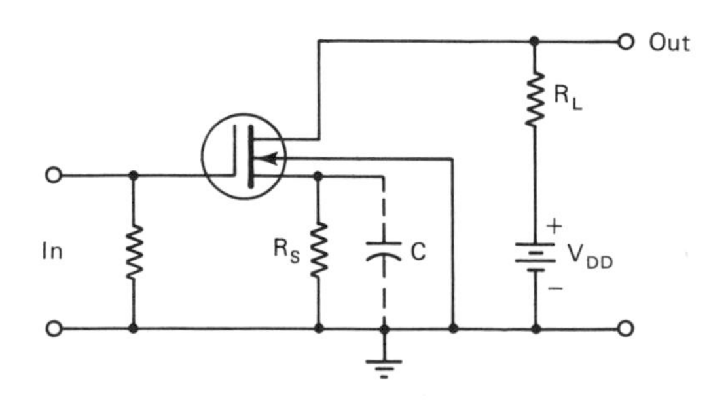

$$A = \frac{g_{fo} r_{os} R_L}{r_{os} + R_L}$$

$$AA = \frac{g_{fs} r_{os} R_L}{r_{os} + (g_{fs} r_{os} + 1) R_s + R_L}$$

Basic Equivalent Circuit
(Bypassed Source Resistor)

```
COMMON-SOURCE AMPLIFIER
Voltage Gain With or Without an Unbypassed Source Resistor
                                ---
Output Resistance With an Unbypassed Source Resistor

.........................................................

Section 1: Gain With Bypassed Source Resistor
Section 2: Gain With Unbypassed Source Resistor
Section 3: Output Resistance With Unbypassed Source Resistor

.........................................................

(Section 1)

Voltage Gain With Bypassed Source Resistor
Gfs (Siemens)= .005
Ros (Ohms)= 20000
RL (Ohms)= 10000
Voltage Gain=   33.33

Voltage Gain of Common-Source Amplifier With Unbypassed Source Resistor
(Section 2)

Gfs (Siemens)= .005
Ros (Ohms)= 20000
RL (Ohms)= 10000
Rs (Ohms)= 1000
Voltage Gain=    7.75


Output Resistance With Unbypassed Source Resistor
(Section 3)

Ros (Ohms)= 20000
Gfs (Siemens)= .005
Rs (Ohms)= 1000
Output Resistance (Ohms)= 121000
```

Figure 8–7. Program for calculating the voltage amplification of a common-source stage with or without an unbypassed source resistor.

```
1 LPRINT"COMMON-SOURCE AMPLIFIER":LPRINT"            ---":A=0:B=0
2 LPRINT"Voltage Gain Versus Load Resistance":C=0:D=0
3 LPRINT"(With Bypassed Source Resistor)":LPRINT"          ---"
4 LPRINT"Voltage Gain Versus Source Resistance"
5 LPRINT"(With Unbypassed Source Resistor)":LPRINT"          ---"
6 LPRINT"Output Resistance Versus Source Resistance"
7 LPRINT"(With Unbypassed Source Resistor)":LPRINT"          ---"
8 LPRINT"* Survey Program *":LPRINT""
9 LPRINT" * * * * * * * * *":LPRINT""
10 INPUT"Gfs (Siemens)=";A
12 INPUT"Ros (Ohms)=";B
14 LPRINT"10000 Ohm Increments for RL":LPRINT"(Other Increments May Be Programme
d)":PRINT"":LPRINT"
15 LPRINT" * * * * * * * * *":LPRINT"":LPRINT"Gfs (Ohms)=";A:LPRINT"Ros (Ohms)="
;B:PRINT"":PRINT"Gfs (Ohms=";A:PRINT"Ros (Ohms)=";B
16 LPRINT"":LPRINT" * * * * * * * * *":LPRINT"":X=0:N=0:PRINT""
17 C=X:M=M+1:P$="####":PRINT"Pass No.";USING P$;M
18 D=A*B*C/(B+C)
19 A$="#######":LPRINT"RL (Ohms)=";USING A$;C:PRINT"RL (Ohms)=";USING A$;C
20 B$="#####.##":LPRINT"Gain=";USING B$;D:LPRINT"":PRINT"Gain=";USING B$;D:PRINT
""
21 X=C+10000
22 IF C>100000! THEN 24
23 GOTO 17
24 LPRINT"End of First Program":LPRINT"":PRINT"End of First Program":PRINT""
25 LPRINT" * * * * * * * * *":LPRINT"":LPRINT"Common-Source Amplifier":E=0:F=0:G
=0:PRINT"Common-Source Amplifier"
26 LPRINT"          ---":LPRINT"Voltage Gain vs. Source Resistance":PRINT"Voltage
Gain vs. Source Resistance"
27 LPRINT"With Unbypassed Source Resistor":LPRINT"          ---":PRINT"With Unbypasse
d Source Resistor":PRINT""
28 LPRINT"(Survey Program)":LPRINT"":PRINT"(Survey Program)"
29 LPRINT" * * * * * * * * *":LPRINT"":PRINT""
30 INPUT"Gfs (Ohms)=";E
32 INPUT"Ros (Ohms)=";F
34 INPUT"RL (Ohms)=";G
36 LPRINT"200 Ohm Increments for Rs":LPRINT"(Other Increments May Be Programmed)
":PRINT "200 Ohm Increments for Rs":PRINT"(Other Increments May Be Programmed)"
37 LPRINT" * * * * * * * * *":LPRINT"":PRINT""
38 C$="##.###":LPRINT"Gfs=";USING C$;E:LPRINT"Ros=";USING A$;F:LPRINT"RL=";USING
 A$;G:PRINT"Gfs=";USING C$;E:PRINT"Ros=";USING A$;F:PRINT"RL=";USING A$;G
39 LPRINT"":PRINT"":LPRINT" * * * * * * * * *":LPRINT"":PRINT"":Y=0:N=0
40 H=Y:N=N+1:PRINT"Pass No.";USING P$;N
41 AA=E*F*G/(F+(E*F+1)*H+G)
42 LPRINT"Rs (Ohms)=";USING A$;H:PRINT"Rs (Ohms)=";USING A$;H
43 LPRINT"Gain=";USING B$;AA:LPRINT"":PRINT"Gain=";USING B$;AA:PRINT""
44 Y=H+200
45 IF Y>2000 THEN 47
46 GOTO 40
47 LPRINT"End of Second Program":LPRINT"":PRINT"End of Second Program":PRINT""
48 LPRINT" * * * * * * * * *":LPRINT"":LPRINT"Common-Source Amplifier":I=0:J=0:K
=0:N=0:PRINT"Common-Source Amplifier"
49 LPRINT"Output Resistance vs. Source Resistance With Unbypassed Source Resisto
r":LPRINT"":PRINT"Output Resistance vs. Source Resistance With Unbypassed Source
 Resistor":PRINT""
50 LPRINT"                    ---"
51 LPRINT"(Survey Program)":LPRINT"":PRINT"(Survey Program)":PRINT""
52 LPRINT" * * * * * * * * *":LPRINT""
53 INPUT"Ros= (Ohms)=";I
55 INPUT"Gfs (Siemens)=";J
57 LPRINT"200 Ohm Increments for Rs":LPRINT"(Other Increments May Be Programmed)
":PRINT"200 Ohm Increments for Rs":PRINT"(Other Increments May Be Programmed)":P
RINT""
58 LPRINT" * * * * * * * * *": LPRINT"":PRINT""
59 D$="#######":LPRINT"Ros=";USING D$;I:E$="##.###":LPRINT"Gfs=";USING E$;J:PRIN
T"Ros=";USING D$;I:PRINT"Gfs=";USING E$;J
```

```
60 LPRINT"":LPRINT" * * * * * * * * *":LPRINT"":T=0:P=0:PRINT""
61 K=T: P=P+1:P$="####":PRINT"Pass No.";USING P$;P
62 N=J*I+1
63 LPRINT"Rs (Ohms)=";USING D$;K:PRINT"Rs (Ohms)=";USING D$;K
64 LPRINT "RO (Ohms)=";USING D$;I+N*K:LPRINT"":PRINT"RO (Ohms)=";USING D$;I+N*K:
PRINT""
65 T=K+200
66 IF T>2000 THEN 68
67 GOTO 61
68 LPRINT"End of Third Program":PRINT"End of Third Program":END
```

```
   COMMON-SOURCE AMPLIFIER
           ---
   Voltage Gain Versus Load Resistance
   (With Bypassed Source Resistor)
           ---
   Voltage Gain Versus Source Resistance
   (With Unbypassed Source Resistor)
           ---
   Output Resistance Versus Source Resistance
   (With Unbypassed Source Resistor)
           ---
   * Survey Program *

    * * * * * * * * *

   10000 Ohm Increments for RL
   (Other Increments May Be Programmed)

 * * * * * * * * *

Gfs (Ohms)= .005
Ros (Ohms)= 20000

 * * * * * * * * *

RL (Ohms)=      0       RL (Ohms)=  40000       RL (Ohms)=  80000
Gain=    0.00           Gain=   66.67           Gain=   80.00

RL (Ohms)=  10000       RL (Ohms)=  50000       RL (Ohms)=  90000
Gain=   33.33           Gain=   71.43           Gain=   81.82

RL (Ohms)=  20000       RL (Ohms)=  60000       RL (Ohms)= 100000
Gain=   50.00           Gain=   75.00           Gain=   83.33

RL (Ohms)=  30000       RL (Ohms)=  70000       RL (Ohms)= 110000
Gain=   60.00           Gain=   77.78           Gain=   84.62

                                                End of First Program

 * * * * * * * * *

Common-Source Amplifier
        ---
Voltage Gain vs. Source Resistance
With Unbypassed Source Resistor
        ---
(Survey Program)

 * * * * * * * * *

200 Ohm Increments for Rs
(Other Increments May Be Programmed)
 * * * * * * * * *
```

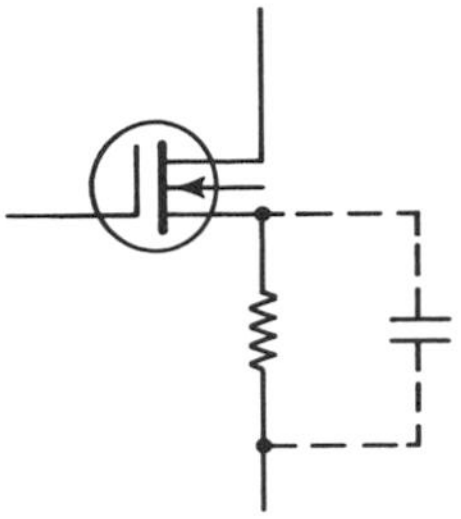

Figure 8–8. (Continues on next page.)

```
Gfs= 0.005
Ros=  20000
RL=  10000

 * * * * * * * * *

Rs (Ohms)=       0        Rs (Ohms)=    800        Rs (Ohms)=   1600
Gain=   33.33             Gain=    9.03            Gain=    5.22

Rs (Ohms)=     200        Rs (Ohms)=   1000        Rs (Ohms)=   1800
Gain=   19.92             Gain=    7.63            Gain=    4.72

Rs (Ohms)=     400        Rs (Ohms)=   1200        Rs (Ohms)=   2000
Gain=   14.20             Gain=    6.61            Gain=    4.31

Rs (Ohms)=     600        Rs (Ohms)=   1400        End of Second Program
Gain=   11.04             Gain=    5.83


 * * * * * * * * *

Common-Source Amplifier
Output Resistance vs. Source Resistance With Unbypassed Source Resistor

                              ---
(Survey Program)

 * * * * * * * * *

200 Ohm Increments for Rs
(Other Increments May Be Programmed)
 * * * * * * * * *

Ros=  20000
Gfs= 0.005

 * * * * * * * * *

Rs (Ohms)=       0        Rs (Ohms)=   1200
RO (Ohms)=   20000        RO (Ohms)= 141200

Rs (Ohms)=     200        Rs (Ohms)=   1400
RO (Ohms)=   40200        RO (Ohms)= 161400

Rs (Ohms)=     400        Rs (Ohms)=   1600
RO (Ohms)=   60400        RO (Ohms)= 181600

Rs (Ohms)=     600        Rs (Ohms)=   1800
RO (Ohms)=   80600        RO (Ohms)= 201800

Rs (Ohms)=     800        Rs (Ohms)=   2000
RO (Ohms)=  100800        RO (Ohms)= 222000

Rs (Ohms)=    1000        End of Third Program
RO (Ohms)=  121000
```

Figure 8–8. Survey program for calculating the voltage-amplification range for a common-source stage with or without an unbypassed source resistor.

```
1 LPRINT"COMMON-SOURCE AMPLIFIER WITH CAPACITIVE LOAD":PRINT"COMMON-SOURCE AMPLI
FIER WITH CAPACITIVE LOAD"
2 LPRINT"":PRINT"":LPRINT"(Voltage Gain and Phase)":PRINT"Voltage Gain and Phase
":LPRINT"":PRINT""
3 INPUT"Gfs (Siemens)=";E
4 LPRINT"Gfs (Siemens)=";E
5 INPUT "Ros (Ohms)=";D
6 LPRINT"Ros (Ohms)=";D
7 INPUT"RL (Ohms)=";G
8 LPRINT"RL (Ohms)=";G
9 INPUT"C (Farads)=";C
10 LPRINT"C (Farads)=";C
11 INPUT "f (Hz)=";F
12 LPRINT"f (Hz)=";F
13 XC=1/(6.283*F*C):FU=D*G:BF=D+G:KA=FU/BF:SH=KA*XC:HS=-90*6.283/360
14 SA=(KA*KA+XC*XC)^.5:AS=-ATN(XC/KA):QU=SH/SA:UQ=HS-AS
15 AN=E*QU:NA=UQ
16 A$="########.##":LPRINT"Voltage Gain (at f Hz)=";USING A$;AN
17 PRINT"Voltage Gain (at f Hz)=";USING A$;AN
18 LPRINT"Phase Angle (Degrees)=";USING A$;-NA*360/6.283
19 PRINT "Phase Angle (Degrees)=";USING A$;-NA*360/6.283
```

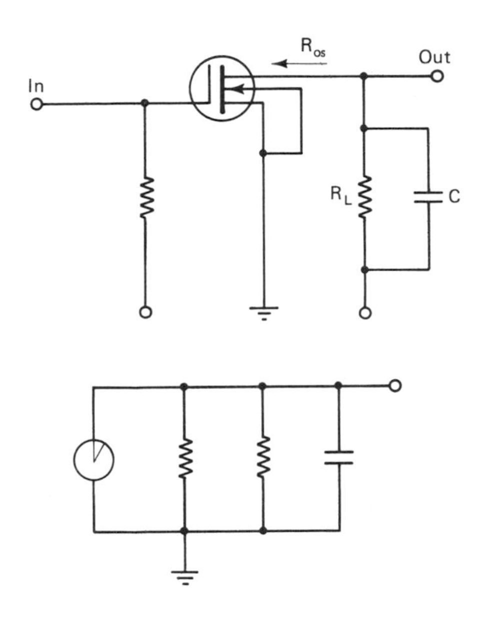

Figure 8–9. Program for calculating the voltage gain, and phase angle, for a common-source amplifier with a capacitive load.

NOTE: When the load is capacitive, it may have significant reactance within the operating frequency range. In turn, the stage will have a falling frequency characteristic, and its phase characteristic will be nonlinear.

fiers with a capacitive output-load impedance have a curved-phase characteristic, or their phase-vs.-frequency plots are curved. As described subsequently, various circuit elaborations are available for improving the linearity (flatness) of a curved-phase characteristic.

DESIGN OPTIMIZATION CONSIDERATIONS

Mixer circuitry can often be optimized by the use of MOSFETs instead of bipolar transistors, inasmuch as cross-modulation activity and spurious-response output are minimized. Device manufacturers have determined that cross-modulation is reduced progressively as a MOSFET approaches cutoff. Dual-gate MOSFETs can be operated in the common-source mode at ultrahigh frequencies and do not require neutralization. Oscillator feedthrough is very small.

Chopper circuitry can usually be optimized by employment of MOSFETs inasmuch as there is no offset voltage with which to contend. A wide dynamic range is available, and the negative temperature coefficient precludes any thermal-runaway hazard. The gate leakage current is very small and is relatively unresponsive to variations in temperature. Their very high input resistance and low input capacitance provide design advantages in product-detector and balanced-modulator circuitry as well as in chopper and gated-amplifier circuits.

9

PRINCIPLES OF OSCILLATOR DESIGN

OSCILLATOR FUNDAMENTALS

An oscillator is an active electronic arrangement that converts dc energy into ac energy. Its output may be sinusoidal or non-sinusoidal, or it may be a combination of both. For example, a shock-excited oscillator generates a damped-sinusoidal waveform in response to an input-trigger pulse. An oscillator may be of the free-running type, monostable type, or bistable type. An oscillator may be unsynchronized, or it may be synchronized (locked) to an input waveform of some type. Specialized types of oscillators generate a pair of output frequencies.

SINUSOIDAL OSCILLATORS

A sinusoidal oscillator is basically a form of positive feedback amplifier. With reference to Figure 9–1, the input voltage v_i is supplied by the feedback voltage $v_f = \beta\ v_o$, whereby we may write:

$$v_i = v_f = \beta\ v_o = (A_v\beta)v_i = 0$$

In turn, the variables may be formulated:

$$(1 - A_v\beta)v_i = 0$$

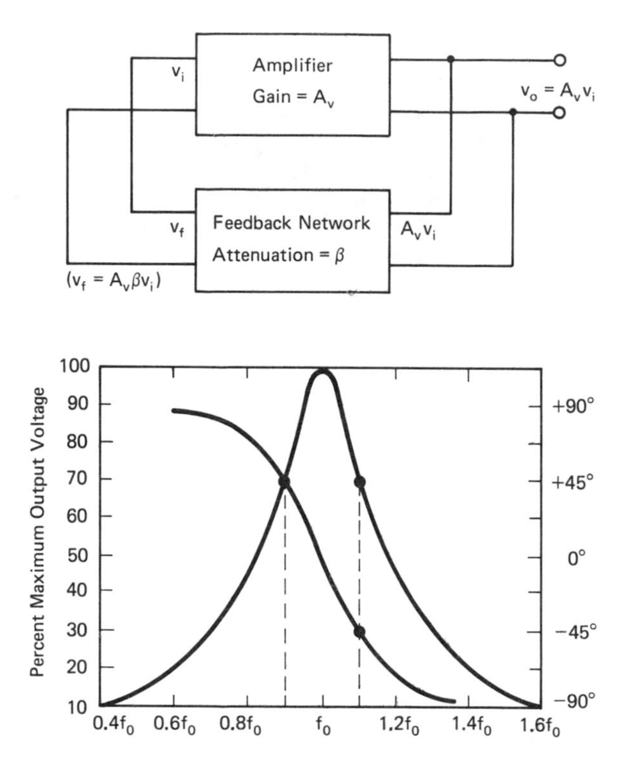

Figure 9–1. Amplifier with positive-feedback network connected in a closed-loop system.

NOTE: Oscillation occurs at a frequency for which the phase at the input is the same as the phase at the output. A parallel-resonant circuit has a leading phase at low frequencies and a lagging phase at high frequencies. Its phase characteristic is 45° at the half-power point on the frequency response (−3 dB down, or 0.707 of maximum response).

To produce an output voltage from the arrangement in Figure 9–1, the input voltage must be greater than zero. Accordingly, the Barkhausen criterion for sustained oscillation may be written:

$$A_v\beta = 1$$

This Barkhausen criterion defines two fundamental requirements for a free-running oscillator:

1. The loop gain (voltage gain around the amplifier and feedback loop) must be unity.
2. The loop-phase shift (phase shift between v_i and v_f) must be zero.

If the foregoing fundamental requirements are met, a sinusoidal-output waveform will be generated by appropriate circuitry. Appropriate circuitry is resonant at a chosen frequency, whereas the amplifier gain is maximum at one frequency. Since only a sine wave contains a single frequency, a sinusoidal output waveform will be generated by a linear (class-A) amplifier.

It is important for the designer to note that if the amplifier and/or feedback network is resonant at two frequencies, oscillation will ensue around the loop that has maximum Q value, and the frequency corresponding to the lower Q value will be suppressed. Or if both loop frequencies happen to be associated with precisely equal Q values, the oscillator will then output a pair of frequencies. Observe that the foregoing remarks assume that the feedback network is arranged so that each frequency is fed back with zero-phase shift to the input.

Note that if the loop gain is less than 1, the oscillator cannot supply its own input for sustained oscillation. (However, if pulsed, the oscillator will generate a damped sine wave.) On the other hand, if the loop gain is greater than 1, the amplitude of oscillation will not be fixed, but will rapidly increase until the oscillator transistor is driven into saturation, into cutoff, or both. In turn, the output waveform will be distorted (the oscillator is operating both as an oscillator and as a clipper).

Automatic gain control may be employed to provide sinusoidal output although the loop gain is greater than 1. Sinusoidal-output waveform (and oscillator stability) are facilitated by utilizing high-Q resonant circuitry in the feedback loop. For example, quartz crystals have very high Q values. Crystals are ground to generate precise frequencies in stipulated oscillator circuits and may be placed in constant-temperature ovens for maximum frequency stability. LCR circuitry provides an oscillatory frequency that is formulated:

$$f_o = 1/(2\pi\sqrt{LC})$$

A parallel-resonant circuit has a frequency response curve as depicted in Figure 9–1. Its half-power bandwidth is approximately formulated:

$$BW = f_o/Q$$

For a parallel-tuned circuit with loss in L only, the resonant frequency for unity power factor is:

$$f_o = \frac{1}{2\pi}\sqrt{\frac{1}{LC} - \frac{R^2}{L^2}}$$

which is the same as for series resonance when L is lossless.

Thus, the effect of resonant-frequency shift in parallel LCR circuits due to R is negligible when the Q value is high.

RF oscillators are generally operated with resonant-circuit loads. A resonant load functions as a wave filter and tends to reject harmonics that may be present in the generated waveform. A skeletonized circuit for a resonant-load oscillator is shown in Figure 9–2. The amplifier voltage gain is formulated:

$$A_v = -A_i Z_L / Z_{in}$$

wherein A_i denotes the common-emitter current gain (h_{fe}) of the transistor, and the minus sign denotes its inherent phase reversal.

If we neglect the comparatively high output resistance of the transistor, the collector-load impedance is formulated:

$$Z_L = \frac{Z2(Z1 + Z3)}{Z1 + Z2 + Z3}$$

The base of the transistor looks into a source impedance that is formulated:

$$Z_s = \frac{Z1(Z2 + Z3)}{Z1 + Z2 + Z3}$$

In turn, the total input impedance of the amplifier is equal to:

$$Z_{in} = Z_s + r_{ie}$$

wherein r_{ie} is the input resistance of the transistor in the CE mode, or r_{ie} is approximately equal to h_{ie}.

Accordingly, the amplifier gain in the absence of feedback is approximately given by:

$$A_v = -h_{fe} Z_L / (Z_s + h_{ie})$$

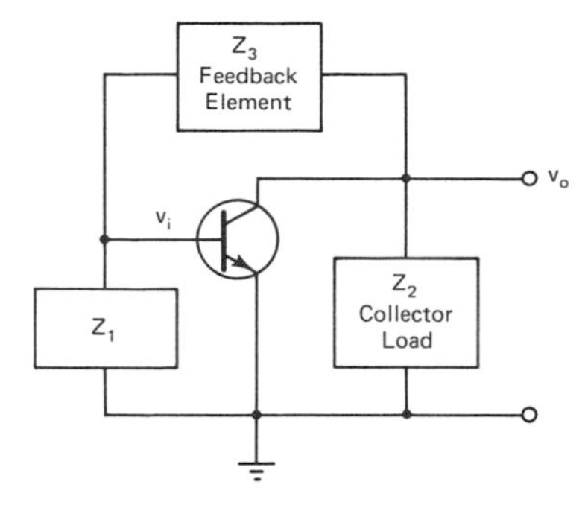

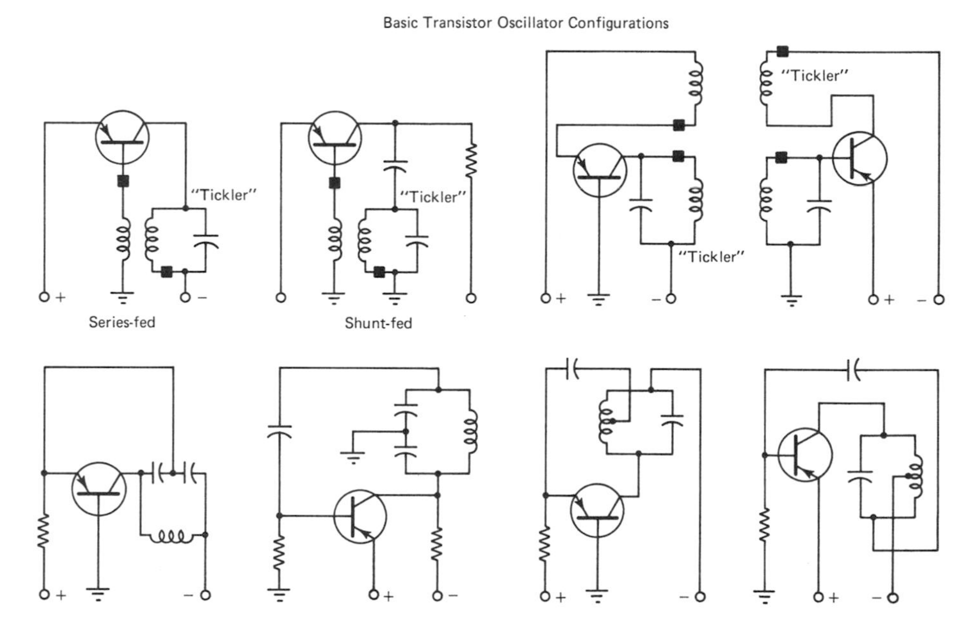

Figure 9–2. Skeletonized resonant-load oscillator arrangements. (Reproduced by special permission of Reston Publishing Company and Campbell Loudoun from *Handbook for Electronic Circuit Design*.)

NOTE: Resonant-load oscillator configurations may be operated in class A, class B, or class C. The mode of operation depends on the amount of positive feedback that is provided from output to input.

With a small amount of positive feedback, the oscillator operates approximately in class A. Ideal class-A operation is impossible, inasmuch as the transistor has a nonlinear transfer characteristic. Alpha crowding limits the amplitude of oscillation and also introduces some harmonic distortion.

With a large amount of positive feedback, the transistor operates in class C. High-frequency resonant-load oscillators are often operated in class C. The transistor is driven substantially into its saturation region, and rectification of the base drive waveform occurs.

A class-C oscillator appears to be reverse-biased when its base-emitter voltage is measured with a dc voltmeter. However, this is merely the average base-emitter bias voltage. An oscilloscope shows that the base conducts strongly in narrow pulses on peaks of the drive waveform.

Class-C oscillators generally employ signal-developed bias with a base-coupling capacitor and base resistor, as exemplified in the lower right-hand diagram (above). The chief advantage of signal-developed bias is that the oscillator is "self-starting."

Oscillator frequency stability depends on a regulated power-supply voltage, on constant ambient temperature, and on a low level of oscillation. (A class-A amplifier with moderate feedback is more stable than a class-C amplifier with substantial feedback.)

Positive feedback is provided by Z3 in Figure 9–2, and the feedback factor is formulated:

$$\beta = Z1/(Z1 + Z3)$$

The loop gain of the oscillator is given by:

$$A_v\beta = h_{fe}X2/X1$$

wherein we regard the impedances in Figure 9–2 as pure reactances at this point.

It follows that to sustain oscillation, the transistor must have a current gain of at least:

$$h_{fe} = X1/X2$$

PHASE-SHIFT (RC) OSCILLATORS

Phase-shift oscillators are a prominent class of low-frequency sine-wave sources. They are designed with respect to an RC network that has a phase shift of an odd multiple of 180° (per stage) at the oscillating frequency. The phase-shift network is connected as a feedback loop from the output to input of an amplifier (often a single transistor). A phase-shift oscillator is often tunable, as depicted in Figure 9–3. Three cascaded RC sections are utilized in this example to develop a 180°-phase shift between collector-output and base-input voltages.

The feedback factor of the RC network is formulated:

$$\beta = V_o/V_i = \frac{1}{1 - 5\alpha^2 - j(6\alpha - \alpha^3)}$$

where $\alpha = 1/(\omega RC)$.

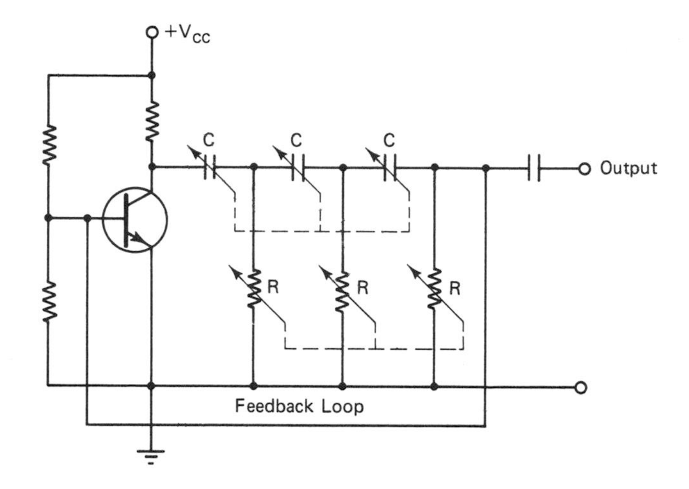

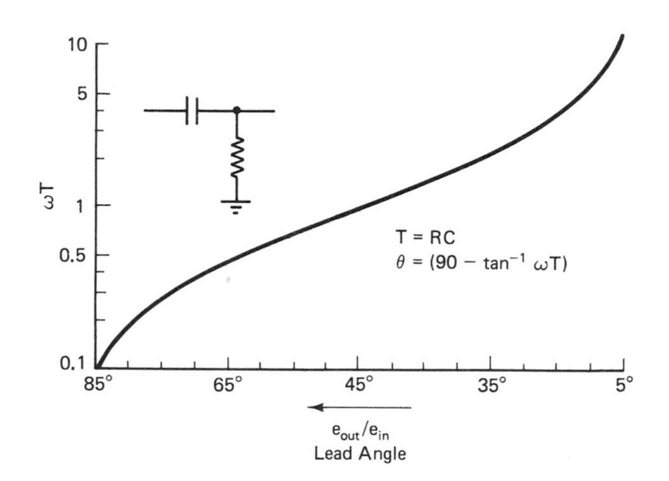

Figure 9–3. Basic RC phase-shift oscillator arrangement with ganged-tuning capacitors and adjustable resistances.

NOTE: A single RC differentiating section has the phase characteristic indicated below. Its characteristics become changed if it is loaded by a following RC section. A three-section network can be designed to provide 60° phase shift through each section. Note that ohms times farads equals seconds.

When the reactive term in the denominator of the foregoing equation is zero, the Barkhausen criterion for self-oscillation is met. In turn, the frequency of oscillation is derived:

$$f = \frac{1}{2\pi RC\sqrt{6}}$$

At this frequency, the feedback factor (β) is −1/29. To meet the requirement that the loop gain ($A_v\beta$) = 1, the amplifier must provide a gain of at least 29 times. Although the basic configuration is best adapted to fixed-frequency applications, it is adequate also for many variable-frequency purposes. The upper-frequency limit is approximately 100 kHz. Although a reasonably good waveform is generated, it does not meet high-fidelity standards.

Lower distortion with simple circuitry can be obtained if a pair of MOSFETs are utilized in a phase-shift oscillator configuration, as exemplified in Figure 9–4. It is essential to use high-gain

DIMENSIONAL UNITS OF MECHANICAL QUANTITIES

Symbol	Physical unit	Dimensional units
F	Force	(F)
L	Length	(L)
t	Time	(T)
M	Mass	(FT^2L^{-1})
W	Energy or work	(FL)
P	Power	(FLT^{-1})
v	Velocity	(LT^{-1})
a	Acceleration	(LT^{-2})

DIMENSIONAL UNITS OF ELECTRICAL QUANTITIES

Symbol	Physical unit	Dimensional units
Q	Charge	(Q)
I	Current	(QT^{-1})
V	Voltage	(FLQ^{-1})
R	Resistance	$(FLTQ^{-2})$
L	Inductance	(FLT^2Q^{-2})
C	Capacitance	$(Q^2L^{-1}F^{-1})$

NOTE: These dimensional units show why the product of ohms and farads is equal to seconds. In other words, resistance has the dimensional units $FLTQ^{-2}$, and capacitance has the dimension units $Q^2L^{-1}F^{-1}$. Accordingly, R times C equals T.

Observe that all derived equations must be dimensionally consistent or there has been some error made in calculation. On the other hand, a few of the equations utilized in electronics are empirical and are not necessarily dimensionally consistent.

MOSFETs. If the three sections of the RC feedback network are identical, the frequency of oscillation is given by the approximate formula:

$$f = 1/(3RC)$$

From a design viewpoint, the arrangement in Figure 9–4 logically has lower distortion than the arrangement in Figure 9–3, inasmuch as the MOSFET configuration employs integrating-circuit phase

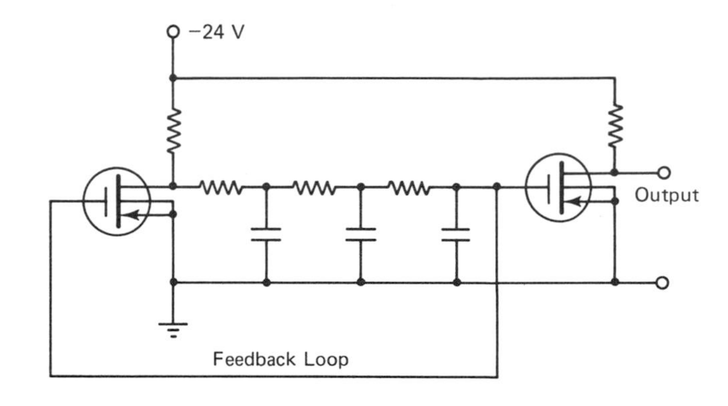

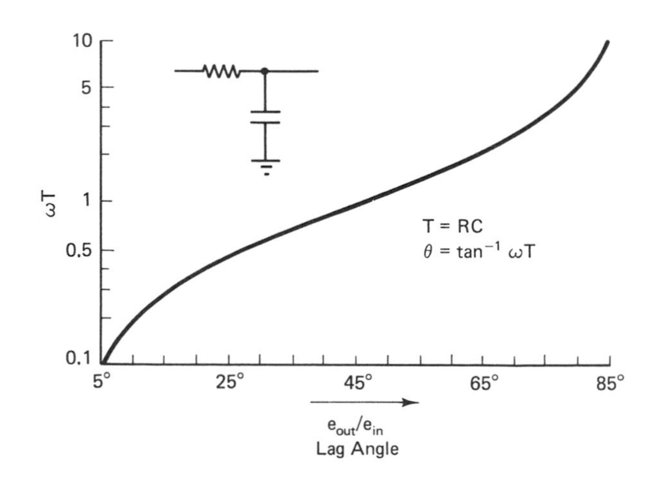

Figure 9–4(A). Basic arrangement for a MOSFET RC phase-shift oscillator.

NOTE: A single RC integrating section has the phase characteristic indicated below. Its characteristics become changed if it is loaded by a following RC section. A three-section network can be designed to provide 60° phase shift through each section. If each RC section has progressively higher impedance, loading will be minimized.

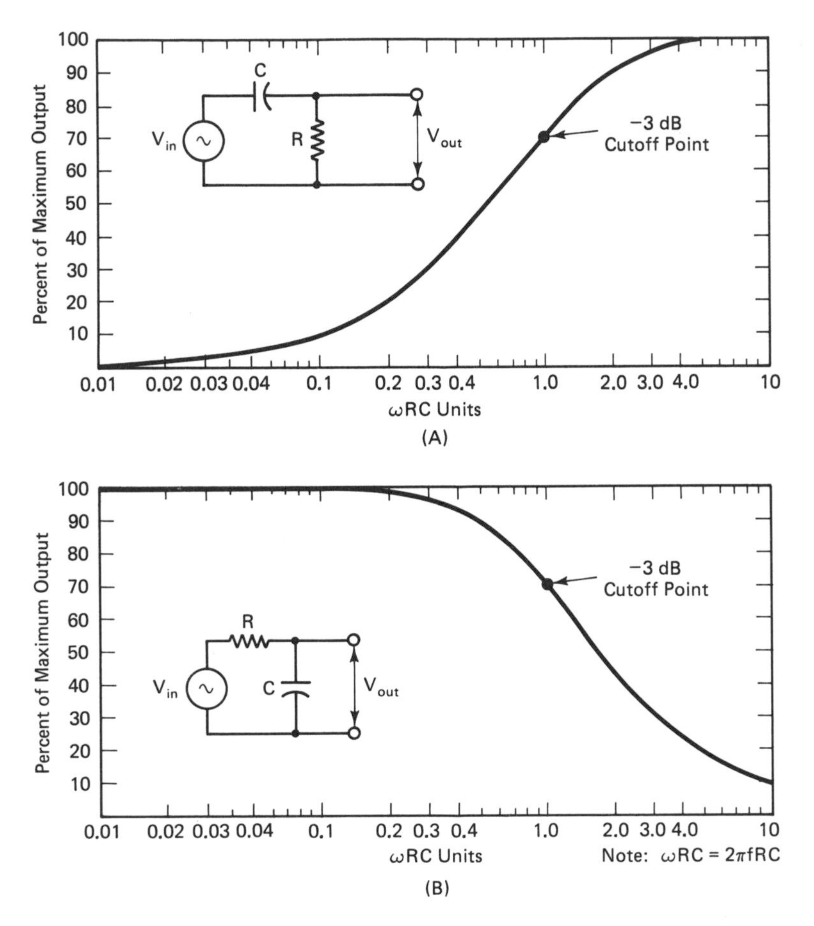

Figure 9–4(B). Universal-frequency response charts. (A) RC differentiating section; (B) RC integrating section. (Reproduced by special permission of Reston Publishing Company and Derek Cameron from *Handbook of Audio Circuit Design*.)

Note: Computer programs for calculating the output/input voltage and phase angle for RC-differentiating and -integrating circuits are provided at the end of the chapter.

shift, whereas the bipolar transistor arrangement employs differentiating-circuit phase shift. In other words, an integrating circuit functions as a low-pass filter and discriminates against harmonic components, whereas a differentiating circuit functions as a high-pass filter and discriminates against the fundamental component.

WIEN-BRIDGE OSCILLATOR

A Wien bridge (Figure 9–5) includes series RC and shunt RC arms. The balance conditions are formulated:

$$Z1Z4 = Z2Z3 \quad \text{or} \quad Z3 = Z1Z4Y2$$

where Y2 = 1/Z2.

The Wien bridge is customarily designed so that R1 = R2 = R and C1 = C2 = C, whereby $R_3/R_4 = 2$, and the balancing frequency is:

$$f = \frac{1}{2\pi RC}$$

Next, a Wien-bridge oscillator is configured as exemplified in Figure 9–6, employing two bipolar transistors in the CE mode. The designer should note that a Wien bridge in an oscillator arrangement does not operate in a balanced condition. Stated otherwise, self-oscillation would be impossible at the balancing frequency, inasmuch as no voltage would be fed back from output to input. Accordingly, appropriate resistance ratios are modified so that an in-phase unbalance voltage is fed back from output to input.

The analysis of this design requirement is as follows:

$$v_a = v_i Z_2/(Z_1 + Z_2) = v_i/3$$

$$v_b = v_i R_2/(R_1 + R_2)$$

In a balanced condition, $R_3/R_4 = 2$, whereby:

$$v_o = v_a - v_b = 0$$

Observe that R_1 and R_2 have no effect upon the resonant frequency, although they determine whether v_o is zero or some other value. Therefore, the designer employs a suitable ratio of R_1 and R_2 for self-oscillation (the value of R_2 will be less than half the value of R_1). In turn, an appropriate unbalance voltage occurs for sustained oscillation; note that this unbalance voltage is in phase with the input voltage.

Coupling capacitors in a Wien-bridge arrangement must be sufficiently large that they introduce negligible phase shift at the lowest frequency of operation. In the example of Figure 9–6, ganged capacitors in the reactive arms function to provide continuous variation of frequency. Various frequency ranges can be obtained by switching the values for resistors R. Observe that R2 does double duty as a bridge arm and also as an emitter resistor for Q1.

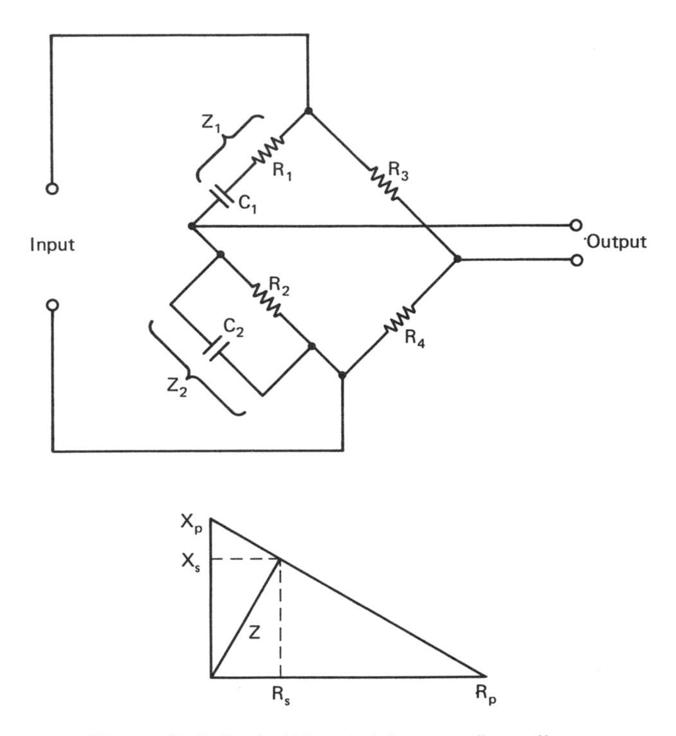

Figure 9–5. Basic Wien-bridge configuration.

NOTE: There is a particular frequency at which Z1 = Z2 = Z, and the bridge will be nulled (output zero) if R3 = R4. If C2 = C1, R2 = R1, and R3 = 2R4, then the null frequency is f = 1/(ωC1R1). In turn, the bridge may be used to measure frequency (over the audio range) by implementing C1 and C2 as ganged variable capacitors and stepping the values of R1 and R2 for various frequency ranges. The ganged capacitors are calibrated in terms of frequency. Observe that the input to the bridge must be a pure sine wave, or it will be difficult to obtain a clear-cut null indication. In other words, the bridge will not be balanced for a harmonic frequency when it is balanced for the fundamental frequency.

PROGRAMMED FORMULAS

A basic overview of the output/input voltage and the output/input phase angle for a single-section RC-differentiating circuit is shown in Figure 9–7. Percentage of maximum response and the leading phase angle are plotted against fRC units. Observe that these curves correspond to an unloaded RC section driven by a constant-voltage generator. In practice, neither of these conditions is met, although the generator might have negligible internal resistance, and the load might have negligible conductance.

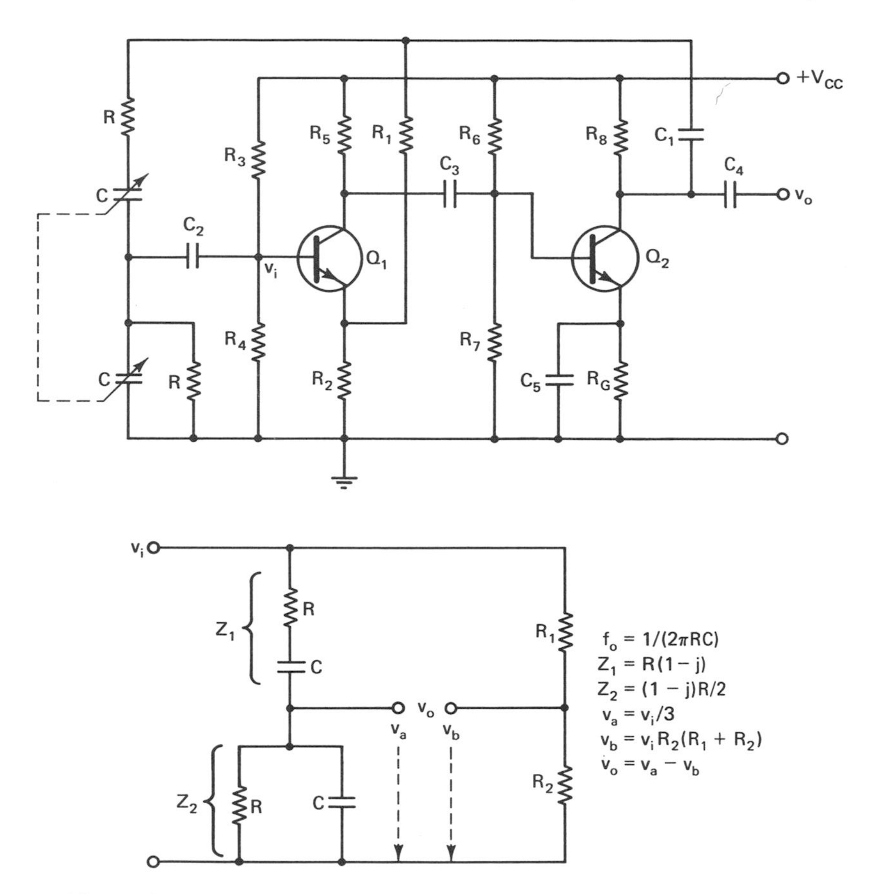

Figure 9–6. Typical two-stage Wien-bridge oscillator arrangement.

It is evident that if the RC section were driven by a constant-current generator, the output voltage from the section would remain constant at any frequency. In practice, the generator is somewhere in between a constant-voltage and a constant-current source, and the designer may need to take the generator internal resistance into account. In other words, a voltage divider is represented by the series combination of the generator internal resistance and the resistance of the RC section.

If the load on the RC section does not have negligible conductance, the capacitor then works into a parallel combination of the load resistance and the resistance of the RC section. Stated otherwise, an RC-differentiating circuit does not necessarily "stand alone" in a system, and its output/input voltage and its output/input

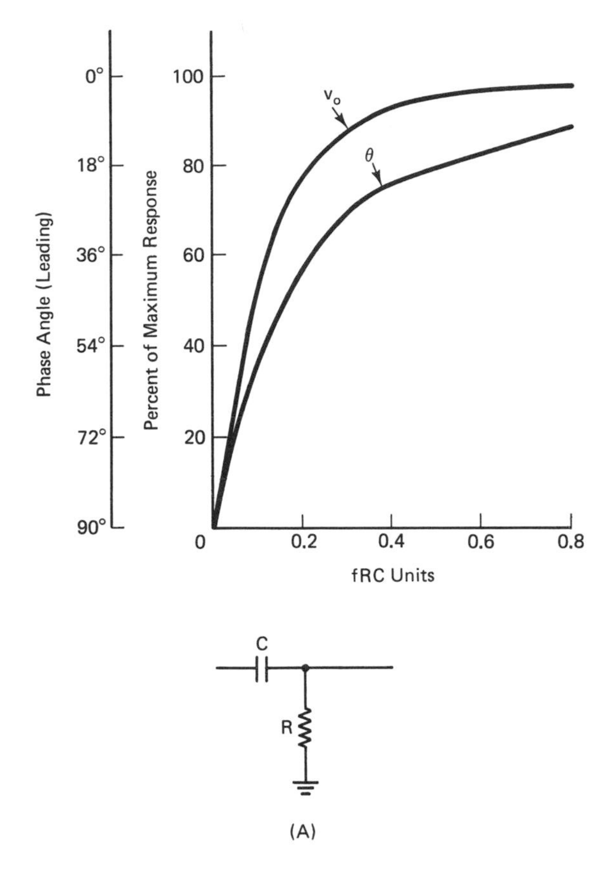

Figure 9–7(A). Percent of maximum-output voltage versus fRC units and phase angle versus fRC units for an RC-differentiating circuit.

phase angle must be considered in context. An RC-differentiating circuit often works into a reactive load, which complicates the circuit analysis, as explained subsequently.

It is sometimes helpful to run a short computer program such as shown in Figure 9–8 for calculating the output voltage and phase angle with respect to the input voltage of a single-section RC-differentiating circuit with chosen f, R, and C values. Observe that this program assumes that the generator internal resistance is negligible and that the load conductance is negligible.

Similarly, a program for calculating the output voltage and phase angle with respect to the input voltage of a single-section RC-integrating circuit with chosen f, R, and C values is sometimes useful, as shown in Figure 9–9. Again, this program assumes that

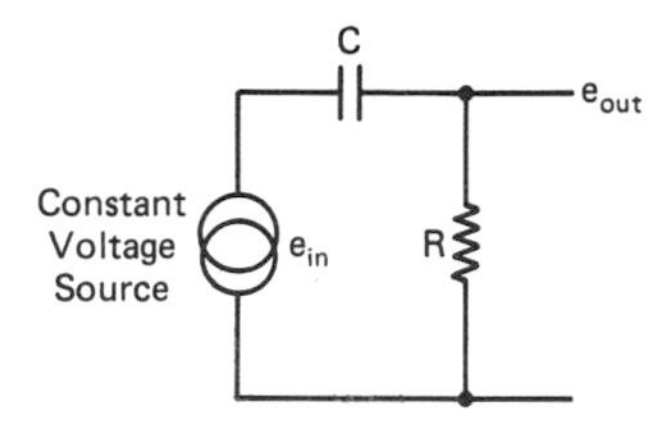

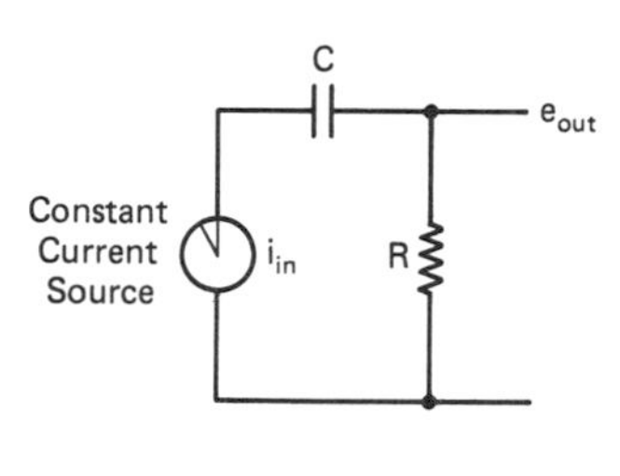

(B)

Figure 9–7(B). Basic distinctions in differentiating-circuit operation from constant-voltage or constant-current sources.

NOTE: With a constant-voltage source, the capacitor-voltage drop approaches e_{in} at low frequencies and approaches zero at high frequencies.

With a constant-current source, the capacitor-voltage drop approaches infinity at low frequencies and approaches zero at high frequencies.

An ideal constant-current source has an infinite open-circuit voltage.

Practical "voltage" sources and practical "current" sources are intermediate to constant-voltage or constant-current sources.

the generator internal resistance is negligible and that the load conductance is negligible.

The designer's task becomes more demanding when two RC sections are to be cascaded. In this situation, the first section is loaded by the first section with the result that calculation of the output/input voltage and phase angle becomes somewhat involved. There are three chief methods available for analyzing this type of network:

1. Mesh analysis.
2. Branch current analysis.
3. Thevenin theorem application.

With reference to Figure 9–10, the basic diagram for mesh analysis of a symmetrical two-section cascaded RC-integrating network is

```
1 LPRINT"OUTPUT VOLTAGE AND PHASE ANGLE FOR RC DIFFERENTIATING CIRCUIT"
2 PRINT"OUTPUT VOLTAGE AND PHASE ANGLE FOR RC DIFFERENTIATING CIRCUIT"
3 LPRINT"**************************************************************"
4 PRINT"**************************************************************"
5 LPRINT"":PRINT""
6 INPUT "f (Hz)=";F
7 LPRINT "f (Hz)=";F
8 INPUT "R (Ohms)=";R
9 LPRINT "R (Ohms)=";R
10 INPUT "C (Farads)=";C
11 LPRINT "C (Farads)=";C
12 N=F*R*C:T=R*C
13 LPRINT"fRC=";N:PRINT"fRC=";N
14 XC=1/(6.283*F*C):AB=(R*R+XC*XC)^.5:BA=-ATN(XC/R)
15 IN=1/AB:NI=-BA:EO=IN*R:OE=NI
16 A$="###.##":LPRINT"Eo/Ei= ";USING A$;EO:PRINT"Eo/Ei=";USING A$;EO
17 LPRINT"Phase Angle (Degrees)=";USING A$;-OE*360/6.283
18 PRINT"Phase Angle (Degrees)=";USING A$;-OE*360/6.283
19 LPRINT"Time Constant (Seconds=";T
20 PRINT"Time Constant (Seconds)=";T
```

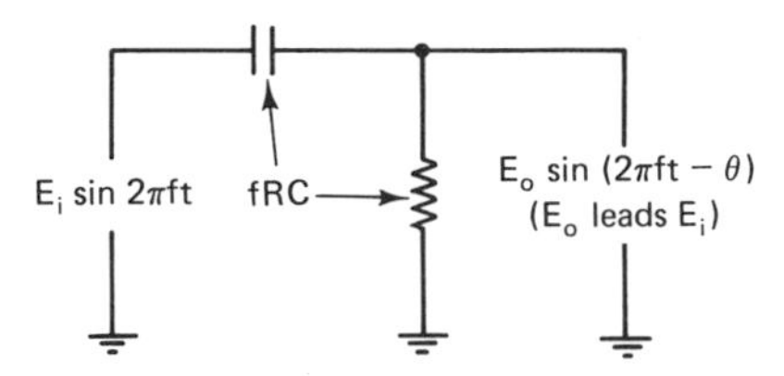

NOTE: If the output of the RC section drives a load resistor, line 8 then requires the value of the load resistor in parallel with the resistor of the RC section.

```
OUTPUT VOLTAGE AND PHASE ANGLE FOR RC DIFFERENTIATING CIRCUIT
**************************************************************

f (Hz)= 500
R (Ohms)= 1000
C (Farads)= .000001
fRC= .5
Eo/Ei=   0.95
Phase Angle (Degrees)=-17.66
Time Constant (Seconds= 9.999999E-04
```

Figure 9–8. Program for calculating the percentage of maximum-output voltage and phase angle versus f, R, and C for an RC-differentiating circuit.

```
1 LPRINT"OUTPUT VOLTAGE AND PHASE ANGLE FOR RC INTEGRATING CIRCUIT"
2 PRINT"OUTPUT VOLTAGE AND PHASE ANGLE FOR RC INTEGRATING CIRCUIT"
3 LPRINT"**********************************************************"
4 PRINT"**********************************************************"
5 LPRINT"":PRINT""
6 INPUT "f (Hz)=";F
7 LPRINT "f (Hz)=";F
8 INPUT "R (Ohms)=";R
9 LPRINT "R (Ohms)=";R
10 INPUT "C (Farads)=";C
11 LPRINT "C (Farads)=";C
12 N=F*R*C:T=R*C
13 LPRINT"fRC=";N:PRINT"fRC=";N
14 XC=1/(6.283*F*C):AB=(R*R+XC*XC)^.5:BA=-ATN(XC/R)
15 IN=1/AB:NI=-BA:EO=IN*XC:OE=NI-90*6.283/360
16 A$="###.##":LPRINT"Eo/Ei= ";USING A$;EO:PRINT"Eo/Ei=";USING A$;EO
17 LPRINT"Phase Angle (Degrees)=";USING A$;-OE*360/6.283
18 PRINT"Phase Angle (Degrees)=";USING A$;-OE*360/6.283
19 LPRINT"Time Constant (Seconds=";T
20 PRINT"Time Constant (Seconds)=";T
```

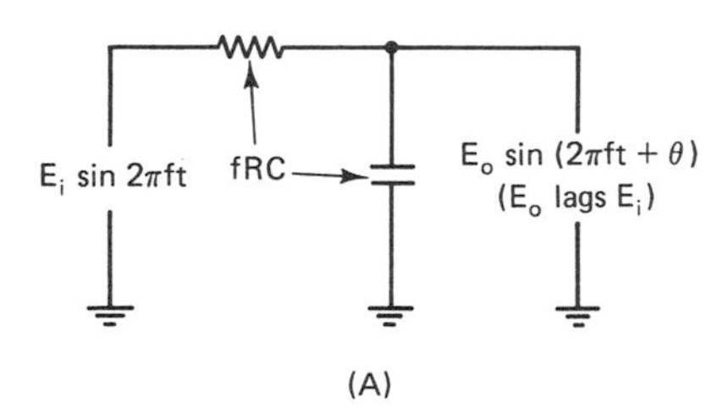

(A)

NOTE: This program assumes that the RC section is driven from a constant-voltage source. If the generator has appreciable internal resistance, this resistance should be added to the resistance in the RC section.

```
OUTPUT VOLTAGE AND PHASE ANGLE FOR RC INTEGRATING CIRCUIT
**********************************************************

f (Hz)= 500
R (Ohms)= 1000
C (Farads)= .000001
fRC= .5
Eo/Ei=    0.30
Phase Angle (Degrees)= 72.34
Time Constant (Seconds= 9.999999E-04
```

Figure 9–9(A). Program for calculating the ratio of output/input voltage and phase angle versus f, R, and C for an RC-integrating circuit.

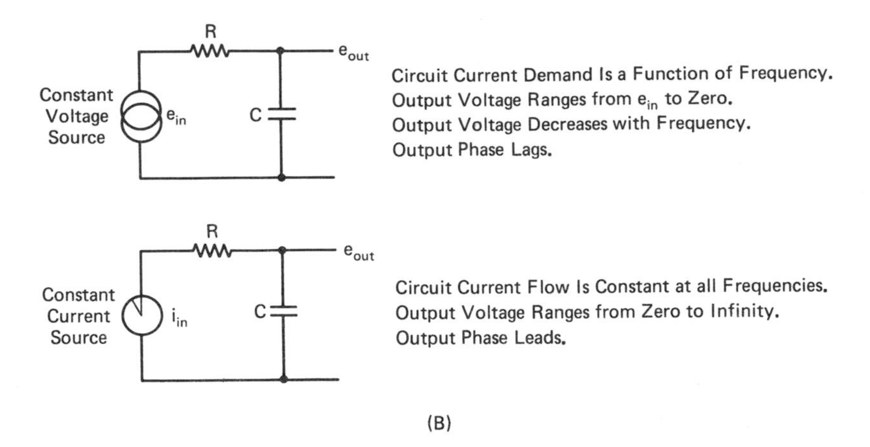

Figure 9–9(B). Basic distinctions in integrating-circuit operation from constant-voltage or constant-current sources.

NOTE: A practical "current" source does not have an infinite open-circuit voltage. In turn, the output voltage from an integrating circuit driven by a practical "current" source ranges from zero to the internal voltage of the "current" source.

shown. A program for mesh analysis of the output/input voltage and phase for a symmetrical two-section cascaded RC-integrating circuit is provided in Figure 9–11. Observe that the R value that is inputted is the series resistance for one section. Similarly, the C value that is inputted is the shunt capacitance for one section.

In some applications the designer wishes to utilize the midpoint voltage from the network, in addition to the output voltage. The midpoint voltage is calculated as shown in Figure 9–12. In other words, the midpoint voltage (and phase) is given by the product of the current and the impedance of R and C in the second mesh. It is evident that the midpoint voltage is greater than the output voltage and that its phase leads the output voltage. Observe that both the midpoint- and the output-voltage phases lag the input-voltage phase.

Branch-current analysis is a variation of mesh-current analysis. It is based on Kirchhoff's current law, which states that the sum of the currents entering and leaving a node is equal to zero. There is no significant distinction in the two methods, apart from details of current and voltage assignments. On the other hand, application of Thevenin's theorem provides both capabilities and limitations insofar as the present example is concerned. In other words, Thevenin's

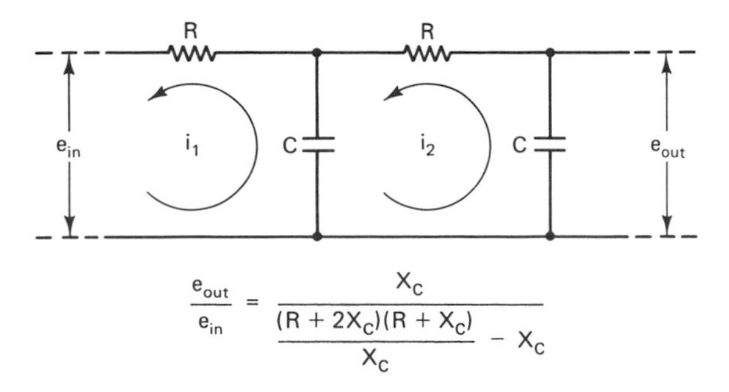

$$\frac{e_{out}}{e_{in}} = \frac{X_C}{\dfrac{(R + 2X_C)(R + X_C)}{X_C} - X_C}$$

Figure 9–10. Basic diagram for mesh analysis of a symmetrical two-section cascaded-RC integrating network.

NOTE: The equation for calculating the ratio of output voltage to input voltage and the output-phase angle with respect to the input-phase angle is derived from the principles that the sum of the voltages around a closed mesh is equal to zero. Thus, the sum of the voltages around the first mesh is equal to the input voltage plus the i_1R drop plus the i_1X_c drop minus the i_2X_C drop. The sum of the voltages around the second mesh is equal to the i_2X_C drop minus the i_1X_C drop plus the i_2XC drop. In turn, the output voltage is equal to the i_2X_C drop. These mesh equations are solved simultaneously (or by substitution) to derive the final e_{out}/e_{in} equation.

theorem simplifies calculation of output voltage and phase to some extent. On the other hand, the theorem provides no immediate conclusion concerning the relation of output voltage and phase to input voltage.

With reference to Figure 9–13, observe that Thevenin's theorem provides simplified calculation of output voltage and phase with respect to the chosen Thevenin terminals in the network. This is just another way of saying that the value of e_{in} is not immediately known; if the black box contains a large number of circuit sections, then the value of e_{in} with respect to e_{out} will be correspondingly large. Note in passing that the Thevenin voltage is defined as the open-circuit output voltage from the black box and that the Thevenin impedance is equal to the output impedance of the black box with e_{in} short-circuited.

To briefly pursue the value of the Thevenin impedance, note that if the black box contains a single-section RC integrating circuit, then the Thevenin impedance consists of the parallel combination of resistance and capacitance in the integrating circuit. It is evident that the value of the Thevenin voltage is not equal to the midpoint voltage previously discussed. In other words, the Thevenin

```
1 LPRINT "SYMMETRICAL TWO-SECTION RC LAG (INTEGRATING) NETWORK"
2 PRINT "SYMMETRICAL TWO-SECTION RC LAG (INTEGRATING) NETWORK"
3 LPRINT "(Computes Eout/Ein, Output/Input Phase Angle, and Phase*Freq. Value>"
4 PRINT "Computes Eout/Ein, Output/Input Phase Angle, and Phase*Freq. Value)"
5 LPRINT"":PRINT"":A$="#######.##"
6 INPUT "R (Ohms)=";R
7 LPRINT "R (Ohms)=";R
8 INPUT "C (Mfd)=";C
9 LPRINT "C (Mfd)=";C
10 INPUT "f (Hz)=";F
11 LPRINT "f (Hz)=";F
12 X=1/(6.2832 *F*C*10^-6):AB=(R^2+4*X^2)^.5:BA=-ATN(2*X/R):AC=(R^2+X^2)^.5
13 CA=-ATN(X/R):BC=AB*AC:CB=BA+CA:AD=BC/X:DA=CB+6.2832/4:BD=AD*COS(DA)
14 DB=AD*SIN(DA):CD=DB+X:DE=(BD^2+CD^2)^.5:ED=-ATN(CD/BD):AE=X/DE
15 EA=-ED-6.2832/4:LPRINT"":PRINT""
16 LPRINT "Eout/Ein=";USING A$;AE:PRINT "Eout/Ein=";USING A$;AE
17 LPRINT "Phase Angle (Degrees)=";USING A$;-EA*360/6.2832
18 PRINT "Phase Angle (Degrees)=";USING A$;-EA*360/6.2832
19 LPRINT "Phase*Frequency Value=";USING A$;EA*F
20 PRINT "Phase*Frequency Value=";USING A$;EA*F
21 END
```

```
SYMMETRICAL TWO-SECTION RC LAG (INTEGRATING) NETWORK
(Computes Eout/Ein, Output/Input Phase Angle, and Phase*Freq. Value>

R (Ohms)= 1000
C (Mfd)= .159145
f (Hz)= 1000

Eout/Ein=          0.33
Phase Angle (Degrees)=      90.00
Phase*Frequency Value=   -1570.84
```

```
SYMMETRICAL TWO-SECTION RC LAG (INTEGRATING) NETWORK
(Computes Eout/Ein, Output/Input Phase Angle, and Phase*Freq. Value>

R (Ohms)= 10000
C (Mfd)= .15
f (Hz)= 400

Eout/Ein=          0.06
Phase Angle (Degrees)=      40.56
Phase*Frequency Value=    -283.19
```

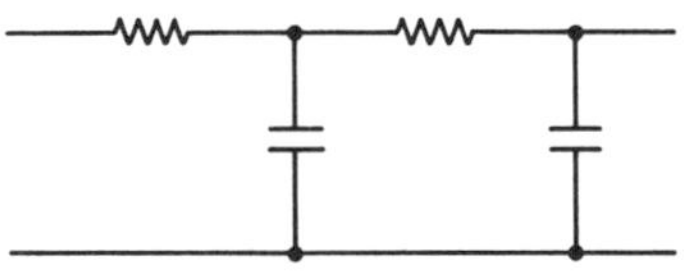

(A)

NOTE: The first RUN *employs a "trick of the trade" wherein the inputted value of C has the same absolute value as R at the operating frequency. In turn, the programmer can easily follow the processing action line by line to observe how the starting equation is being solved step by step.*

Figure 9–11(A). Program for mesh analysis of output/input voltage and phase for a symmetrical two-section cascaded RC-integrating network.

```
1 LPRINT "SYMMETRICAL TWO-SECTION RC LAG (INTEGRATING) NETWORK"
2 PRINT "SYMMETRICAL TWO-SECTION RC LAG (INTEGRATING) NETWORK"
3 LPRINT "(Computes Eout/Ein, Output/Input Phase Angle, and Phase*Freq. Value>"
4 PRINT "Computes Eout/Ein, Output/Input Phase Angle, and Phase*Freq. Value)"
5 A$="#######.##":LPRINT "** Survey Program **":PRINT "** Survey Program **"
6 LPRINT"":PRINT"":INPUT "R (Ohms)=";R
7 LPRINT "R (Ohms)=";R
8 INPUT "C (Mfd)=";C
9 LPRINT "C (Mfd)=";C"LPRINT"":PRINT""
10 INPUT "f (Hz)=";H
11 LPRINT "f (Hz)=";H:N=0:G=H
12 F=G+N:B$="########":LPRINT"f (Hz)=";USING B$;F:PRINT"f (Hz)=";USING B$;F
13 X=1/(6.2832 *F*C*10^-6):AB=(R^2+4*X^2)^.5:BA=-ATN(2*X/R):AC=(R^2+X^2)^.5
14 CA=-ATN(X/R):BC=AB*AC:CB=BA+CA:AD=BC/X:DA=CB+6.2832/4:BD=AD*COS(DA)
15 DB=AD*SIN(DA):CD=DB+X:DE=(BD^2+CD^2)^.5:ED=-ATN(CD/BD):AE=X/DE
16 EA=-ED-6.2832/4:LPRINT"":PRINT""
17 LPRINT "Eout/Ein=";USING A$;AE:PRINT "Eout/Ein=";USING A$;AE
18 LPRINT "Phase Angle (Degrees)=";USING A$;-EA*360/6.2832
19 PRINT "Phase Angle (Degrees)=";USING A$;-EA*360/6.2832
20 LPRINT "Phase*Frequency Value=";USING A$;EA*F:N=N+100
21 PRINT "Phase*Frequency Value=";USING A$;EA*F:LPRINT"":PRINT""
22 IF F>5000 THEN 24
23 GOTO 12
24 END
```

```
SYMMETRICAL TWO-SECTION RC LAG (INTEGRATING) NETWORK
(Computes Eout/Ein, Output/Input Phase Angle, and Phase*Freq. Value>
** Survey Program **

R (Ohms)= 1000
C (Mfd)= 1 LPRINT:PRINT
f (Hz)= 1
f (Hz)=          1

Eout/Ein=         1.00
Phase Angle (Degrees)=     178.92
Phase*Frequency Value=      -3.12

f (Hz)=        101

Eout/Ein=         0.50
Phase Angle (Degrees)=     107.42
Phase*Frequency Value=    -189.35

f (Hz)=        201

Eout/Ein=         0.26
Phase Angle (Degrees)=      81.08
Phase*Frequency Value=    -284.42

f (Hz)=        301

Eout/Ein=         0.16
Phase Angle (Degrees)=      65.57
Phase*Frequency Value=    -344.49

f (Hz)=        401

Eout/Ein=         0.11
Phase Angle (Degrees)=      54.72
Phase*Frequency Value=    -382.96

f (Hz)=        501

Eout/Ein=         0.08
Phase Angle (Degrees)=      46.67
Phase*Frequency Value=    -408.07

f (Hz)=        601

Eout/Ein=         0.06
Phase Angle (Degrees)=      40.51
Phase*Frequency Value=    -424.92

f (Hz)=        701
```

(B)

NOTE: A lower limit for the value of f is inputted. In turn, the program increments f upward in 100-Hz steps.

Figure 9–11(B). Survey program for mesh analysis of the output/input voltage and phase for a symmetrical two-section cascaded RC-integrating network.

```
1 LPRINT "SYMMETRICAL TWO-SECTION RC LAG (INTEGRATING) NETWORK"
2 PRINT "SYMMETRICAL TWO-SECTION RC LAG (INTEGRATING) NETWORK"
3 LPRINT "(Computes Output Impedance and Phase Angle)"
4 PRINT "(Computes Output Impedance and Phase Angle)"
5 A$="#######.##":LPRINT "** Survey Program **":PRINT "** Survey Program **"
6 LPRINT"":PRINT"":INPUT "R (Ohms)=";R
7 LPRINT "R (Ohms)=";R
8 INPUT "C (Mfd)=";C
9 LPRINT "C (Mfd)=";C"
10 INPUT "f (Hz)=";H
11 LPRINT "f (Hz)=";H:N=0:G=H
12 F=G+N:B$="########":LPRINT"f (Hz)=";USING B$;F:PRINT"f (Hz)=";USING B$;F
13 X=1/(6.2832 *F*C*10^-6):AB=R*X:BA=-6.2832/4:AC=(R^2+X^2)^.5:CA=-ATN(X/R)
14 AD=AB/AC:DA=BA-CA:AE=AD*COS(DA):EA=AD*SIN(DA):AF=AE+R:N=N+100
15 AG=(AF^2+EA^2)^.5:GA=-ATN(EA/AF):AH=AG*X:HA=GA-6.2832/4:AJ=EA-X
16 AK=(AF^2+AJ^2)^.5:KA=-ATN(AJ/AF):AN=AH/AK:NA=HA-KA:LPRINT"":PRINT""
17 LPRINT "Zout (Ohms)=";USING A$;AN:PRINT "Zout (Ohms)=";USING A$;AN
18 LPRINT "Phase Angle (Degrees)=";USING A$;-NA*360/6.2832
19 PRINT "Phase Angle (Degrees)=";USING A$;-NA*360/6.2832
20 IF F>5000 THEN 22
21 LPRINT"":PRINT"":GOTO 12
22 END
```

```
SYMMETRICAL TWO-SECTION RC LAG (INTEGRATING) NETWORK
(Computes Output Impedance and Phase Angle)
** Survey Program **

R (Ohms)= 1000
C (Mfd)= 1
f (Hz)= 1
f (Hz)=        1

Zout (Ohms)=     1999.74
Phase Angle (Degrees)=     179.10

f (Hz)=      101

Zout (Ohms)=     1051.60
Phase Angle (Degrees)=     125.02

f (Hz)=      201

Zout (Ohms)=      616.75
Phase Angle (Degrees)=     113.35

f (Hz)=      301

Zout (Ohms)=      441.73
Phase Angle (Degrees)=     108.97

f (Hz)=      401

Zout (Ohms)=      347.41
Phase Angle (Degrees)=     106.28

f (Hz)=      501

Zout (Ohms)=      287.26
Phase Angle (Degrees)=     104.24

f (Hz)=      601
```

R R C C Z_{out}

$$Z_{out} = \frac{\left[\frac{R \times X_C}{R + X_C} + R\right] X_C}{\frac{R \times X_C}{R + X_C} + R + X_C}$$

(C)

Figure 9–11(C). Program for calculating the output impedance of a symmetrical two-section cascaded RC-integrating network.

NOTE: When angles greater than 90 degrees are being processed, there is a possibility of a 180-degree ambiguity in the final answer. To check for 180-degree ambiguity, draw a rough vector diagram of the network. (Vector diagramming is explained in Chapter 3.)

```
1 LPRINT "UNSYMMETRICAL TWO-SECTION RC LAG (INTEGRATING) NETWORK"
2 PRINT "UNSYMMETRICAL TWO-SECTION RC LAG (INTEGRATING) NETWORK"
3 LPRINT "(Computes Output Impedance and Phase Angle)"
4 PRINT "(Computes Output Impedance and Phase Angle)"
5 A$="#######.##":LPRINT"":PRINT"":INPUT "f (Hz)=";F
6 LPRINT "f (Hz)=";F
7 INPUT "R1 (Ohms)=";RO
8 LPRINT "R1 (Ohms)=";RO
9 INPUT "C1 (Mfd)=";CO
10 LPRINT "C1 (Mfd)=";CO
11 INPUT "R2 (Ohms)=";RT
12 LPRINT "R2 (Ohms)=";RT
13 INPUT "C2 (Mfd)=";CT
14 LPRINT "C2 (Mfd)=";CT
15 XO=1/(6.2832 *F*CO*10^-6):XT=1/(6.2832*F*CT*10^-6):AB=RO*XO:BA=-6.2832/4
16 AC=(RO^2+XO^2)^.5:CA=-ATN(XO/RO):AD=AB/AC:DA=BA-CA:AE=AD*COS(DA)
17 AF=AD*SIN(DA):AG=AE+RT:AH=(AG^2+AF^2)^.5:HA=-ATN(AF/AG):AJ=AH*XT
18 JA=HA-6.2832/4:AK=AF-XT:KA=FA-6.2832/4:AL=(AG^2+AK^2)^.5:LA=-ATN(AK/AG)
19 AM=AJ/AL:MA=JA-LA:AN=AM:NA=MA:LPRINT"":PRINT""
20 LPRINT "Zout (Ohms)=";USING A$;AN
21 PRINT "Zout (Ohms)=";USING A$;AN
22 LPRINT "Phase Angle (Degrees)=";USING A$;-NA*360/6.2832
23 PRINT "Phase Angle (Degrees)=";USING A$;-NA*360/6.2832
24 END
```

```
UNSYMMETRICAL TWO-SECTION RC LAG (INTEGRATING) NETWORK
(Computes Output Impedance and Phase Angle)

f (Hz)= 1000
R1 (Ohms)= 1000
C1 (Mfd)= .15
R2 (Ohms)= 10000
C2 (Mfd)= .015

Zout (Ohms)=    7307.15
Phase Angle (Degrees)=     133.82
```

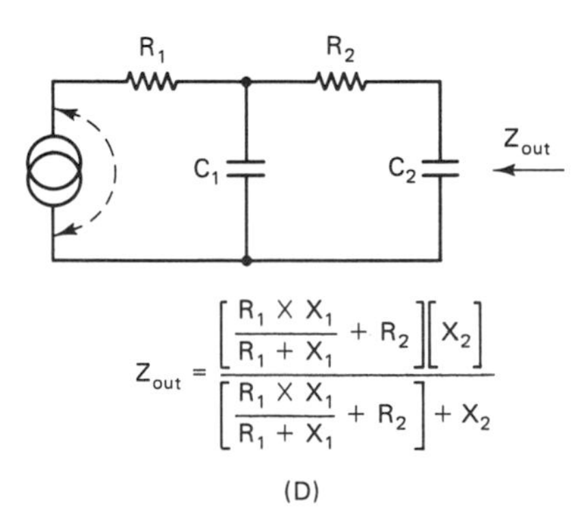

$$Z_{out} = \frac{\left[\dfrac{R_1 \times X_1}{R_1 + X_1} + R_2\right]\left[X_2\right]}{\left[\dfrac{R_1 \times X_1}{R_1 + X_1} + R_2\right] + X_2}$$

(D)

Figure 9–11(D). Program for calculating the output impedance of an unsymmetrical two-section cascaded RC-integrating network.

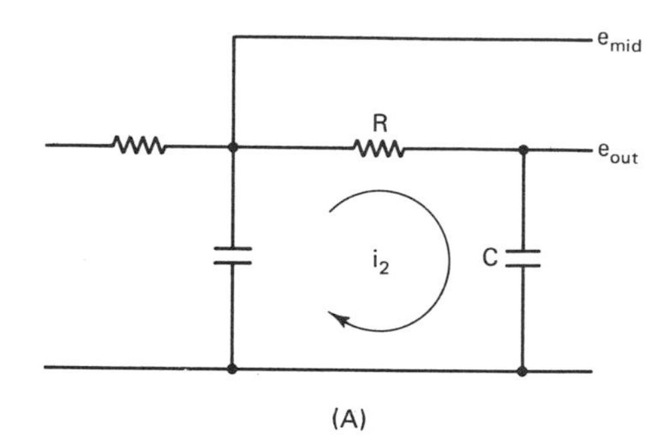

Figure 9–12(A). Calculation of midpoint voltage and phase angle with respect to the input voltage for a symmetrical two-section RC-integrating network.

NOTE: The output voltage e_{out} is equal to the product of i_2 and the reactance of C. The midpoint voltage e_{mid} is equal to the product of i_2 and the impedance of R and C. For example, if the value of i_2 is 0.5 mA and the reactance of C is 1000 ohms, the value of e_{out} will be 0.5 volt. In turn, if the value of R is 1000 ohms, the value of e_{mid} will be .707 volt. The phase of e_{out} lags the phase of e_{mid} by 45 degrees in this example.

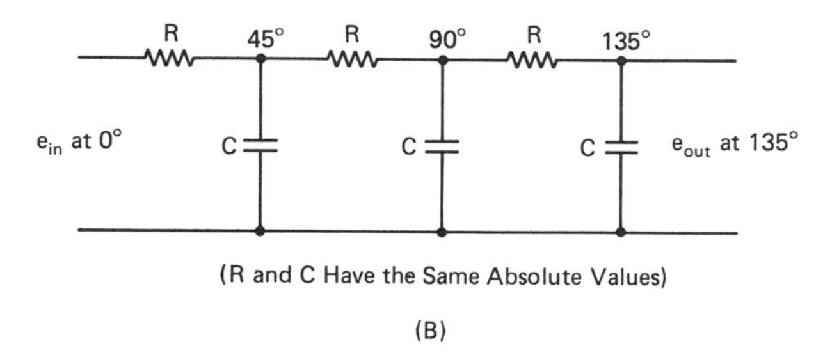

Figure 9–12(B). A useful special case for a symmetrical three-section cascaded RC-integrating network wherein the absolute values of resistance and reactance are equal.

NOTE: This special case for a symmetrical three-section cascaded RC-integrating network can be solved by inspection with reference to the network depicted in Figure 9–12(A). Each successive RC section shifts the phase of the output by an additional 45°.

voltage is an open-circuit value that is reduced by the current demand of the output section in Figure 9–13.

It should also be noted that the current demand of the output section in Figure 9–13 is not directly calculated with respect to R and C only. Calculation of the current demand with respect to the Thevenin voltage entails application of the Thevenin voltage to an output circuit comprising the sum of the Thevenin impedance and the impedance of the RC section. In summary, when the designer

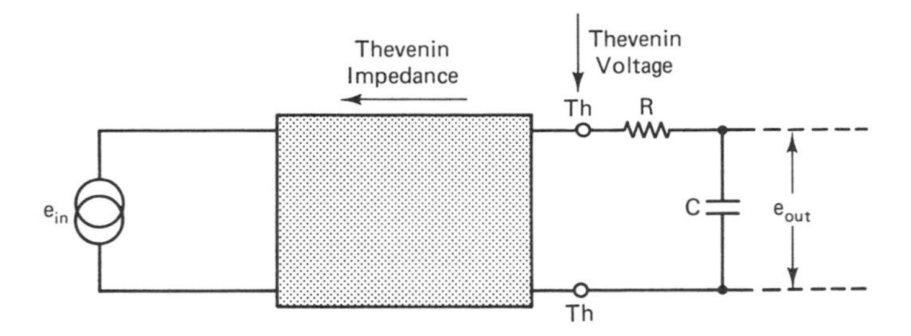

NOTE: The Thevenin theorem provides the capability of output-voltage and phase calculation from the voltage and impedance values at any intermediate point in a network. On the other hand, the Thevenin theorem imposes the limitation that the calculated output-voltage and phase values relate only to the chosen intermediate point (Th). Stated otherwise, the calculated output-voltage and phase values have no immediate relation to e_{in}. This relation depends on the network that is present in the black box. Accordingly, it is generally advisable for the designer to employ mesh analysis or branch-current analysis when an e_{out}/e_{in} ratio is to be calculated.

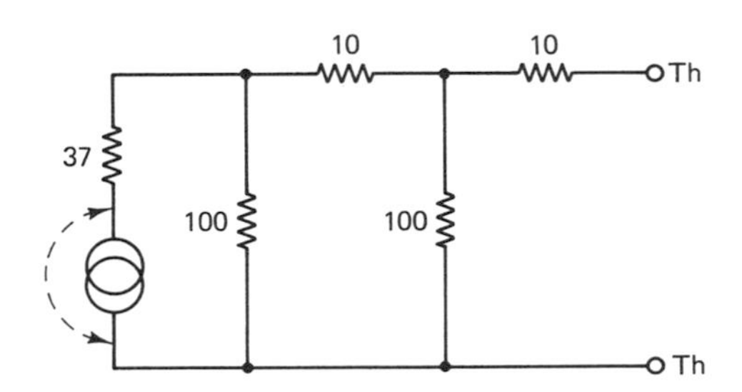

Figure 9–13. Visualization of the capabilities and limitations of Thevenin's theorem.

NOTE: If the Thevenin resistance is measured with an ohmmeter in this example, a value of 37 ohms will be obtained. Observe that if all of the resistors are removed, with the exception of the 37-ohm resistor in series with the generator, an ohmmeter will still indicate a Thevenin resistance of 37 ohms. This is a practical example of the need for mesh analysis when an $e_{out}e_{in}$ ratio is to be determined.

wishes to calculate the output voltage and phase relation to the input voltage, it is generally more economical to utilize mesh analysis or branch-current analysis.

The designer is frequently concerned with the output/input voltage and phase relation for two-section unsymmetrical cascaded RC-integrating circuits. In turn, it can be helpful to have a computer program available for rapid calculation. With reference to Figure 9–14, R2 may have a larger or smaller value than R1, and C2 may have a larger or smaller value than C1. As exemplified in the RUN, R1 and R2 may have the same value, and C1 and C2 may have the same value. When the R and C values are equal, the same answer is obtained as when the previous program for a symmetrical network is utilized.

Unsymmetrical two-section cascaded RC-differentiating networks are extensively used in various areas of electronic design. In turn, it is helpful to have a program available for rapid calculation of the output voltage and phase relation to the input voltage. A survey program for this purpose is provided in Figure 9–15.

DESIGN OPTIMIZATION CONSIDERATIONS

In many applications, it is necessary for the oscillatory frequency to remain as constant as may be feasible from a production cost standpoint. Temperature stability is a prime consideration, and transistor junction capacitances require close attention. The designer endeavors to make the junction capacitances only a small fraction of the total tuned-circuit capacitance. Changes in transistor-terminal resistance and variations in supply voltages can be minimized by employment of high-Q tuned circuits. Regulated

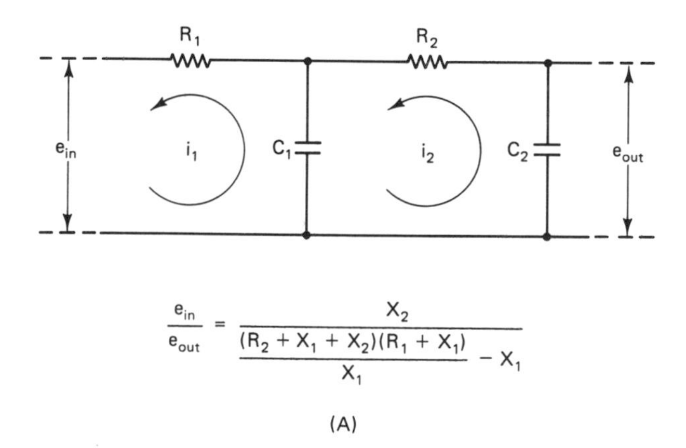

$$\frac{e_{in}}{e_{out}} = \frac{X_2}{\dfrac{(R_2 + X_1 + X_2)(R_1 + X_1)}{X_1} - X_1}$$

(A)

Figure 9–14(A). Basic diagram for mesh analysis of an unsymmetrical two-section cascaded RC-integrating network.

NOTE: This is a widely used network in oscillator circuitry and various other applications. Observe that if R1 = R2 and C1 = C2, the above equation reduces to the equation that was previously shown for a symmetrical two-section cascaded RC-integrating network.

```
1 LPRINT "UNSYMMETRICAL TWO-SECTION CASCADED INTEGRATING NETWORK":PRINT"UNSYMMET
RICAL TWO-SECTION CASCADED INTEGRATING NETWORK"
2 LPRINT"Eout/Ein, Phase Angle, and Phase Slope at Specified Frequency":PRINT"Eo
ut/Ein, Phase Angle, and Phase Slope at Specified Frequency"
3 LPRINT""
4 PRINT""
5 INPUT"R1 (Ohms)=";RO
6 LPRINT"R1 (Ohms)=";RO
7 INPUT"R2 (Ohms)=";RT
8 LPRINT"R2 (Ohms)=";RT
9 INPUT"C1 (Mfd)=";CO
10 LPRINT"C1 (Mfd)=";CO
11 INPUT"C2 (Mfd)=";CT
12 LPRINT"C2 (Mfd)=";CT
13 INPUT"f (Hz)=";F
14 LPRINT"f (Hz)";F
15 XO=1/(6.283*F*CO*10^-6):XT=1/(6.283*F*CT*10^-6)
16 AB=(RO*RO+XO*XO)^.5:BA=-ATN(XO/RO):LPRINT"":PRINT""
17 LK=XO+XT:BC=(RT*RT+LK*LK)^.5:CB=-ATN(LK/RT):AC=BC*AB:CA=CB+BA
18 AD=AC/XO:DA=CA+6.283/4:AE=AD*COS(DA):EA=AD*SIN(DA)
19 BE=EA+XO:CE=(AE*AE+BE*BE)^.5:EC=-ATN(BE/AE):AN=XT/CE:NA=-EC-6.283/4
20 A$="####.##":LPRINT"Eout/Ein=";USING A$;AN
21 LPRINT"Phase Angle (Degrees)=";USING A$;180+NA*360/6.283
22 PRINT"Eout/Ein=";USING A$;AN:PRINT"Phase Angle (Degrees)=";USING A$;180+NA*36
0/6.283
23 H=F+10
24 XP=1/(6.283*H*CO*10^-6):XU=1/(6.283*H*CT*10^-6)
25 AZ=(RO*RO+XP*XP)^.5:ZA=-ATN(XP/RO)
26 KL=XP+XU:BZ=(RT*RT+KL*KL)^.5:CY=-ATN(KL/RT):AQ=BZ*AZ:CQ=CY+ZA
27 FG=AQ/XP:GF=CQ+6.283/4:BF=FG*COS(GF):FB=FG*SIN(GF)
28 LM=FB+XP:MN=(BF*BF+LM*LM)^.5:NM=-ATN(LM/BF):BL=XU/MN:LB=-NM-6.283/4
31 PS=(NA-LB)*360/62.83
32 C$="#####.####":LPRINT"Phase Slope (Deg/Hz at f)=";USING C$;-PS:PRINT"Phase S
lope (Deg/Hz at f)=";USING C$;-PS
```

NOTE: This program provides computation of the output/input voltage ratio, output/input phase angle, and phase slope for an unsymmetrical two-section cascaded RC-integrating network at a specified frequency. As exemplified in the first RUN, symmetrical values are also accommodated. This RUN also illustrates equal reactance and resistance values at the operating frequency.

Observe that the phase slope (tangent to the phase characteristic at the operating frequency) is comparatively large at low frequencies and comparatively small at high frequencies. In other words, the phase characteristic for this network is nonlinear. (If the phase characteristic were linear, the phase slope would have the same value at any frequency within the operating range.)

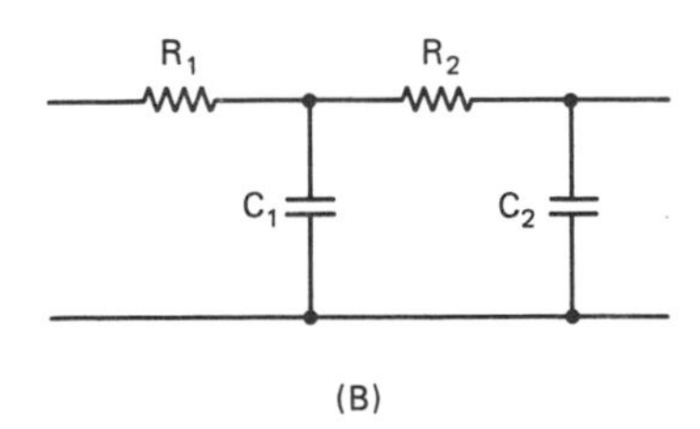

(B)

Figure 9–14(B). (Continues on next pages.)

```
UNSYMMETRICAL TWO-SECTION CASCADED INTEGRATING NETWORK
Eout/Ein, Phase Angle, and Phase Slope at Specified Frequency

R1 (Ohms)= 1000
R2 (Ohms)= 1000
C1 (Mfd)= .15915
C2 (Mfd)= .15915
f (Hz) 1000

Eout/Ein=   0.33
Phase Angle (Degrees)=  90.00
Phase Slope (Deg/Hz at f)=    0.0380
```

```
UNSYMMETRICAL TWO-SECTION CASCADED INTEGRATING NETWORK
Eout/Ein, Phase Angle, and Phase Slope at Specified Frequency

R1 (Ohms)= 1000
R2 (Ohms)= 1500
C1 (Mfd)= .2
C2 (Mfd)= .1
f (Hz) 50

Eout/Ein=   0.99
Phase Angle (Degrees)=   8.06
Phase Slope (Deg/Hz at f)=    0.1599
```

```
UNSYMMETRICAL TWO-SECTION CASCADED INTEGRATING NETWORK
Eout/Ein, Phase Angle, and Phase Slope at Specified Frequency

R1 (Ohms)= 1000
R2 (Ohms)= 1500
C1 (Mfd)= .2
C2 (Mfd)= .1
f (Hz) 500

Eout/Ein=   0.63
Phase Angle (Degrees)=  63.53
Phase Slope (Deg/Hz at f)=    0.0835
```

```
UNSYMMETRICAL TWO-SECTION CASCADED INTEGRATING NETWORK
Eout/Ein, Phase Angle, Slope, and Delay at Specified Frequency

R1 (Ohms)= 1000
R2 (Ohms)= 1500
C1 (Mfd)= .2
C2 (Mfd)= .1
f (Hz) 5000

Eout/Ein=   0.03
Phase Angle (Degrees)= 153.70
Phase Slope (Deg/Hz at f)=    0.0049
Phase Delay (Seconds)= 0.000015
```

```
UNSYMMETRICAL TWO-SECTION CASCADED INTEGRATING NETWORK
Eout/Ein, Phase Angle, Slope, and Delay at Specified Frequency

R1 (Ohms)= 1000
R2 (Ohms)= 1500
C1 (Mfd)= .2
C2 (Mfd)= .1
f (Hz) 15000

Eout/Ein=   0.00
Phase Angle (Degrees)= 170.92
Phase Slope (Deg/Hz at f)=    0.0006
Phase Delay (Seconds)= 0.000002
```

Figure 9–14(B). Program for mesh analysis of an unsymmetrical two-section cascaded RC-integrating network.

NOTE: The $E_{out}E_{in}$ value in the final RUN is printed out as zero, although its actual value is greater than zero. (The print format in this example rejects values smaller than 0.005.)

```
1 LPRINT "UNSYMMETRICAL TWO-SECTION CASCADED INTEGRATING NETWORK":PRINT"UNSYMMET
RICAL TWO-SECTION CASCADED INTEGRATING NETWORK"
2 LPRINT"Eout/Ein, Phase Angle, Slope, and Delay at Specified Frequency":PRINT"E
out/Ein, Phase Angle, Slope, and Delay at Specified Frequency"
3 LPRINT""
4 PRINT""
5 INPUT"R1 (Ohms)=";RO
6 LPRINT"R1 (Ohms)=";RO
7 INPUT"R2 (Ohms)=";RT
8 LPRINT"R2 (Ohms)=";RT
9 INPUT"C1 (Mfd)=";CO
10 LPRINT"C1 (Mfd)=";CO
11 INPUT"C2 (Mfd)=";CT
12 LPRINT"C2 (Mfd)=";CT
13 INPUT"f (Hz)=";F
14 LPRINT"f (Hz)";F
15 XO=1/(6.283*F*CO*10^-6):XT=1/(6.283*F*CT*10^-6)
16 AB=(RO*RO+XO*XO)^.5:BA=-ATN(XO/RO):LPRINT"":PRINT""
17 LK=XO+XT:BC=(RT*RT+LK*LK)^.5:CB=-ATN(LK/RT):AC=BC*AB:CA=CB+BA
18 AD=AC/XO:DA=CA+6.283/4:AE=AD*COS(DA):EA=AD*SIN(DA)
19 BE=EA+XO:CE=(AE*AE+BE*BE)^.5:EC=-ATN(BE/AE):AN=XT/CE:NA=-EC-6.283/4
20 A$="####.##":LPRINT"Eout/Ein=";USING A$;AN
21 LPRINT"Phase Angle (Degrees)=";USING A$;180+NA*360/6.283
22 PRINT"Eout/Ein=";USING A$;AN:PRINT"Phase Angle (Degrees)=";USING A$;180+NA*36
0/6.283
23 H=F+10
24 XP=1/(6.283*H*CO*10^-6):XU=1/(6.283*H*CT*10^-6)
25 AZ=(RO*RO+XP*XP)^.5:ZA=-ATN(XP/RO)
26 KL=XP+XU:BZ=(RT*RT+KL*KL)^.5:CY=-ATN(KL/RT):AQ=BZ*AZ:CQ=CY+ZA
27 FG=AQ/XP:GF=CQ+6.283/4:BF=FG*COS(GF):FB=FG*SIN(GF)
28 LM=FB+XP:MN=(BF*BF+LM*LM)^.5:NM=-ATN(LM/BF):BL=XU/MN:LB=-NM-6.283/4
31 PS=(NA-LB)*360/62.83
32 C$="#####.####":LPRINT"Phase Slope (Deg/Hz at f)=";USING C$;-PS:PRINT"Phase S
lope (Deg/Hz at f)=";USING C$;-PS
33 PD=NA/(6.283*F)
34 D$="##.######":LPRINT"Phase Delay (Seconds)=";USING D$;-PD:PRINT"Phase Delay
(Seconds)=";USING D$;-PD
```

Figure 9–14(C). (Continues on next pages.)

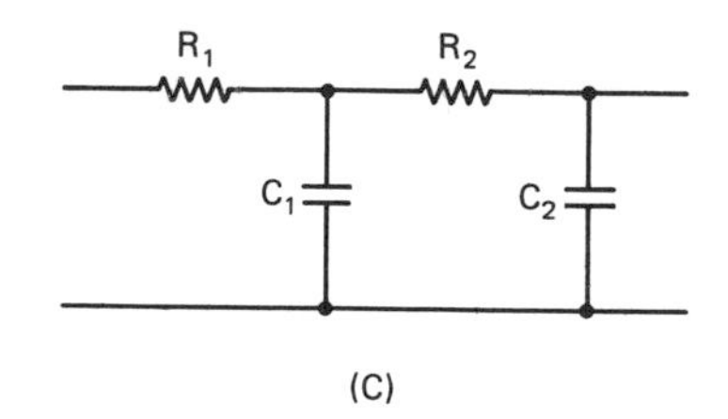

(C)

```
UNSYMMETRICAL TWO-SECTION CASCADED INTEGRATING NETWORK
Eout/Ein, Phase Angle, Slope, and Delay at Specified Frequency

R1 (Ohms)= 1000
R2 (Ohms)= 1500
C1 (Mfd)= .2
C2 (Mfd)= .1
f (Hz) 5

Eout/Ein=   1.00
Phase Angle (Degrees)=   0.80
Phase Slope (Deg/Hz at f)=    0.1619
Phase Delay (Seconds)= 0.099554

UNSYMMETRICAL TWO-SECTION CASCADED INTEGRATING NETWORK
Eout/Ein, Phase Angle, Slope, and Delay at Specified Frequency

R1 (Ohms)= 1000
R2 (Ohms)= 1500
C1 (Mfd)= .2
C2 (Mfd)= .1
f (Hz) 50

Eout/Ein=   0.99
Phase Angle (Degrees)=   8.06
Phase Slope (Deg/Hz at f)=    0.1599
Phase Delay (Seconds)= 0.009552

UNSYMMETRICAL TWO-SECTION CASCADED INTEGRATING NETWORK
Eout/Ein, Phase Angle, Slope, and Delay at Specified Frequency

R1 (Ohms)= 1000
R2 (Ohms)= 1500
C1 (Mfd)= .2
C2 (Mfd)= .1
f (Hz) 500

Eout/Ein=   0.63
Phase Angle (Degrees)=  63.53
Phase Slope (Deg/Hz at f)=    0.0835
Phase Delay (Seconds)= 0.000647

UNSYMMETRICAL TWO-SECTION CASCADED INTEGRATING NETWORK
Eout/Ein, Phase Angle, and Phase Slope at Specified Frequency

R1 (Ohms)= 1000
R2 (Ohms)= 1500
C1 (Mfd)= .2
C2 (Mfd)= .1
f (Hz) 5000

Eout/Ein=   0.03
Phase Angle (Degrees)= 153.70
Phase Slope (Deg/Hz at f)=    0.0049
```

```
UNSYMMETRICAL TWO-SECTION CASCADED INTEGRATING NETWORK
Eout/Ein, Phase Angle, and Phase Slope at Specified Frequency

R1 (Ohms)= 1000
R2 (Ohms)= 1500
C1 (Mfd)= .2
C2 (Mfd)= .1
f (Hz) 15000

Eout/Ein=   0.00
Phase Angle (Degrees)= 170.92
Phase Slope (Deg/Hz at f)=    0.0006
```

Figure 9–14(C). Program for computation of E_{out}/E_{in} voltage, phase angle, phase slope, and phase delay for an unsymmetrical two-section RC-integrating network.

Note: This program exemplifies computation of the phase delay in addition to the phase slope for an RC-integrating network. The phase-delay value is often of concern to the designer, inasmuch as it shows the difference in time of arrival for various signal-frequency components at the output of the network. In turn, the designer can decide whether the difference in delay times is significant and whether phase-compensation circuitry is required.

```
1 LPRINT "UNSYMMETRICAL TWO-SECTION CASCADED RC LEAD NETWORK"
2 PRINT "UNSYMMETRICAL TWO-SECTION CASCADED RC LEAD NETWORK"
3 LPRINT "Eout/Ein, Phase Angle, Phase Slope, Envelope Delay, Input Impedance,
         Output Impedance, Midpoint Eout/Ein, and Phase Angle"
4 PRINT "Eout/Ein, Phase Angle, Phase Slope, Envelope Delay, Input Impedance,
        Output Impedance, Midpoint Eout/Ein, and Phase Angle"
5 LPRINT "":PRINT"":INPUT "R1 (Ohms)=";RO
6 LPRINT "R1 (Ohms)=";RO:INPUT "R2 (Ohms)=";RT
7 LPRINT "R2 (Ohms)=";RT:INPUT "C1 (Mfd)=";CO
8 LPRINT "C1 (Mfd)=";CO:INPUT "C2 (Mfd)=";CT
9 LPRINT "C2 (Mfd)=";CT:INPUT "f (Hz)=";F
10 LPRINT"f (Hz)=";F:LPRINT"":PRINT"":A$="########":B$="####.##":C$="####.####"
11 XO=1/(6.2832*F*CO*10^-6):XT=1/(6.2832*F*CT*10^-6):D$="##.######"
12 RS=RO+RT:AB=(RS^2+XT^2)^.5:BA=ATN(XT/RS):AC=(RO^2+XO^2)^.5:CA=ATN(XO/RO)
13 AD=AB*AC:DA=BA+CA:AE=AD/RO:EA=DA:AF=AE*COS(EA):AG=AE*SIN(EA):AH=AF-RO
14 AJ=(AH^2+AG^2)^.5:JA=ATN(AG/AH):AK=RT/AJ:KA=-JA:KA=KA*360/6.2832:N=N+1000
15 LPRINT "Eout/Ein=";USING B$;AK:PRINT "Eout/Ein=";USING B$;AK:KA=KA-180
16 IF ABS(KA)>180 THEN KA=KA+180
17 LPRINT"Phase Angle (Degrees)=";USING B$;KA:H=F+10:AAK=AK:KKA=KA
18 PRINT "Phase Angle (Degrees)=";USING B$;KA:XP=1/(6.2832*H*CO*10^-6)
19 XU=1/(6.2832*H*CT*10^-6):RS=RO+RT:AB=(RS^2+XU^2)^.5:BA=ATN(XU/RS)
20 AC=(RO^2+XP^2)^.5:CA=ATN(XP/RO):AD=AB*AC:DA=BA+CA:AE=AD/RO:EA=DA
21 AF=AE*COS(EA):AG=AE*SIN(EA):AH=AF-RO:AJ=(AH^2+AG^2)^.5:JA=ATN(AG/AH)
22 HK=R1/AJ:KH=-JÅ:KH=KH*360/6.2832:DG=KA-KH:PS=.1*DG:KA=KA*6.2832/360
23 LPRINT "Phase Slope (Deg/Hz at f)=";USING C$;-PS
24 PRINT "Phase Slope (Deg/Hz at f)=";USING C$;-PS:PD=KA/(6.2832*F)
25 LPRINT "Phase Delay (Sec)=";USING D$;-PD
26 PRINT "Phase Delay (Sec)=";USING D$;-PD
27 AB=(RT^2+XT^2)^.5:BA=ATN(XT/RT):AC=XO*AB:CA=BA-6.2832/4:AD=XO+XT
28 AE=(RT^2+AD^2)^.5:EA=ATN(AD/RT):AF=AC/AE:FA=CA-EA:AG=AF*COS(FA)
29 AH=AF*SIN(FA):AJ=RO+AG:AK=(AJ^2+AH^2)^.5:KA=ATN(AH/AJ):KA=KA*360/6.2832
30 LPRINT "Input Impedance (Ohms)=";USING A$;AK:KA=KA+180
31 PRINT "Input Impedance (Ohms)=";USING A$;AK
32 LPRINT "Phase Angle (Degrees)=";USING B$;-KA
33 PRINT "Phase Angle (Degrees)=";USING B$;-KA
34 AB=RO*XO:BA=-6.2832/4:AC=(RO^2+XO^2)^.5:CA=ATN(XO/RO):AD=AB/AC:DA=BA-CA
```

Figure 9–15(A). Basic diagram for mesh analysis of an unsymmetrical two-section cascaded RC-differentiating network.

```
35 AE=AD*COS(DA):AF=AD*SIN(DA):AG=AF+XT:AH=(AE^2+AG^2)^.5:HA=ATN(AG/AE)
36 AJ=RT*AH:JA=HA:AK=(RT^2+AH^2)^.5:KA=ATN(AH/RT):AL=AJ/AK:LA=-KA
37 LA=LA*360/6.2832:LA=LA+90:LA=-LA
38 LPRINT "Output Impedance (Ohms)=";USING A$;AL
39 PRINT "Output Impedance (Ohms)=";USING A$;AL
40 LPRINT "Phase Angle (Degrees)=";USING B$;LA
41 PRINT "Phase Angle (Degrees)-";USING B$;LA
42 TH=ATN(XT/RT):EM=AAK/COS(TH):TH=TH*360/6.2832:SH=TH+KKA
43 LPRINT "Midpoint Eout/Ein=";USING B$;EM
44 PRINT "Midpoint Eout/Ein=";USING B$;EM
45 LPRINT "Phase Angle (Degrees)=";USING B$;SH
46 PRINT "Phase Angle (Degrees)=";USING B$;SH
47 END
```

```
UNSYMMETRICAL TWO-SECTION CASCADED RC LEAD NETWORK
Eout/Ein, Phase Angle, Phase Slope, Envelope Delay, Input Impedance,
Output Impedance, Midpoint Eout/Ein, and Phase Angle

R1 (Ohms)= 1000
R2 (Ohms)= 10000
C1 (Mfd)= .15
C2 (Mfd)= .015
f (Hz)= 1000

Eout/Ein=   0.45
Phase Angle (Degrees)= -93.23
Phase Slope (Deg/Hz at f)=  18.0542
Phase Delay (Sec)= 0.000259
Input Impedance (Ohms)=     1385
Phase Angle (Degrees)=-133.45
Output Impedance (Ohms)=     7115
Phase Angle (Degrees)= -44.64
Midpoint Eout/Ein=    0.65
Phase Angle (Degrees)= -46.54
```

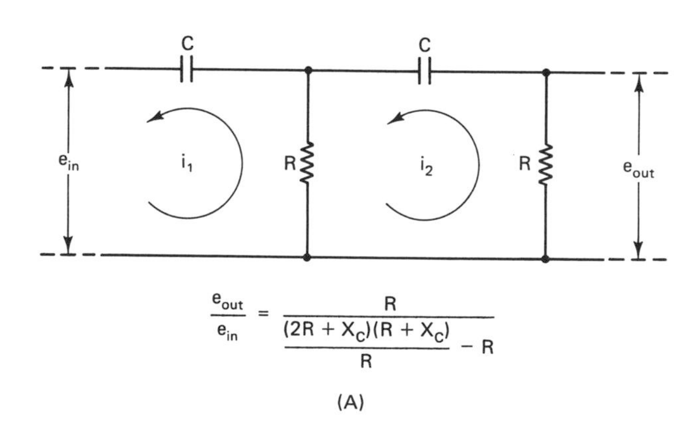

$$\frac{e_{out}}{e_{in}} = \frac{R}{\frac{(2R + X_C)(R + X_C)}{R} - R}$$

(A)

NOTE: This is a widely used RC network in oscillator circuitry and various other applications. Observe that the output voltage leads the input voltage, whereas the analogous integrating network has a lagging output voltage. Observe that this program processes a symmetrical network of R1 = R2 and C1 = C2.

```
1 LPRINT "UNSYMMETRICAL TWO-SECTION CASCADED RC LEAD NETWORK"
2 PRINT "UNSYMMETRICAL TWO-SECTION CASCADED RC LEAD NETWORK"
3 LPRINT "Eout/Ein, Phase Angle, Phase Slope, Envelope Delay, Input Impedance,
         Output Impedance, Midpoint Eout/Ein, and Phase Angle"
4 PRINT "Eout/Ein, Phase Angle, Phase Slope, Envelope Delay, Input Impedance,
         Output Impedance, Midpoint Eout/Ein, and Phase Angle"
5 LPRINT "* Survey Program *":PRINT "* Survey Program *":LPRINT:PRINT
6 A$="########":B$="####.##":C$="#####.####":INPUT "R1 (Ohms)=";RO
7 LPRINT "R1 (Ohms)=";RO:INPUT "R2 (Ohms)=";RT
8 LPRINT "R2 (Ohms)=";RT:INPUT "C1 (Mfd)=";CO
9 LPRINT "C1 (Mfd)=";CO:INPUT "C2 (Mfd)=";CT
10 LPRINT "C2 (Mfd)=";CT:INPUT "f (Hz)=";H
11 LPRINT "f (Hz)=";H:N=0:G=H:LPRINT"":PRINT""
12 F=G+N:LPRINT "f (Hz)=";USING A$;F:PRINT "f (Hz)=";USING A$;F
13 XO=1/(6.2832*F*CO*10^-6):XT=1/(6.2832*F*CT*10^-6):D$="##.######"
14 RS=RO+RT:AB=(RS^2+XT^2)^.5:BA=ATN(XT/RS):AC=(RO^2+XO^2)^.5:CA=ATN(XO/RO)
15 AD=AB*AC:DA=BA+CA:AE=AD/RO:EA=DA:AF=AE*COS(EA):AG=AE*SIN(EA):AH=AF-RO
16 AJ=(AH^2+AG^2)^.5:JA=ATN(AG/AH):AK=RT/AJ:KA=-JA:KA=KA*360/6.2832:N=N+1000
17 LPRINT "Eout/Ein=";USING B$;AK:PRINT "Eout/Ein=";USING B$;AK:KA=KA-180
18 IF ABS(KA)>180 THEN KA=KA+180
19 LPRINT"Phase Angle (Degrees)=";USING B$;KA:H=F+10:AAK=AK:KKA=KA
20 PRINT "Phase Angle (Degrees)=";USING B$;KA:XP=1/(6.2832*H*CO*10^-6)
21 XU=1/(6.2832*H*CT*10^-6):RS=RO+RT:AB=(RS^2+XU^2)^.5:BA=ATN(XU/RS)
22 AC=(RO^2+XP^2)^.5:CA=ATN(XP/RO):AD=AB*AC:DA=BA+CA:AE=AD/RO:EA=DA
23 AF=AE*COS(EA):AG=AE*SIN(EA):AH=AF-RO:AJ=(AH^2+AG^2)^.5:JA=ATN(AG/AH)
24 HK=RT/AJ:KH=-JA:KH=KH*360/6.2832:DG=KA-KH:PS=.1*DG:KA=KA*6.2832/360
25 LPRINT "Phase Slope (Deg/Hz at f)=";USING C$;-PS
26 PRINT "Phase Slope (Deg/Hz at f)=";USING C$;-PS:PD=KA/(6.2832*F)
27 LPRINT "Phase Delay (Sec)=";USING D$;-PD
28 PRINT "Phase Delay (Sec)=";USING D$;-PD
29 AB=(RT^2+XT^2)^.5:BA=ATN(XT/RT):AC=XO*AB:CA=BA-6.2832/4:AD=XO+XT
30 AE=(RT^2+AD^2)^.5:EA=ATN(AD/RT):AF=AC/AE:FA=CA-EA:AG=AF*COS(FA)
31 AH=AF*SIN(FA):AJ=RO+AG:AK=(AJ^2+AH^2)^.5:KA=ATN(AH/AJ):KA=KA*360/6.2832
32 LPRINT "Input Impedance (ohms)=";USING A$;AK:KA=KA+180
33 PRINT "Input Impedance (Ohms)=";USING A$;AK
34 LPRINT "Phase Angle (Degrees)=";USING B$;-KA
35 PRINT "Phase Angle (Degrees)=";USING B$;-KA
36 AB=RO*XO:BA=-6.2832/4:AC=(RO^2+XO^2)^.5:CA=ATN(XO/RO):AD=AB/AC:DA=BA-CA
37 AE=AD*COS(DA):AF=AD*SIN(DA):AG=AF+XT:AH=(AE^2+AG^2)^.5:HA=ATN(AG/AE)
38 AJ=RT*AH:JA=HA:AK=(RT^2+AH^2)^.5:KA=ATN(AH/RT):AL=AJ/AK:LA=-KA
39 LA=LA*360/6.2832:LA=LA+90:LA=-LA
40 LPRINT "Output Impedance (Ohms)=";USING A$;AL
41 PRINT "Output Impedance (Ohms)=";USING A$;AL
42 LPRINT "Phase Angle (Degrees)=";USING B$;LA
43 PRINT "Phase Angle (Degrees)-";USING B$;LA
44 TH=ATN(XT/RT):EM=AAK/COS(TH):TH=TH*360/6.2832:SH=TH+KKA
45 LPRINT "Midpoint Eout/Ein=";USING B$;EM
46 PRINT "Midpoint Eout/Ein=";USING B$;EM
47 LPRINT "Phase Angle (Degrees)=";USING B$;SH
48 PRINT "Phase Angle (Degrees)=";USING B$;SH
49 IF F>20000 THEN 51
50 LPRINT:PRINT:GOTO 12
51 END
```

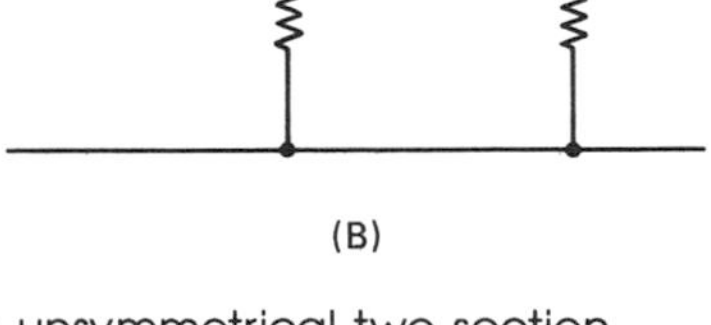

(B)

Figure 9–15(B). Survey program for mesh analysis of an unsymmetrical two-section cascaded RC-differentiating network.

power-supply voltages are desirable, but may be ruled out on the basis of cost.

Changes in oscillator loading are accompanied by variation in oscillatory frequency. In turn, a buffer amplifier is often employed between the oscillator section and the load. If a buffer cannot be utilized, it is helpful to impose a minimum power-output demand on the oscillator. Automatic-frequency control (AFC) is an attractive approach to oscillator-frequency-stability problems. However, AFC is feasible only in appropriate systems, such as in superheterodyne receiver circuitry.

Another aspect of oscillator design optimization concerns device and component-production tolerances in variable-frequency oscillator circuitry. Typical variable-frequency oscillators operate over a wide frequency range and are tuned with a calibrated control dial. Production tolerances affect the "spread" on the calibrated scale and reduce the accuracy rating of the oscillator. This problem can be handled by the availability of several types of calibrated control dials that provide different "spreads," for selection at the test station in production.

Frequency stability versus temperature variation in resonant-load oscillators such as depicted in Figure 9–2 can usually be improved by designing the tank circuit with one or more fixed capacitors with a suitable temperature coefficient. Thus, a fixed capacitor with a positive temperature coefficient or with a negative temperature coefficient or with a zero temperature coefficient may be employed at an appropriate point in a tuned circuit.

10

CONTEXTUAL CIRCUIT ANALYSIS

NETWORK SECTIONAL INTERACTION

Network analysis starts with the characteristics of components such as resistors, capacitors, and inductors, and devices such as diodes, bipolar, and unipolar transistors. These are network building blocks from which circuit sections such as RC-differentiating and -integrating sections are designed, and LCR-resonant circuits are designed. It was previously noted that when RC-integrating or -differentiating sections are cascaded, the first section is loaded by the second section, and so on. Stated otherwise, the characteristics of the first section, considered in isolation, are changed by loading interaction of the second section.

Furthermore, a cascaded integrating circuit, for example, often works into a load that has a significant current demand. In such a case, the characteristics of the cascaded-integrating circuit, considered in isolation, are modified by the load-current demand. Again, the cascaded-integrating circuit is frequently energized by a voltage source with significant internal resistance. In turn, the characteristics of the cascaded integrating circuit, considered with respect to the load-current demand, are further modified by the internal resistance of the source.

Again, consider the design of an LCR parallel-resonant load circuit. A parallel-resonant circuit has a certain resonant frequency, impedance, and Q value. Its selectivity (bandwidth) is given by the quotient of its resonant frequency and Q value. If the LCR circuit is energized by a constant-current source, the foregoing parameters remain unchanged. However, an LCR circuit is frequently energized from a voltage source with the significant internal resistance. Accordingly, its impedance and its Q value are reduced. If the LCR circuit works into a load with a significant current demand, its impedance and its Q value are further reduced.

A simple LCR parallel-resonant load circuit is often unsuitable for direct application in bipolar-transistor circuitry, inasmuch as source and load interaction results in poor selectivity (excessive bandwidth). In turn, the designer typically taps the inductor at one or two points in order to reduce the source and/or load interaction and thereby obtain an optimum bandwidth. These design approaches exemplify basic contextual-circuit analysis. Typical contextual-circuit analyses are comparatively involved.

LOADED RC-INTEGRATING CIRCUIT

A basic example of loaded RC integrating circuitry is shown in Figure 10–1. The RC section is an elementary low-pass filter; it is also employed as a phase-shifting arrangement, as previously noted. It is apparent that if the integrating circuit works into a very high load resistance, its frequency response and phase characteristic will approximate that of an unloaded integrating circuit. However, if the load resistance has a comparatively small value, the capacitive reactance (which determines frequency response and phase shift) will tend to become masked by the shunt resistance.

An input/output equation for a resistively loaded RC-integrating circuit is shown in Figure 10–2. The e_{out}/e_{in} ratio is associated with a phase angle that relates the phase of the output voltage to the phase of the input voltage. This RC configuration is sufficiently involved that it is helpful to have a computer program available for initial evaluation. Since the output voltage amplitude and its phase angle with respect to frequency are of basic interest, a survey program such as provided in Figure 10–3 is most appropriate.

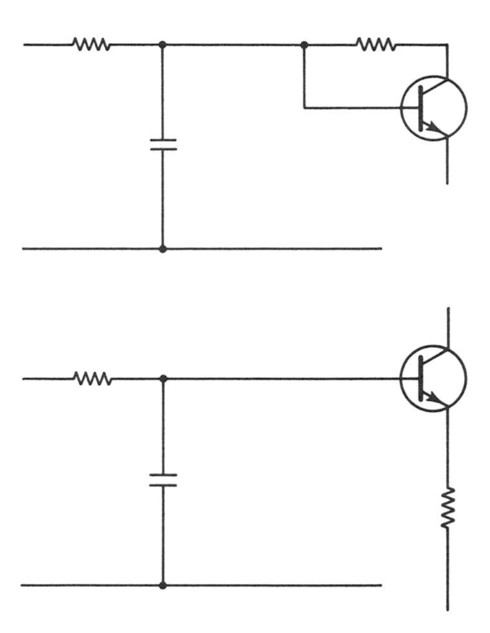

Figure 10–1. Basic-load circuitry for an RC-integrating section.

NOTE: An RC-integrating section often drives a transistor, which in turn presents a resistive load to the section. As previously noted, the input resistance to a bipolar transistor may have a low value or a comparatively high value, depending upon stage circuitry details. The design engineer is usually interested in the frequency and/or phase response of the loaded RC section and also its insertion loss.

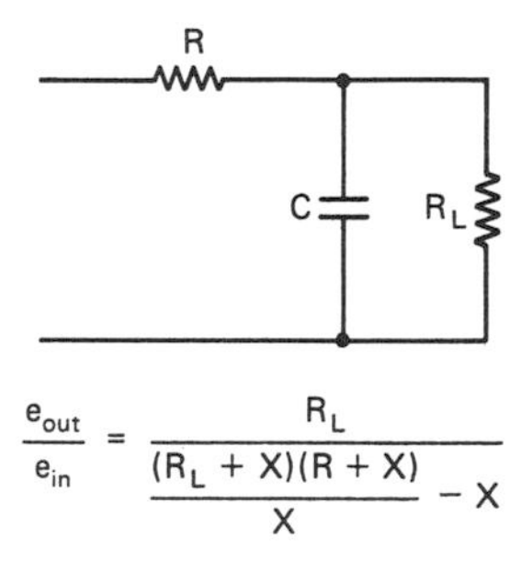

$$\frac{e_{out}}{e_{in}} = \frac{R_L}{\dfrac{(R_L + X)(R + X)}{X} - X}$$

Figure 10–2. Basic configuration and output/input equation for analysis of an RC-integrating section with a resistive load.

NOTE: This equation follows from a mesh analysis of the loaded RC-integrating section. It is evident that the load will reduce the output voltage in accordance with its conductance. Also, the load will shift the output phase to a lesser or greater extent, depending on its value. The output voltage lags the input voltage.

```
1 LPRINT "RESISTIVE LOADED RC LAG CIRCUIT"
2 PRINT "RESISTIVE LOADED RC LAG CIRCUIT"
3 LPRINT "Computes Ein/Eout and Phase Angle"
4 PRINT "Computes Ein/Eout and Phase Angle"
5 LPRINT "** Constant (Controlled) Voltage Source **"
6 PRINT "** Constant (Controlled) Voltage Source **"
7 LPRINT "(Survey Program)":PRINT "(Survey Program)"
8 LPRINT"":PRINT"":A$="########":B$="####.##"
9 INPUT "R (Ohms)=";R
10 LPRINT "R (Ohms)=";R
11 INPUT "RL (Ohms)=";RL
12 LPRINT "RL (Ohms)=";RL
13 INPUT "C (Mfd)=";C
14 LPRINT "C (Mfd)=";C
15 INPUT "f (Hz)=";H
16 N=0:G=H:LPRINT"":PRINT""
17 F=G+N:LPRINT "f (Hz)=";USING A$;F
18 PRINT "f (Hz)=";USING A$;F
19 X=1/(6.2832*F*C*10^-6):AB=(RL^2+X^2)^.5
20 AC=RL*X:CA=-1.3216:AD=AC/AB:DA=CA-BA:AE=AD*COS(DA)
21 AF=AD*SIN(DA):AG=R+AE:AH=(AG^2+AF^2)^.5:HA=ATN(AF/AG)
22 AJ=AD/AH:JA=DA-HA:EO=AJ:OE=JA:OE=-OE
23 PRINT "Eout/Ein=";USING B$;EO
24 PRINT "Phase Angle (Degrees)=";USING B$;OE*360/6.2832
25 LPRINT "Eout/Ein=";USING B$;EO
26 LPRINT "Phase Angle (Degrees)=";USING B$;OE*360/6.2832
27 N=N+100:LPRINT"":PRINT""
28 IF F>20000 THEN 30
29 GOTO 17
30 END
```

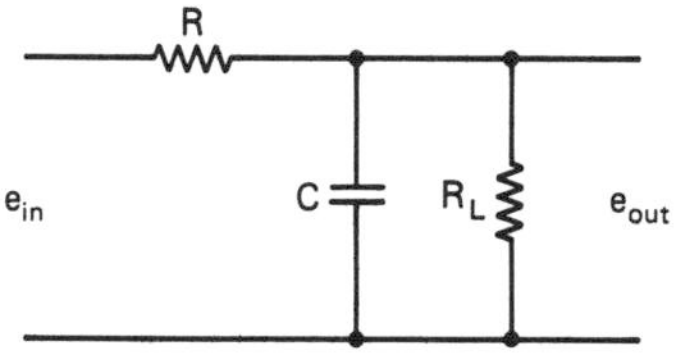

R
e_{in}
C
R_L
e_{out}

```
RESISTIVE LOADED RC LAG CIRCUIT
Computes Ein/Eout and Phase Angle
** Constant (Controlled) Voltage Source **
(Survey Program)

R (Ohms)= 1000
RL (Ohms)= 10000
C (Mfd)= .15

f (Hz)=      100
Eout/Ein=    0.96
Phase Angle (Degrees)=    7.34

f (Hz)=      200
Eout/Ein=    0.93
Phase Angle (Degrees)=   11.11

f (Hz)=      300
Eout/Ein=    0.90
Phase Angle (Degrees)=   15.14

f (Hz)=      400
Eout/Ein=    0.86
Phase Angle (Degrees)=   19.02

f (Hz)=      500
Eout/Ein=    0.82
Phase Angle (Degrees)=   22.65

f (Hz)=      600
Eout/Ein=    0.79
Phase Angle (Degrees)=   25.99

f (Hz)=      700
Eout/Ein=    0.75
Phase Angle (Degrees)=   29.04

f (Hz)=      800
Eout/Ein=    0.72
Phase Angle (Degrees)=   31.83

f (Hz)=      900
Eout/Ein=    0.68
Phase Angle (Degrees)=   34.36

f (Hz)=     1000
Eout/Ein=    0.65
Phase Angle (Degrees)=   36.67
```

Figure 10–3. A survey program for analysis of an RC-integrating section with a resistive load.

LOADED TWO-SECTION CASCADED RC-INTEGRATING NETWORK

With reference to Figure 10–4, typical load arrangements for a two-section cascaded RC-integrating network are shown. In the first example, the RC network outputs its current demand into a 250-ohm effective-load resistance. In the second example, the network outputs its current demand into a 2500-ohm effective resistance. Observe that this is a bilateral network—in other words, it can conduct current in either direction. This is just another way of saying that the load-resistance value will be reflected back to the input and that the input-impedance value and phase is a function of the load resistance.

The unsymmetrical two-section cascaded RC-integrating network is a linear bilateral network because its output current in response to constant-voltage input is the same if the generator and the ammeter are interchanged, as depicted in Figure 10–5. Note that a constant-voltage generator has zero internal resistance and that an ideal ammeter has zero internal resistance. Accordingly, the

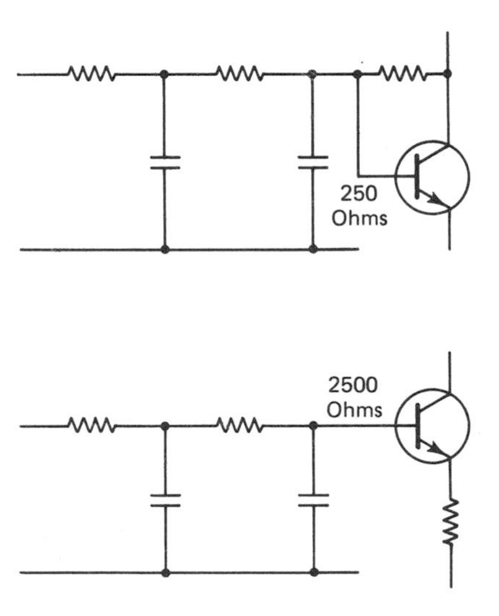

Figure 10–4. Typical load arrangements for an integrating network.

NOTE: The load into which an integrating network supplies its current demand may have a low value in some applications, or it may have a high value in other applications. As shown by the computer program in Figure 10–6, the characteristics of the integrating network are modified by the value of the load resistance that is utilized.

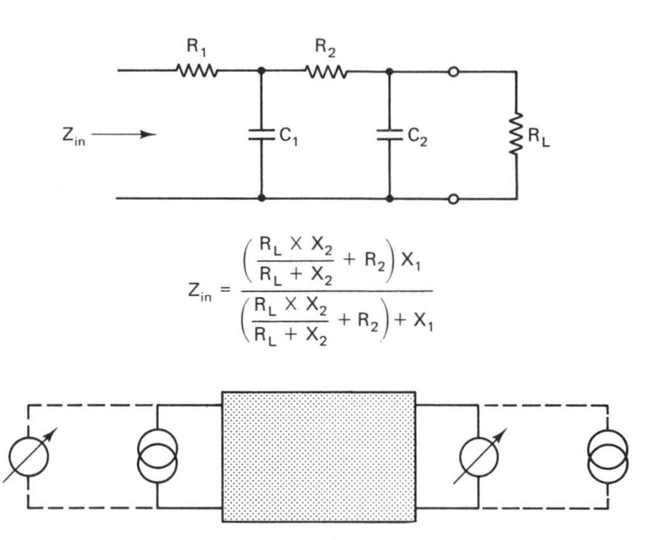

$$Z_{in} = \frac{\left(\frac{R_L \times X_2}{R_L + X_2} + R_2\right) X_1}{\left(\frac{R_L \times X_2}{R_L + X_2} + R_2\right) + X_1}$$

Figure 10–5. Basic diagram for calculating the input impedance of an unsymmetrical two-section cascaded RC-integrating circuit with resistive load.

NOTE: A resistive load reduces the input impedance of the network compared with the input impedance of an unloaded network. A resistive load also reduces the phase angle of the input impedance. The input current leads the input voltage. Observe that when RL has a comparatively high value, the input impedance approximates the input impedance of an unloaded network.

The unsymmetrical two-section cascaded integrating circuit is a linear bilateral network because its input/output E-I relation is the same as its output/input E-I relation.

black box reduces to a resistor connected in series with a capacitor, and the current flow is necessarily the same in either direction.

Mesh analysis of the two-section cascaded RC-integrating network leads to the input-impedance equation shown in Figure 10–5. Since this is a somewhat involved expression, it is frequently helpful in preliminary design procedures to employ a computer program such as the routine provided in Figure 10–6(A). It yields the magnitude and phase angle of the input impedance at a stipulated frequency. In various circumstances, the designer needs to determine the input-impedance variation of a tentative network versus frequency, and this is easily accomplished by means of the survey program provided in Figure 10–6(B).

```
1 LPRINT"UNSYMMETRICAL TWO-SECTION CASCADED RC INTEGRATING CIRCUIT (With Resisti
ve Load)" :PRINT"TWO-SECTION CASCADED RC INTEGRATING CIRCUIT (With Resistive Loa
d)"
2 LPRINT"* Input Impedance and Phase Angle *":PRINT"* Input Impedance and Phase
Angle *":LPRINT"":PRINT""
3 INPUT "R1 (Ohms)=";RO
4 LPRINT"R1 (Ohms)=";RO
5 INPUT"R2 (Ohms)=";RT
6 LPRINT"R2 (Ohms)=";RT
7 INPUT"C1 (Mfd)=";CO
8 LPRINT"C1 (Mfd)=";CO
9 INPUT"C2 (Mfd)=";CT
10 LPRINT"C2 (Mfd)=";CT
11 INPUT"RL (Ohms)=";RL
12 LPRINT"RL (Ohms)=";RL
13 INPUT "f (Hz)=";F
14 LPRINT"F (Hz)=";F:LPRINT"":PRINT""
15 XO=1/(6.283*F*CO*10^-6):XT=1/(6.283*F*CT*10^-6):AB=RL*XT:BA=-6.283/4:AC=(RL*R
L+XT*XT)^.5
16 CA=-ATN(XT/RL):AD=AB/AC:DA=BA-CA:AE=AD*COS(DA):AF=AD*SIN(DA)
17 AG=AE+RT:AH=(AG*AG+AF*AF)^.5:HA=-ATN(AF/AG)
18 AJ=AH*XO:JA=HA-6.283/4:AK=AF+XO:KA=-6.283/4
19 AL=(AG*AG+AK*AK)^.5:LA=-ATN(AK/AG):AM=AJ/AL:MA=JA-LA
20 AN=AM*COS(MA):AO=AM*SIN(MA):AP=RO+AN
21 AQ=(AP*AP+AO*AO)^.5:QA=-ATN(AO/AP):PF=F*QA:LPRINT"":PRINT""
22 A$="#######":LPRINT"Zin (Ohms)=";USING A$;AQ:PRINT"Zin (Ohms)=";USING A$;AQ
23 LPRINT"Phase Angle (Degrees)=";USING A$;-QA*360/6.283
24 PRINT"Phase Angle (Degrees)=";USING A$;-QA*360/6.283
```

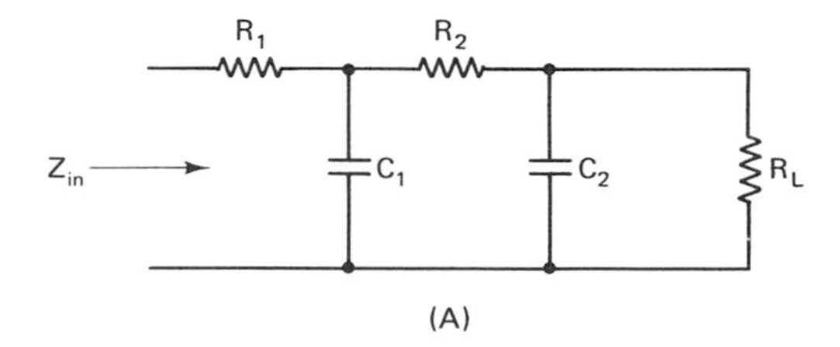


```
UNSYMMETRICAL TWO-SECTION CASCADED RC INTEGRATING CIRCUIT (With Resistive Load)
* Input Impedance and Phase Angle *

R1 (Ohms)= 1000
R2 (Ohms)= 2500
C1 (Mfd)= .15
C2 (Mfd)= .1
RL (Ohms)= 15000
F (Hz)= 1200


Zin (Ohms)=    1592
Phase Angle (Degrees)=     -36
```

Figure 10–6(A). Program for calculating the input impedance of an unsymmetrical two-section cascaded RC-integrating circuit with resistive load.

```
1 LPRINT"UNSYMMETRICAL TWO-SECTION CASCADED RC INTEGRATING CIRCUIT (With Resisti
ve Load)" :PRINT"TWO-SECTION CASCADED RC INTEGRATING CIRCUIT (With Resistive Loa
d)"
2 LPRINT"* Survey: Input Impedance and Phase Angle":PRINT"* Survey: Input Impeda
nce and Phase Angle":LPRINT"":PRINT""
3 INPUT"R1 (Ohms)=";RO
4 LPRINT"R1 (Ohms)=";RO
5 INPUT"R2 (Ohms)=";RT
6 LPRINT"R2 (OHMS)';RT
7 INPUT "C1 (Mfd)=";CO
8 LPRINT"C1 (Mfd)=";CO
9 INPUT"C2 (Mfd)=";CT
10 LPRINT"C2 (Mfd)=";CT
11 INPUT"RL (Ohms)=";RL
12 LPRINT"RL(Ohms)=";RL
13 INPUT"f (Hz)=";H
14 N=0:G=H:LPRINT"":PRINT""
15 F=G+N:A$="########":LPRINT"f (Hz)=";USING A$;F:PRINT"f (Hz)=";USING A$;F
16 XO=1/(6.283*F*CO*10^-6):XT=1/(6.283*F*CT*10^-6):AB=RL*XT:BA=-6.283/4:AC=(RL*R
L+XT*XT)^.5
17 CA=-ATN(XT/RL):AD=AB/AC:DA=BA-CA:AE=AD*COS(DA):AF=AD*SIN(DA)
18 AG=AE+RT:AH=(AG*AG+AF*AF)^.5:HA=-ATN(AF/AG)
19 AJ=AH*XO:JA=HA-6.283/4:AK=AF+XO:KA=-6.283/4
20 AL=(AG*AG+AK*AK)^.5:LA=-ATN(AK/AG):AM=AJ/AL:MA=JA-LA
21 AN=AM*COS(MA):AO=AM*SIN(MA):AP=RO+AN
22 AQ=(AP*AP+AO*AO)^.5:QA=-ATN(AO/AP)
23 A$="#######":LPRINT"Zin (Ohms)=";USING A$;AQ:PRINT"Zin (Ohms)=";USING A$;AQ
24 LPRINT"Phase Angle (Degrees)=";USING A$;-QA*360/6.283
25 PRINT"Phase Angle (Degrees)=";USING A$;-QA*360/6.283:LPRINT"":PRINT""
26 N=N+100
27 IF F>2500 THEN 29
28 GOTO 15
29 END
```

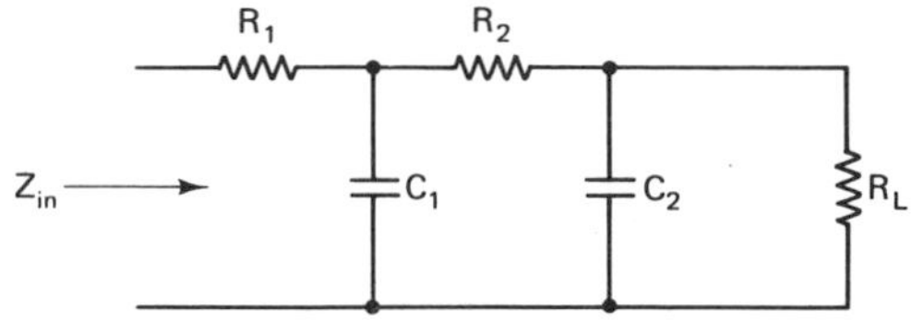

R_1 R_2 Z_{in} C_1 C_2 R_L

(B)

```
UNSYMMETRICAL TWO-SECTION CASCADED RC INTEGRATING CIRCUIT (With Resistive Load)
* Survey: Input Impedance and Phase Angle

R1 (Ohms)= 1000
R2 (OHMS)';RT
C1 (Mfd)= .15
C2 (Mfd)= .1
RL(Ohms)= 15000

f (Hz)=        600
Zin (Ohms)=    2844
Phase Angle (Degrees)=     -46

f (Hz)=        700
Zin (Ohms)=    2453
Phase Angle (Degrees)=     -44

f (Hz)=        800
Zin (Ohms)=    2173
Phase Angle (Degrees)=     -42

f (Hz)=        900
Zin (Ohms)=    1967
Phase Angle (Degrees)=     -41

f (Hz)=       1000
Zin (Ohms)=    1811
Phase Angle (Degrees)=     -39

f (Hz)=       1100
Zin (Ohms)=    1689
Phase Angle (Degrees)=     -37
```

Figure 10–6(B). Survey program for calculating the input impedance of an unsymmetrical two-section cascaded RC-integrating circuit with resistive load.

```
1 LPRINT "UNSYMMETRICAL TWO-SECTION CASCADED RC INTEGRATING CIRCUIT (With or Wit
hout Resistive Load)":PRINT"UNSYMMETRICAL TWO-SECTION CASCADED RC INTEGRATING CI
RCUIT (With or Without Resistive Load)"
2 LPRINT"* Unloaded Output Impedance, Phase Angle *":PRINT"* Unloaded Output Imp
edance, Phase Angle *"
3 LPRINT"* Unloaded Eout/Ein, Thevenin Impedance, and Loaded Eout/Ein *":PRINT"*
 Unloaded Eout/Ein, Thevenin Impedance, and Loaded Eout/Ein *":LPRINT"":PRINT""
4 INPUT "R1 (ohms)=";RO
5 LPRINT "R1 (ohms)=";RO
6 INPUT "R2 (ohms)=";RT
7 LPRINT "R2 (ohms)=";RT
8 INPUT "C1 (mfd)=";CO
9 LPRINT "C1 (mfd)=";CO
10 INPUT "C2 (mfd)=";CT
11 LPRINT "C2 (mfd)=";CT
12 INPUT "RL (ohms)=";RL
13 LPRINT "RL (ohms)=";RL
14 INPUT "f (hz)=";F
15 LPRINT "f (hz)=";F:LPRINT "":PRINT ""
16 XO=1/(6.283*F*CO*10^-6):XT=1/(6.283*F*CT*10^-6):AB=RO*XO:BA=-6.283/4
17 AC=(RO*RO+XO*XO)^.5:CA=-ATN(XO/RO):AD=AB/AC:DA=BA-CA
18 AE=AD*COS(DA):AF=AD*SIN(DA):AG=AE+RT
19 AH=(AG*AG+AF*AF)^.5:HA=-ATN(AF/AG):AJ=AH*XT:JA=HA-6.283/4
20 AK=AF-XT:KA=FA-6.283/4:AL=(AG*AG+AK*AK)^.5:LA=-ATN(AK/AG)
21 AM=AJ/AL:MA=JA-LA:AN=AM:NA=MA:FF=F*NA
22 A$="########":LPRINT "Zout (ohms)=";USING A$;AN:PRINT "Zout (ohms)=";USING A$
;AN
23 LPRINT "Phase angle, degrees (RL open)=";USING A$;-NA*360/6.283:PRINT "Phase
angle, degrees (RL open)=";USING A$;-NA*360/6.283
25 YZ=(RO*RO+XO*XO)^.5:ZY=-ATN(XO/RO)
26 XY=XO+XT:WX=(RT*RT+XY*XY)^.5:XW=-ATN(XY/RT):VW=WX*YZ:WV=XW+ZY
27 UV=VW/XO:VU=WV+6.283/4:TU=UV*COS(VU):UT=UV*SIN(VU)
28 ST=UT+XO:RS=(TU*TU+ST*ST)^.5:SR=-ATN(ST/TU):PU=XT/RS:UP=-SR-6.283/4
29 B$="####.##":LPRINT "Eout/Ein (RL open)=";USING B$;PU:PRINT "Eout/Ein (RL ope
n)=";USING B$;PU
30 GU=PU*RL:HU=AN*COS(NA):JU=AN*SIN(NA):UG=UP
31 KU=HU-RL:LU=(KU*KU+JU*JU)^.5:UL=-ATN(JU/KU):NU=GU/LU:UN=UG-UL
32 C$="######":LPRINT "Thevenin impedance (ohms)=";USING C$;AN:PRINT "Thevenin i
mpedance (ohms)=";USING C$;AN
33 D$="##.##":LPRINT "Eout/Ein (loaded)=";USING D$;NU:PRINT "Eout/Ein (loaded)="
;USING D$;NU
```

R1 R2 ein C1 C2 RL eout

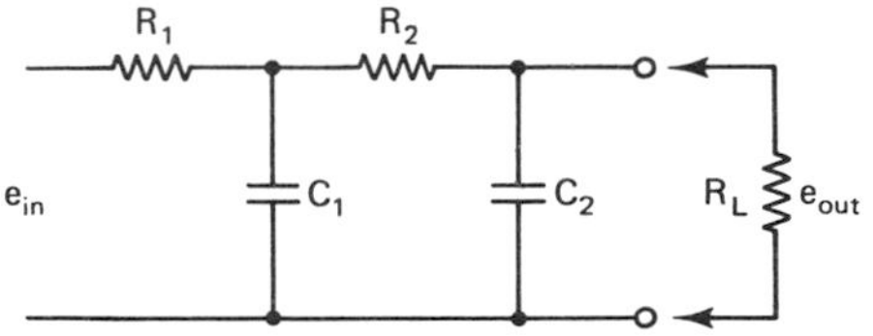

(C)

```
UNSYMMETRICAL TWO-SECTION CASCADED RC INTEGRATING CIRCUIT (With or Without Resis
tive Load)
* Unloaded Output Impedance, Phase Angle *
* Unloaded Eout/Ein, Thevenin Impedance, and Loaded Eout/Ein *

R1 (ohms)= 1500
R2 (ohms)= 2000
C1 (mfd)= .15
C2 (mfd)= .1
RL (ohms)= 3000
f (hz)= 1000

Zout (ohms)=     1218
Phase angle, degrees (RL open)=      117
Eout/Ein (RL open)=   0.27
Thevenin impedance (ohms)=  1218
Eout/Ein (loaded)= 0.22
```

Figure 10–6(C). Program for calculating the output/input voltage for an unsymmetrical two-section cascaded RC-integrating network with resistive load.

OUTPUT IMPEDANCE OF RC NETWORK WITH SIGNIFICANT SOURCE RESISTANCE

The designer is often concerned also with the output impedance of the unsymmetrical two-section cascaded RC-integrating network when energized by a voltage source with significant internal resistance. With reference to Figure 10–7, typical sources may have an internal resistance as low as 500 ohms or as high as 50,000 ohms. Inasmuch as this is a linear bilateral network, its output impedance is a function of the source resistance. The magnitude and phase angle of the output impedance are given by the equation shown in Figure 10–8.

Observe that the output impedance of the RC network is derived with respect to a short-circuited generator in Figure 10–8. The resulting equation is comparatively involved, and preliminary design procedures are facilitated by availability of an appropriate computer program such as provided in Figure 10–9. Since this is a

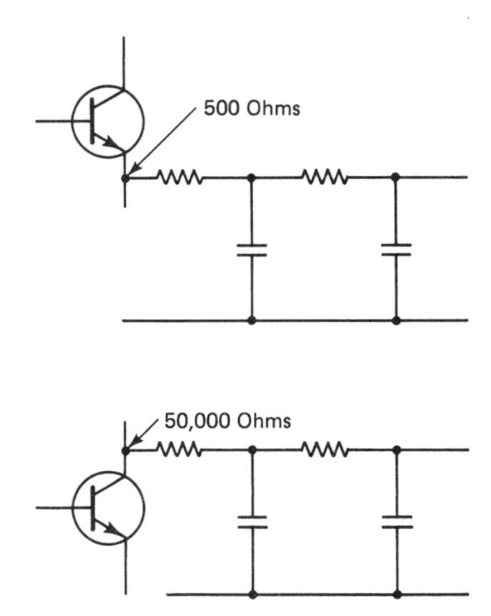

Figure 10–7. Typical source arrangements for an integrating network.

NOTE: The source resistance from which an integrating network obtains its current demand may have a low value in some applications or it may have a high value in other applications. As shown by the computer program in Figure 10–9, the output impedance of the network and the phase of its output voltage are modified by the value of source resistance that is employed.

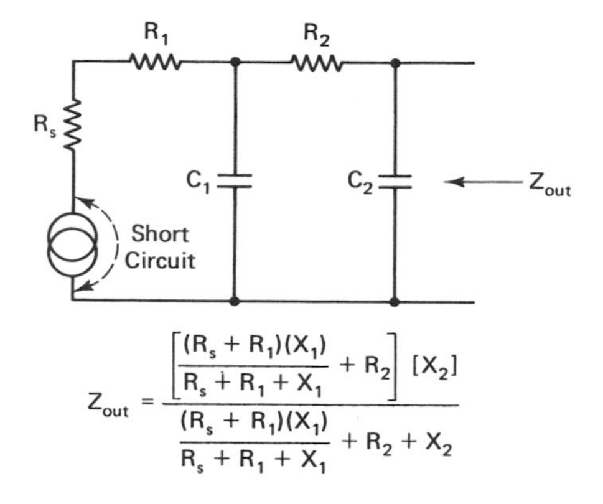

$$Z_{out} = \frac{\left[\frac{(R_s + R_1)(X_1)}{R_s + R_1 + X_1} + R_2\right][X_2]}{\frac{(R_s + R_1)(X_1)}{R_s + R_1 + X_1} + R_2 + X_2}$$

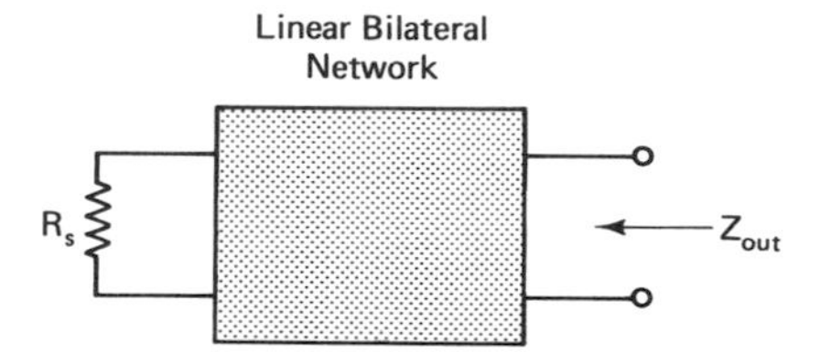

Figure 10–8. Basic diagram for calculation of the output impedance and phase angle for an unsymmetrical two-section cascaded RC-integrating network with source resistance.

NOTE: The output impedance of the network may or may not increase as the source resistance increases, depending on the values of the capacitors and the frequency of operation. The output impedance is calculated with respect to a short-circuited generator and is essentially a Thevenin impedance.

capacitive network, its output impedance has a leading phase angle. As the source-resistance value increases, the output impedance increases and its phase angle also increases.

ANALYSIS OF RESONANT CIRCUITRY

As previously discussed, parallel-resonant circuitry is extensively utilized in tuned-amplifier arrangements. A resonant circuit does not "stand alone," but functions instead in combination with a source resistance and a load resistance. In some situations, the source resistance may be so high that it can be disregarded. Similarly, in some situations the load resistance may be so high that

```
1 LPRINT"UNSYMMETRICAL TWO-SECTION CASCADED RC INTEGRATING CIRCUIT":PRINT"UNSYMM
ETRICAL TWO-SECTION INTEGRATING CIRCUIT"
2 LPRINT"* With Source Resistance *":PRINT"* With Source Resistance *"
3 LPRINT"(Output Resistance and Phase Angle)":PRINT"(Output Resistance and Phase
 Angle)":LPRINT"":PRINT""
4 INPUT "R1 (Ohms)=";RO
5 LPRINT"R1 (Ohms)=";RO
6 INPUT"R2 (Ohms)=";RT
7 LPRINT"R2 (Ohms)=";RT
8 INPUT"C1 (Mfd)=";CO
9 LPRINT"C1 (Mfd)=";CO
10 INPUT "C2 (Mfd)=";CT
11 LPRINT"C2 (Mfd)=";CT
12 INPUT "Rs (Ohms)=";RS
13 LPRINT"Rs (Ohms)=";RS
14 INPUT "f (Hz)=";F
15 LPRINT"f (Hz)=";F:LPRINT"":PRINT""
16 XO=1/(6.283*F*CO*10^-6):XT=1/(6.283*F*CT*10^-6):AB=(RS+RO)*XO:BA=-6.283/4
17 AC=((RS+RO)*(RS+RO)+XO*XO)^.5:N=RS+RO:CA=-ATN(XO/N):AD=AB/AC:DA=BA-CA
18 AE=AD*COS(DA):AF=AD*SIN(DA):AG=AE+RT
19 AH=(AG*AG+AF*AF)^.5:HA=-ATN(AF/AG):AJ=AH*XT:JA=HA-6.283/4
20 AK=AF-XT:KA=FA-6.283/4:AL=(AG*AG+AK*AK)^.5:LA=-ATN(AK/AG)
21 AM=AJ/AL:MA=JA-LA:AN=AM:NA=MA
22 A$="########":LPRINT"Zout (Ohms)=";USING A$;AN:PRINT"Zout (Ohms)=";USING A$;A
N
23 LPRINT"Phase Angle, Degrees=";USING A$;-NA*360/6.283:PRINT"Phase Angle, Degre
es=";USING A$;-NA*360/6.283
```

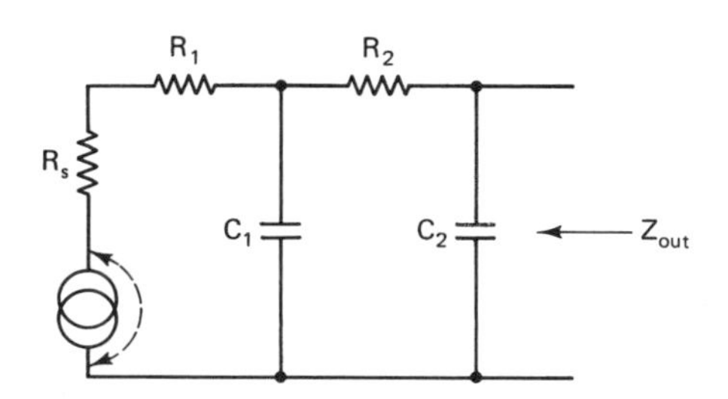


```
UNSYMMETRICAL TWO-SECTION CASCADED RC INTEGRATING CIRCUIT
* With Source Resistance *
(Output Resistance and Phase Angle)

R1 (Ohms)= 1500
R2 (Ohms)= 2000
C1 (Mfd)= .16
C2 (Mfd)= .2
Rs (Ohms)= 950
f (Hz)= 1000

Zout (Ohms)=       693
Phase Angle, Degrees=      105
```

Figure 10–9. Program for calculation of the output impedance and phase angle for an unsymmetrical two-section cascaded RC-integrating network with source resistance.

it can be disregarded. However, the designer usually needs to take both the source resistance and the load resistance into consideration.

With reference to Figure 10–10, the characteristics of the LCR tank are affected by the value of R_s, which may have a value as low as 3 kilohms or as high as 100 kilohms. The resonant frequency of the tank is essentially the same as for an LCR series circuit, and the value of the source resistance has little effect on the resonant frequency. However, the value of the source resistance has a significant effect both on the impedance of the tank at resonance and on

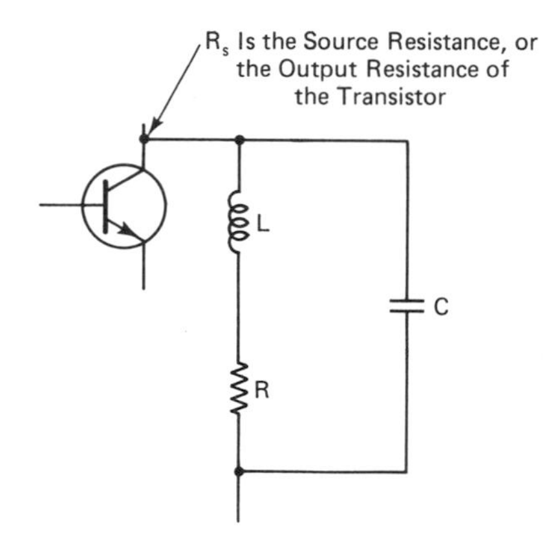

Resonant frequency $f_o = 1/(2\pi FC)$, approximately
Z at resonance $= L/R_C$, approximately
Q at resonance $= X_L/R$

When R_s is sufficiently low that it must be taken into consideration, the resonant frequency is essentially unchanged. However, Z becomes less, and Q becomes less. The effect of R_s on Z is the same as if R_s shunts the LCR circuit. The effect of R_s on Q is the same as if a resistance equal to $(X_C)^2/R_s$ were connected in series with C. In other words, Q becomes equal to XC divided by $(X_C)^2/R_s$.

Figure 10–10. Basic diagram for calculation of parallel-resonant frequency, with impedance and Q values for constant-current source and for voltage source with series resistance.

NOTE: R is the ac resistance of the inductor; it may be somewhat greater than the dc resistance of the winding.

This configuration is in very wide use in many areas of electronic design. The following program provides rapid calculation of f_o, Z, and Q, both for a constant-current source and for a voltage source with finite internal resistance.

Observe that the bandwidth of the tuned circuit is approximately equal to f_o/Q in any case.

its Q value. In other words, the magnitude of the load impedance and its selectivity (bandwidth) are functions of the source resistance.

Observe in Figure 10–10 that the source resistance R_s is effectively connected in shunt to the tank. In turn, the magnitude of the load impedance is given by the parallel combination of R_s and L/(RC) at resonance. Note also that the effect of R_s on the Q value of the LCR parallel circuit is equivalent to the insertion of a resistance $(XC)^2/R_s$ in series with C. Since XL = XC at resonance, the modified Q value becomes equal to R_s/XC. As previously noted, a reactance-squared value has a zero-degree phase angle (it is a resistive value).

Preliminary design procedures are facilitated by the availability of a suitable computer program for calculation of Z and Q values both for a constant-current source and a voltage source with finite internal resistance. With reference to Figure 10–11, an appropriate routine is provided. As exemplified in the RUN, a source resistance as high as 25 kilohms is nevertheless quite significant in modifying the Z and Q values of a typical LCR tank. Thus, at a resonant frequency of 1061 kHz, a 25-kilohm source resistance reduces the tank Z from 40 kilohms to approximately 15 kilohms and reduces the tank Q from 37 to approximately 23.5.

As previously noted, bipolar-transistor circuitry commonly employs tapped-tank arrangements, such as shown in Figure 10–12. In this basic example, the inductor is center-tapped. The designer is concerned with the reduction in the tank Z and Q values resulting from the connection of the collector lead to the center tap on the inductor. Analysis of this configuration is facilitated by noting that the tank Z at resonance is twice the value of Z at the center tap with respect to a constant-current source. In turn, the effect of R_s on the impedance of the center-tapped tank is equivalent to R_s connected in shunt to Z/2.

As anticipated, the resonant frequency of the tank is virtually unchanged by connection of the collector lead to the center tap on the inductor. However, the loading action of R_s reduces the Z and Q values of the tank; these reductions are less severe than if the collector lead is connected to the top of the tank circuit. Insofar as reduction of the Q value is concerned when the collector lead is connected to the center tap on the inductor, it is helpful to note that if the collector lead were connected to only one turn at the bottom of the inductor, the resulting Q value of the tank would be essentially the same as for a constant-current source.

```
1 LPRINT"PARALLEL RESONANT CIRCUIT":PRINT"PARALLEL RESONANT CIRCUIT"
2 LPRINT"Impedance and Q at Resonance":PRINT"Impedance and Q at Resonance"
3 LPRINT"* With Constant-Current Source *":PRINT"* With Constant-Current Source
*"
4 LPRINT"* With Source Resistance *":PRINT"* With Source Resistance *"
5 LPRINT"":PRINT""
6 INPUT"L (mH)=";L
7 LPRINT"L (mH)=";L
8 INPUT "C (pF)=";C
9 LPRINT"C (pF)=";C
10 INPUT "R (Ohms)=";R
11 LPRINT"R (Ohms)=";R
12 INPUT"Rs (Ohms)=";RS
13 LPRINT"Rs (Ohms)=";RS:LPRINT""
14 A=(L*.001*C*10^-12)^.5:FO=1/(6.283*A):LPRINT"":PRINT""
15 A$="########":LPRINT"Fo (Hz, approx.)=";USING A$;FO:PRINT"Fo (Hz, approx.)=";
USING A$;FO:LPRINT"":PRINT""
16 Z=L*.001/(R*C*10^-12)
17 B$="######":LPRINT"Z (Ohms, approx.)=";USING B$;Z:PRINT"Z (Ohms, approx.)=";U
SING B$;Z
18 LPRINT "(With Constant-Current Source)":PRINT"(With Constant-Current Source":
LPRINT"":PRINT""
19 Q=6.283*FO*L*.001/R
20 C$="####":LPRINT"Q (approx.)=";USING C$;Q:PRINT"Q (approx.)=";USING C$;Q
21 LPRINT"(With Constant-Current Source)":PRINT"(With Constant-Current Source":L
PRINT"":PRINT""
22 ZL=RS*Z/(RS+Z)
23 D$="######":LPRINT"Z (Ohms, approx.)=";USING D$;ZL:PRINT"Z (Ohms, approx.)=";
USING D$;ZL
24 LPRINT"(With Source Resistance Rs)":PRINT"(With Source Resistance Rs)":LPRINT
"":PRINT""
25 B=1/(6.283*FO*C*10^-12):T=B*B/RS:U=B/T
26 E$="###.#":LPRINT"Q (approx.)=";USING E$;U:PRINT"Q (approx.)=";USING E$;U
27 LPRINT"(With Source Resistance Rs)":PRINT"(With Source Resistance Rs)"
```

```
PARALLEL RESONANT CIRCUIT
Impedance and Q at Resonance
* With Constant-Current Source *
* With Source Resistance *

L (mH)= .15
C (pF)= 150
R (Ohms)= 25
Rs (Ohms)= 25000


Fo (Hz, approx.)= 1061064

Z (Ohms, approx.)= 40000
(With Constant-Current Source)

Q (approx.)=  40
(With Constant-Current Source)

Z (Ohms, approx.)= 15385
(With Source Resistance Rs)

Q (approx.)= 25.0
(With Source Resistance Rs)
```

R_s
L
C
R

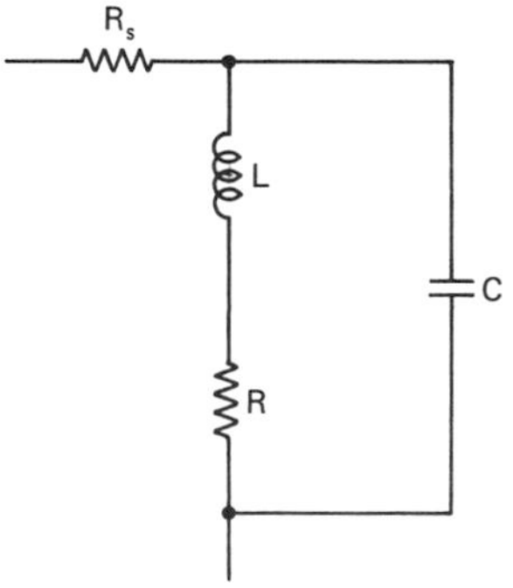

Figure 10–11. Program for calculating the parallel-resonant frequency with Z and Q values for a constant-current source and for a voltage source with finite internal resistance.

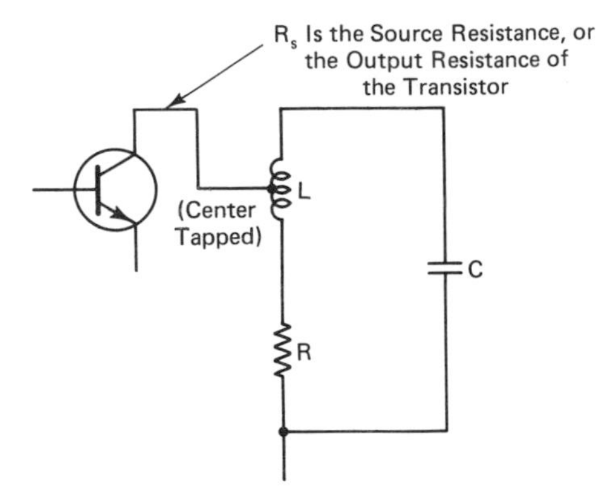

Resonant frequency $f_o = 1/(2\pi FC)$, approximately

Z at resonance = 1/2 the value of Z when the input voltage is applied at the top of the inductor

Q at resonance is the same as when the input voltage is applied at the top of the inductor

} (Constant-Current Source)

When R_s is sufficiently low that it must be taken into consideration, the resonant frequency is essentially unchanged. However, Z becomes less, and Q becomes less. The effect of R_s on Z is the same as if R_s shunts one-half of the LCR circuit. The effect of R_s on Q is the same as the average value of the unloaded Q value of the fully loaded Q value (Q value with the input voltage applied at the top of the inductor).

Figure 10–12. Basic diagram for calculation of f_o, Z, and Q values for a parallel-resonant circuit with center-tapped inductor.

NOTE: *This arrangement is in extensive use in tuned amplifier circuitry because it provides a higher Q value than if the input voltage were applied at the top of the inductor.*

As exemplified by the program in Figure 10–15, the bandwidth for the center-tapped configuration is approximately 83 percent of the bandwidth with the input voltage applied at the top of the inductor.

It follows that when the collector lead is connected to the top of the tank, the modified Q value is given by the routine in Figure 10–11 and that when the collector lead is connected to a single turn at the bottom of the tank, the Q value is given by XL/R. Accordingly, when the collector lead is connected to the center tap on the inductor, the modified Q value is then equal to the average of the unloaded Q value and the modified Q value with the collector lead connected at the top of the tank.

Since the foregoing calculations are somewhat tedious, it is helpful in preliminary design procedures to have a suitable computer program available, such as provided in Figure 10–13. Observe that in the exemplified RUN, a source resistance of 25 kilohms

```
1 LPRINT"PARALLEL RESONANT CIRCUIT":PRINT"PARALLEL RESONANT CIRCUIT"
2 LPRINT"CENTER-TAPPED INDUCTOR":PRINT"CENTER-TAPPED INDUCTOR":LPRINT"(Impedance
 and Q at Resonance)":PRINT"(Impedance and Q at Resonance"
3 LPRINT"* With Constant-Current Source *":PRINT"* With Constant-Current Source
*"
4 LPRINT"* With Source Resistance *":PRINT"* With Source Resistance *"
5 LPRINT"":PRINT""
6 INPUT"L (mH)=";L
7 LPRINT"L (mH)=";L
8 INPUT "C (pF)=";C
9 LPRINT"C (pF)=";C
10 INPUT "R (Ohms)=";R
11 LPRINT"R (Ohms)=";R
12 INPUT"Rs (Ohms)=";RS
13 LPRINT"Rs (Ohms)=";RS:LPRINT""
14 A=(L*.001*C*10^-12)^.5:FO=1/(6.283*A):LPRINT"":PRINT""
15 A$="########":LPRINT"Fo (Hz, approx.)=";USING A$;FO:PRINT"Fo (Hz, approx.)=";
USING A$;FO:LPRINT"":PRINT""
16 Z=L*.001/(R*C*10^-12)
17 B$="######":LPRINT"Z (Ohms, approx.)=";USING B$;Z:PRINT"Z (Ohms, approx.)=";U
SING B$;Z
18 LPRINT "(With Constant-Current Source)":PRINT"(With Constant-Current Source":
LPRINT"":PRINT""
19 Q=6.283*FO*L*.001/R
20 C$="####":LPRINT"Q (approx.)=";USING C$;Q:PRINT"Q (approx.)=";USING C$;Q
21 LPRINT"(With Constant-Current Source)":PRINT"(With Constant-Current Source":L
PRINT"":PRINT""
22 Y=Z/2:ZL=RS*Y/(RS+Y)
23 D$="######":LPRINT"Z (Ohms, approx.)=";USING D$;ZL:PRINT"Z (Ohms, approx.)=";
USING D$;ZL
24 LPRINT"(With Source Resistance Rs)":PRINT"(With Source Resistance Rs)":LPRINT
"":PRINT""
25 B=1/(6.283*FO*C*10^-12):T=B*B/RS:U=B/T:QL=(U+Q)/2
26 E$="###.#":LPRINT"Q (approx.)=";USING E$;QL:PRINT"Q (approx.)=";USING E$;QL
27 LPRINT"(With Source Resistance Rs)":PRINT"(With Source Resistance Rs)"
```

```
PARALLEL RESONANT CIRCUIT
CENTER-TAPPED INDUCTOR
(Impedance and Q at Resonance)
* With Constant-Current Source *
* With Source Resistance *

L (mH)= .15
C (pF)= 150
R (Ohms)= 25
Rs (Ohms)= 25000


Fo (Hz, approx.)= 1061064

Z (Ohms, approx.)= 40000
(With Constant-Current Source)

Q (approx.)=  40
(With Constant-Current Source)

Z (Ohms, approx.)= 11111
(With Source Resistance Rs)

Q (approx.)= 32.5
(With Source Resistance Rs)
```

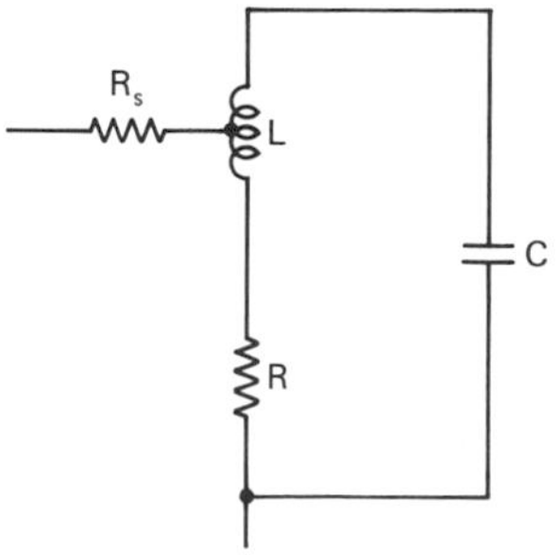

Figure 10–13. Program for calculating f_o, Z, and Q values for a parallel-resonant circuit with center-tapped inductor.

connected at center tap on the tank inductor results in a reduction of tank input impedance from 40 kilohms to approximately 11 kilohms. Also, the Q value of the tank becomes approximately 32 with the drive voltage applied at the center tap. This corresponds to a reduction of approximately 17 percent in bandwidth. In other words, the tank circuit has higher selectivity when the drive voltage is applied at the center tap instead of the top of the inductor.

Designers may prefer to tap the tank inductor off-center to obtain a specified bandwidth. In turn, the more generalized approach shown in Figure 10–14 is employed. It is evident that the tank bandwidth will approach its unloaded value as the tap point is moved nearer the lower end of the winding. Of course, the input impedance to the tap point also becomes progressively smaller. Preliminary design procedures are facilitated with the aid of an appropriate computer program, such as provided in Figure 10–15. As seen in the exemplified RUNS, the bandwidth at a 25-percent tap point is approximately 29 kHz, whereas the bandwidth at a 75-

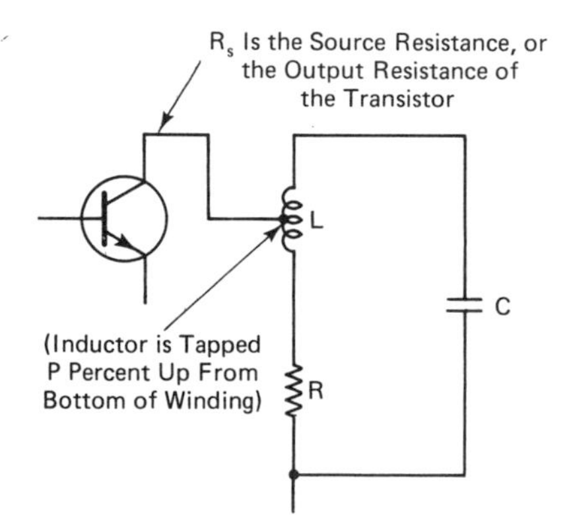

Resonant frequency $f_o = 1/(2\pi\sqrt{LC})$, approximately
Z at resonance = P percent of the value of Z when the input voltage is applied at the top of the inductor
Q at resonance is the same as when the input voltage is applied at the top of the inductor
(Constant-Current Source)

When the source resistance R_s is sufficiently low that it must be taken into consideration, the resonant frequency is essentially unchanged. However, Z becomes less and Q becomes less. The effect of R_s on Z is the same as if R_s shunts P percent of the LCR circuit. The effect of R_s on Q is the same as the unloaded Q value minus the percentage P of the difference between the unloaded and the fully loaded Q values.

Figure 10–14. Basic diagram for calculation of f_o, Z, Q, and BW values for a parallel-resonant circuit with inductor-tapped P percent up from lower end of winding.

```
1 LPRINT"PARALLEL RESONANT CIRCUIT":PRINT"PARALLEL RESONANT CIRCUIT"
2 LPRINT"INDUCTOR TAPPED P% UP FROM LOWER END OF WINDING":PRINT"INDUCTOR TAPPED
P% UP FRONM LOWER END OF WINDING":LPRINT"(Impedance and Q at Resonance)":PRINT"(
Impedance and Q at Resonance)"
3 LPRINT"* With Constant-Current Source *":PRINT"* With Constant-Current Source
*"
4 LPRINT"* With Source Resistance *":PRINT"* With Source Resistance *"
5 LPRINT"":PRINT""
6 INPUT"L (mH)=";L
7 LPRINT"L (mH)=";L
8 INPUT "C (pF)=";C
9 LPRINT"C (pF)=";C
10 INPUT "R (Ohms)=";R
11 LPRINT"R (Ohms)=";R
12 INPUT"Rs (Ohms)=";RS
13 LPRINT"Rs (Ohms)=";RS
14 INPUT"P (%)=";P
15 LPRINT"P (%)=";P:PRINT""
16 A=(L*.001*C*10^-12)^.5:FO=1/(6.283*A):LPRINT"":PRINT""
17 A$="########":LPRINT"Fo (Hz, approx.)=";USING A$;FO:PRINT"Fo (Hz, approx.)=";
USING A$;FO:LPRINT"":PRINT""
18 Z=L*.001/(R*C*10^-12)
19 B$="######":LPRINT"Z (Ohms, approx.)=";USING B$;Z:PRINT"Z (Ohms, approx.)=";U
SING B$;Z
20 LPRINT "(With Constant-Current Source)":PRINT"(With Constant-Current Source":
LPRINT"":PRINT""
21 Q=6.283*FO*L*.001/R
22 C$="####":LPRINT"Q (approx.)=";USING C$;Q:PRINT"Q (approx.)=";USING C$;Q
23 LPRINT"(With Constant-Current Source)":PRINT"(With Constant-Current Source":L
PRINT"":PRINT""
24 Y=Z*P*.01:ZL=RS*Y/(RS+Y)
25 D$="######":LPRINT"Z (Ohms, approx.)=";USING D$;ZL:PRINT"Z (Ohms, approx.)=";
USING D$;ZL
26 LPRINT"(With Source Resistance Rs)":PRINT"(With Source Resistance Rs)":LPRINT
"":PRINT""
27 XC=1/(6.283*FO*C*10^-12):RN=XC*XC/RS:QL=XC/RN
28 BW=FO/Q
29 E$="######":LPRINT"BW (Hz)=";USING E$;BW:PRINT"BW (Hz)=";USING E$;BW
30 LPRINT"(With Constant-Current Source)":PRINT"(With Constant-Current Source)":
LPRINT"":PRINT""
31 QM=Q-P*.01*(Q-QL):WB=FO/QM
32 F$="####":LPRINT"Q=";USING F$;QM:PRINT"Q=";USING F$;QM
33 LPRINT"(With Source Resistance Rs)":PRINT"(With Source Resistance Rs)":LPRINT
"":PRINT""
34 G$="######":LPRINT"BW (Hz)=";USING G$;WB:PRINT"BW (Hz)=";USING G$;WB
35 LPRINT"(With Source Resistance Rs)":PRINT"(With Source Resistance Rs)"
```

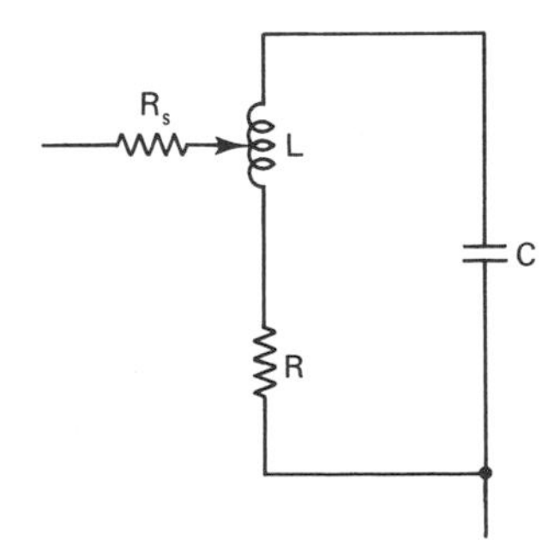

Figure 10–15. (Continues on next page.)

```
PARALLEL RESONANT CIRCUIT
INDUCTOR TAPPED P% UP FROM LOWER END OF WINDING
(Impedance and Q at Resonance)
* With Constant-Current Source *
* With Source Resistance *

L (mH)= .15
C (pF)= 150
R (Ohms)= 25
Rs (Ohms)= 25000
P (%)= 25

Fo (Hz, approx.)= 1061064

Z (Ohms, approx.)= 40000              BW (Hz)= 26527
(With Constant-Current Source)        (With Constant-Current Source)

Q (approx.)=  40                      Q=  36
(With Constant-Current Source)        (With Source Resistance Rs)

Z (Ohms, approx.)=  7143              BW (Hz)= 29271
(With Source Resistance Rs)           (With Source Resistance Rs)
```

Figure 10–15. Program for calculation of f_o, Z, Q, and BW values for a parallel-resonant circuit with inductor-tapped P percent up from lower end of winding.

percent tap point is 36 kHz, approximately, for the specified L, C, and R values.

In some applications, two taps are taken on the tank inductor, as shown in Figure 10–16. Thus, R_s may denote the output resistance of a driver transistor and R_L may denote the base-input resistance of the following transistor. Since the collector output resistance is generally greater than the base-input resistance, a doubly tapped tank inductor assists in improvement of power transfer. At the same time, tapping down provides improvement of selectivity. As noted in Figure 10–16, the doubly tapped tank can be analyzed in two steps.

Observe that the program provided in Figure 10–15 can be employed for analysis of a doubly tapped tank configuration. In other words, the program is run twice. On the first run, R_L is disregarded, and the reduced Q value due to R_s is computed. On the second run, the R_L tap point is inputted, along with a modified value for R that yields the reduced Q value that has been computed. In turn, the modified Q value due to connection of R_L is computed. This modified Q value determines the bandwidth of the doubly tapped tank arrangement.

To briefly recapitulate, consider that the unloaded Q value of the tank circuit is 35 and that the value of XL is 1000 ohms at the resonant frequency of 1 MHz. This corresponds to an R value of approximately 28 ohms. Now, if connection of R_s to the inductor results in a reduction of the Q value to 30, this corresponds to a modified R value of 33 ohms, approximately. The 33-ohm value is

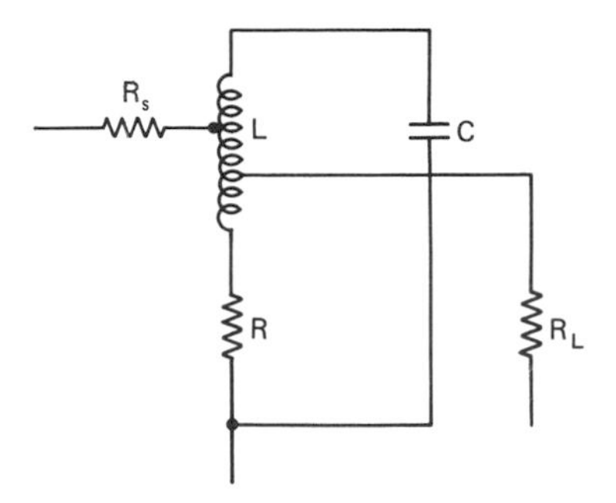

Figure 10–16. A doubly tapped LCR tank configuration.

NOTE: This arrangement is readily analyzed in two steps. First, the Q value is determined with respect to R_s. This is, of course, less than the unloaded Q value of the circuit. This reduced Q value is then taken as the reference Q value, and a modified Q value is then calculated with respect to R_L. This modified Q value determines the bandwidth of the tank circuit with both R_s and R_L connected.

then inputted as the R value on the second run, and the R_L value is inputted as the R_s value. In turn, the modified value of Q is printed out, along with the bandwidth resulting from the loading action of both R_s and R_L. Observe that P corresponds to the R_L tap point on the second run.

The Z value that is printed out on the second run denotes the output impedance of the tank at the R_L tap point. Although seldom realized precisely, maximum power transfer (highest efficiency) will occur if the tank-output impedance is equal to R_L. Since R_L draws tank current, it affects the input impedance at the R_s tap point. If it is desired to compute this reduction in Z at the R_s tap point, the program must be run a third time. On the third run, a repetition of the first run is inputted, except that the value of R is adjusted to reflect the modified Q value. In turn, the value of Z that is printed out corresponds to the input impedance at the R_L tap point on the tank inductor.

Again, although it is seldom precisely realized, maximum power transfer will occur if the tank input impedance is equal to R_s. Inasmuch as bandwidth is usually the primary concern, the designer ordinarily accepts moderate impedance mismatches as a necessary compromise. Note in passing that a suitable impedance mismatch is occasionally specified in order to stabilize rf amplifier operation without resorting to neutralizing circuitry. This is a form of "losser" stabilization. Again, a suitable impedance mismatch is sometimes specified in a low-level stage to optimize the noise factor.

11

BASIC FILTER DESIGN PROCEDURES

GENERAL CONSIDERATIONS

A filter is a frequency-selective arrangement. RC filters are used in a very wide spectrum of electronic equipment. It was previously noted that a differentiating circuit functions as a high-pass filter and that an integrating circuit functions as a low-pass filter. When a differentiating circuit and an integrating circuit are cascaded as shown in Figure 11–1, bandpass action is obtained with suitable component values.

With reference to Figure 11–2, the exemplified component values provide a peak frequency of approximately 1500 Hz and half-power cutoff frequencies of approximately 500 Hz and 4000 Hz. The maximum e_{out}/e_{in} ratio is 0.74 because the first RC section is noticeably loaded by the second section. The output is very low at very low input frequencies and rises with comparative rapidity at frequencies in the order of a few hundred Hz. However, the output falls comparatively slowly through the high-frequency interval of the pass band.

CUTOFF CHARACTERISTICS

The output from an RC bandpass filter will fall more rapidly through the high-frequency interval of the pass band if a two-section low-pass filter is employed, as depicted in Figure 11–3. In this simple example, it is seen that the second harmonic of the input frequency (15,750 Hz) is attenuated to 75 percent with respect to the fundamental when one section is utilized. However, the second harmonic of the input frequency is attenuated to 52 percent with respect to the fundamental when two sections are utilized.

The attenuation of an RC low-pass filter in terms of dB per octave approaches a limit as the number of sections is increased. In the limit, the cutoff characteristic has the shape of the standard probability curve (Gaussian cutoff characteristic). It is customary to specify the dB-per-octave rolloff value of a filter as the slope of the tangent (in dB units) to the characteristic at the half-power point.

RC FILTER TONE CONTROLS

Most tone controls and frequency equalizers are designed as RC-filter networks. With reference to Figure 11–4, a typical losser-type bass-treble tone-control arrangement is shown. This is termed a losser network inasmuch as it consists of passive components and has no active device to compensate for I^2R losses. Next, observe the response of the network when the component values listed in Figure 11–5 are utilized, with the bass control set for maximum boost and

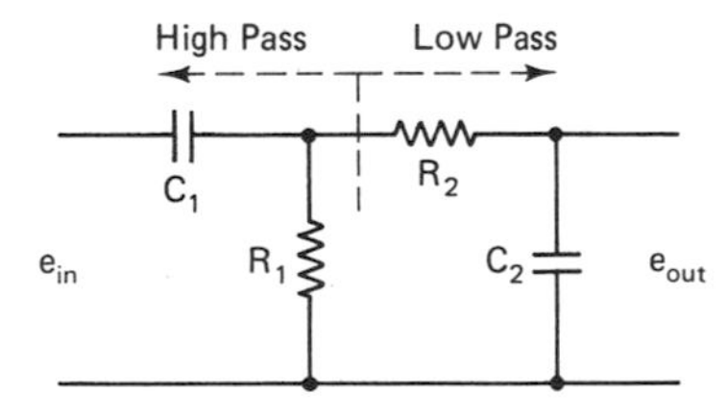

Figure 11–1. Basic RC bandpass filter configuration.

NOTE: This cascaded high-pass/low-pass RC filter arrangement provides bandpass action. The input section has a specified low-frequency cutoff value, and the output section has a specified high-frequency cutoff value. Although the input section is loaded by the output section, this factor can be minimized by employing a comparatively high-impedance output section.

With a constant-voltage source the e_{out}/e_{in} relationship is formulated:

$$\frac{e_{out}}{e_{in}} = \frac{X2}{\dfrac{(R1 + R2 + X2)(R1 + X1)}{R1} - R1}$$

```
1 LPRINT"RC BANDPASS FILTER NETWORK":PRINT"RC BANDPASS FILTER NETWORK"
2 LPRINT"* Eout/Ein, Phase *":PRINT"* Eout/Ein, Phase *"
3 LPRINT"":PRINT""
4 INPUT"C1 (Pf)=";CO
5 LPRINT"C1 (pF)=";CO
6 INPUT "R1 (Ohms)=";RO
7 LPRINT"R1 (Ohms)=";RO
8 INPUT "C2 (pF)=";CT
9 LPRINT"C2 (pF)=";CT
10 INPUT"R2 (Ohms)=";RT
11 LPRINT"R2 (Ohms)=";RT
12 INPUT "f (Hz)=";F
13 LPRINT"f (Hz)=";F:LPRINT"":PRINT""
14 XO=1/(6.283*F*CO*10^-12):XT=1/(6.283*F*CT*10^-12)
15 AB=RO+RT:AC=(AB*AB+XT*XT)^.5:CA=-ATN(XT/AB)
16 AD=(RO*RO+XO*XO)^.5:DA=-ATN(XO/RO)
17 AE=AC*AD:EA=CA+DA:AF=AE/RO:FA=EA
18 AG=AF*COS(FA):AH=AF*SIN(FA):AJ=AG-RO
19 AK=(AJ*AJ+AH*AH)^.5:KA=-ATN(AH/AJ)
20 AL=XT/AK:LA=-KA-6.283/4:PF=-F*LA:LPRINT"":PRINT""
21 A$="####.##":LPRINT"Eout/Ein=";USING A$;AL:PRINT"Eout/Ein=";USING A$;AL
22 LPRINT"Phase Angle (Degrees)=";USING A$;-LA*360/6.283:PRINT"Phase Angle (Degr
ees)=";USING A$;-LA*360/6.283
```

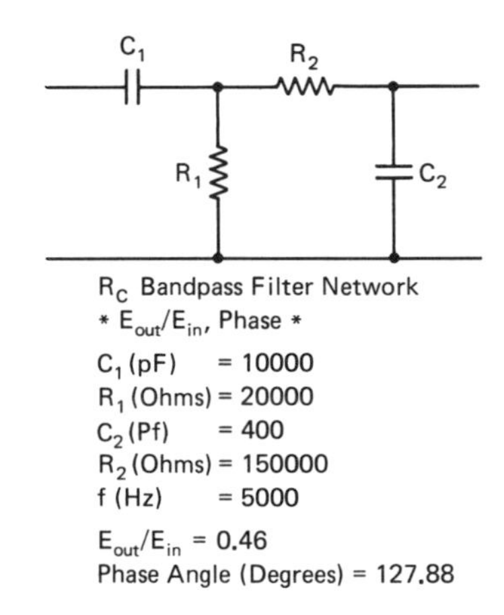


```
RC BANDPASS FILTER NETWORK
* Eout/Ein, Phase *

C1 (pF)= 10000
R1 (Ohms)= 20000
C2 (Pf)= 400
R2 (Ohms)= 150000
f (Hz)= 5000


Eout/Ein=    0.46
Phase Angle (Degrees)= 127.88
```

Figure 11–2. Program for calculating E_{out}/E_{in} and phase angle for an RC bandpass filter network.

```
1 LPRINT"RC LOW-PASS ONE- AND TWO-SECTION FILTER NETWORKS":PRINT"RC LOW-PASS ONE
- AND TWO-SECTION FILTER NETWORKS"
2 LPRINT"* Eout/Ein, Phase *":PRINT"* Eout/Ein, Phase *"
3 LPRINT"":PRINT""
4 INPUT"f (Hz)=";F
5 LPRINT"f (Hz)=";F:XC=0:AB=0:BA=0:IN=0:NI=0:EO=0:OE=0
6 INPUT"R (Ohms)=";R
7 LPRINT"R (OHMS)=";R
8 INPUT"C (pF)=";C
9 LPRINT"C (pF)=";C
10 XC=1/(6.283*F*C*10^-12):AB=(R*R+XC*XC)^.5:BA=-ATN(XC/R)
11 IN=1/AB:NI=-BA:EO=IN*XC:OE=NI-6.283/4:LPRINT"":PRINT""
12 A$="####.##":LPRINT"Eout/Ein=";USING A$;EO:PRINT"Eout/Ein=";USING A$;EO
13 LPRINT"Phase Angle (Degrees)=";USING A$;-OE*360/6.283:PRINT"Phase Angle (Degr
ees)=";USING A$;-OE*360/6.283:B$="#####"
14 LPRINT"(For One Section)":PRINT"(For One Section)":LPRINT"":PRINT""
15 AB=(R*R+4*XC*XC)^.5:BA=-ATN(2*XC/R):AC=(R*R+XC*XC)^.5:CA=-ATN(XC/R)
16 BC=AB*AC:CB=BA+CA:AD=BC/XC:DA=CB+6.283/4:BD=AD*COS(DA):DB=AD*SIN(DA):CD=DB+XC

17 DE=(BD*BD+CD*CD)^.5:ED=-ATN(CD/AB):AE=XC/DE:EA=-ED-6.283/4
18 LPRINT"Eout/Ein=";USING A$;AE:PRINT"Eout/Ein=";USING A$;AE
19 LPRINT"Phase Angle (Degrees)=";USING A$;-EA*360/6.283:PRINT"Phase Angle (Degr
ees)=";USING A$;-EA*360/6.283
20 LPRINT"(For Two Sections)":PRINT"(For Two Sections)"
```

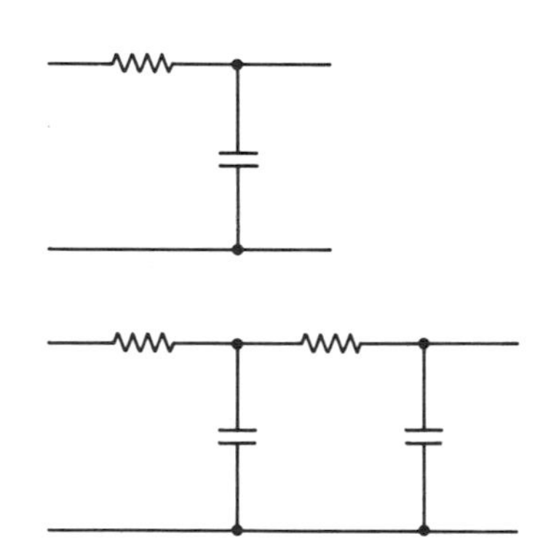

```
RC LOW-PASS ONE- AND TWO-SECTION
FILTER NETWORKS
* Eout/Ein, Phase *

f (Hz)= 15750
R (OHMS)= 150000
C (pF)= 40

Eout/Ein=    0.86
Phase Angle (Degrees)=   30.70

(For One Section)

Eout/Ein=    0.53
Phase Angle (Degrees)= 107.24

(For Two Sections)
```

```
RC LOW-PASS ONE- AND TWO-SECTION
FILTER NETWORKS
* Eout/Ein, Phase *

f (Hz)= 31500
R (OHMS)= 150000
C (pF)= 40

Eout/Ein=    0.64
Phase Angle (Degrees)=   49.90

(For One Section)

Eout/Ein=    0.28
Phase Angle (Degrees)=   80.00

(For Two Sections)
```

Figure 11–3. Comparison program for E_{out}/E_{in} and phase relations for one-section and two-section RC low-pass filter networks.

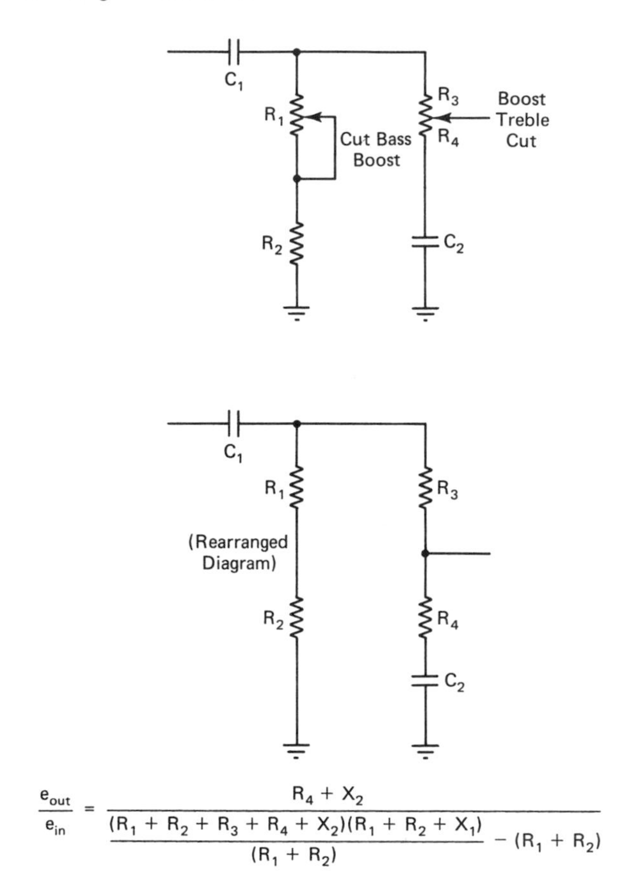

$$\frac{e_{out}}{e_{in}} = \frac{R_4 + X_2}{\dfrac{(R_1 + R_2 + R_3 + R_4 + X_2)(R_1 + R_2 + X_1)}{(R_1 + R_2)} - (R_1 + R_2)}$$

Figure 11–4. Basic diagram for RC bass-treble losser-type tone control.

Note: Component values are typically specified so that response is essentially uniform over the audio-frequency range when the controls are set to their midpoints. In turn, bass response is comparatively reduced when the bass control is set to the top of R1 and is conversely greater when set to the bottom of R1. Similarly, treble response is comparatively boosted when the treble control is set to the top of R3-R4 and vice versa.

the treble control set to its midpoint. In turn, the output is 94 percent of maximum at 100 Hz, but 51 percent of maximum at 10 kHz.

Next, with reference to Figure 11–6, a survey program is provided for the same analysis. The same component values are used as before, but with the bass control set near its maximum cut position and the treble control set near its maximum boost position. It is evident in the RUN that the frequency response increases in the range from 500 Hz to 4 kHz. (This response is subsequently analyzed in greater detail.) Next, when the bass control is set for maximum boost and the treble control is set for maximum cut, the output is 91 percent at 500 Hz, but is only 8 percent at 19 kHz.

The low bass response is evaluated to good advantage with the modified survey program shown in Figure 11–7. Here, the analysis starts at 50 Hz and is incremented in 50-Hz steps to 500 Hz. With the bass control set for maximum cut and the treble control set for maximum boost, the output is only 20 percent at 50 Hz, but increases to 93 percent at 300 Hz. A survey program is helpful to rapidly determine the effect of various component values and to calculate the result of various component tolerances.

AMPLIFIED (ACTIVE) TONE CONTROLS

An amplified or active tone control comprises a frequency-selective network supplemented by one or more amplifying devices. In this

```
1 LPRINT"RC BASS-TREBLE LOSSER-TYPE TONE CONTROL":PRINT"RC BASS-TREBLE LOSSER TY
PE TONE CONTROL"
2 LPRINT"* Eout/Ein, Phase *":PRINT"* Eout/Ein, Phase *"
3 LPRINT"":PRINT""
4 INPUT"f (Hz)=";F
5 LPRINT"f (Hz)=";F
6 INPUT"R1 (Ohms)=";RO
7 LPRINT"R1 (Ohms)=";RO
8 INPUT"R2 (Ohms)=";RT
9 LPRINT"R2 (Ohms)=";RT
10 INPUT"R3 (Ohms)=";RH
11 LPRINT"R3 (Ohms)=";RH
12 INPUT"R4 (Ohms)=";RF
13 LPRINT"R4 (Ohms)=";RF
14 INPUT"C1 (Mfd)=";CO
15 LPRINT"C1 (Mfd)=";CO
16 INPUT"C2 (Mfd)=";CT
17 LPRINT"C2 (Mfd)=";CT
18 LPRINT"":PRINT""
19 A=RO+RT:B=RO+RT+RH+RF:XO=1/(6.283*F*CO*10^-6): XT=1/(6.283*F*CT*10^-6)
20 AB=(B*B+XT*XT)^.5:BA=-ATN(XT/B)
21 AC=(A*A+XO*XO)^.5:CA=-ATN(XO/A)
22 AD=(RF*RF+XT*XT)^.5:DA=-ATN(XT/RF)
23 AE=AB*AC:EA=BA+CA:AF=AE/A:FA=EA
24 AG=AF*COS(FA):AH=AF*SIN(FA):AJ=AG-A
25 AK=(AJ*AJ+AH*AH)^.5:KA=-ATN(AH/AJ)
26 AN=AD/AK:NA=DA-KA
27 A$="####.##":LPRINT"Eout/Ein=";USING A$;AN:PRINT"Eout/Ein=";USING A$;AN
28 LPRINT"Phase Angle (Degrees)=";USING A$;-NA*360/6.283:PRINT"Phase Angle (Degr
ees)=";USING A$;-NA*360/6.283
```

Figure 11–5. (Continues on next page.)

```
RC BASS-TREBLE LOSSER-TYPE TONE CONTROL
* Eout/Ein, Phase *

f (Hz)= 100
R1 (Ohms)= 1500000
R2 (Ohms)= 33000
R3 (Ohms)= 50000
R4 (Ohms)= 50000
C1 (Mfd)= .02
C2 (Mfd)= .001

Eout/Ein=    0.95
Phase Angle (Degrees)= 177.61
```

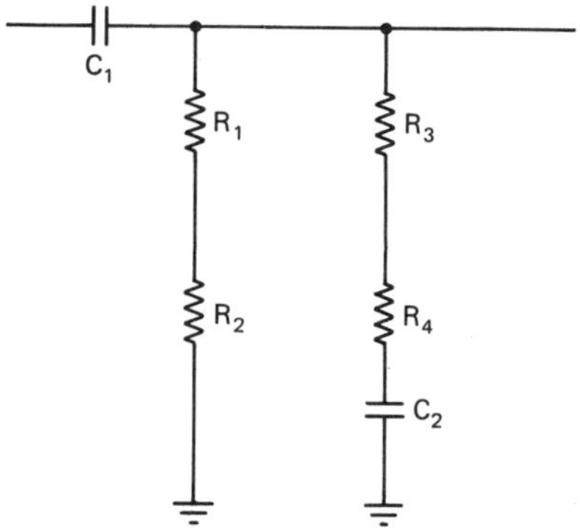

```
RC BASS-TREBLE LOSSER-TYPE TONE CONTROL
* Eout/Ein, Phase *

f (Hz)= 10000
R1 (Ohms)= 1500000
R2 (Ohms)= 33000
R3 (Ohms)= 50000
R4 (Ohms)= 50000
C1 (Mfd)= .02
C2 (Mfd)= .001

Eout/Ein=    0.52
Phase Angle (Degrees)=  27.18
```

Figure 11–5. Program for calculating frequency response of RC bass-treble losser-type tone control.

```
1 LPRINT"RC BASS-TREBLE LOSSER-TYPE TONE CONTROL":PRINT"RC BASS-TREBLE LOSSER TY
PE TONE CONTROL"
2 LPRINT"* Eout/Ein, Phase *":PRINT"* Eout/Ein, Phase *":F=0:N=0
3 LPRINT "(Survey Program)"
4 PRINT "(Survey Program)"
5 LPRINT"":PRINT""
6 INPUT"R1 (Ohms)=";RO
7 LPRINT"R1 (Ohms)=";RO
8 INPUT"R2 (Ohms)=";RT
9 LPRINT"R2 (Ohms)=";RT
10 INPUT"R3 (Ohms)=";RH
11 LPRINT"R3 (Ohms)=";RH
12 INPUT"R4 (Ohms)=";RF
13 LPRINT"R4 (Ohms)=";RF
14 INPUT"C1 (Mfd)=";CO
15 LPRINT"C1 (Mfd)=";CO
16 INPUT"C2 (Mfd)=";CT
17 LPRINT"C2 (Mfd)=";CT:F=0:G=0
18 LPRINT"":PRINT""
19 G=G+1:F=G*50:A=RO+RT:B=RO+RT+RH+RF:XO=1/(6.283*F*CO*10^-6):XT=1/(6.283*F*CT*1
0^-6)
20 AB=(B*B+XT*XT)^.5:BA=-ATN(XT/B)
21 AC=(A*A+XO*XO)^.5:CA=-ATN(XO/A)
22 AD=(RF*RF+XT*XT)^.5:DA=-ATN(XT/RF)
23 AE=AB*AC:EA=BA+CA:AF=AE/A:FA=EA
24 AG=AF*COS(FA):AH=AF*SIN(FA):AJ=AG-A
25 AK=(AJ*AJ+AH*AH)^.5:KA=-ATN(AH/AJ)
26 AN=AD/AK:NA=DA-KA
```

```
27 B$="######":LPRINT"f (Hz)=";USING B$;F:PRINT"f (Hz)=";USING B$;F
28 A$="#####.#":LPRINT"Eout/Ein=";USING A$;AN:PRINT"Eout/Ein=";USING A$;AN
29 LPRINT"Phase Angle (Degrees)=";USING A$;-NA*360/6.283:PRINT"Phase Angle (Degr
ees)=";USING A$;-NA*360/6.283:LPRINT "":PRINT ""
31 IF F>1000 THEN 33
32 GOTO 18
33 END
```

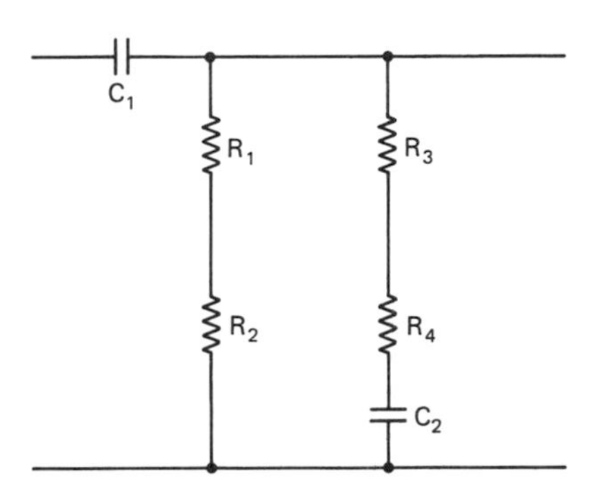


```
RC BASS-TREBLE LOSSER-TYPE TONE CONTROL
* Eout/Ein, Phase *
(Survey Program)

R1 (Ohms)= 10
R2 (Ohms)= 33000
R3 (Ohms)= 10
R4 (Ohms)= 99990
C1 (Mfd)= .02
C2 (Mfd)= .001

f (Hz)=   500
Eout/Ein=     0.9
Phase Angle (Degrees)=   -9.5

f (Hz)=  1000
Eout/Ein=     0.9
Phase Angle (Degrees)=  130.0

f (Hz)=  1500
Eout/Ein=     1.0
Phase Angle (Degrees)=  103.7

f (Hz)=  2000
Eout/Ein=     1.0
Phase Angle (Degrees)=   85.1

f (Hz)=  2500
Eout/Ein=     1.0
Phase Angle (Degrees)=   71.7

f (Hz)=  3000
Eout/Ein=     1.0
Phase Angle (Degrees)=   61.6

f (Hz)=  3500
Eout/Ein=     1.0
Phase Angle (Degrees)=   53.9

f (Hz)=  4000
Eout/Ein=     1.0
Phase Angle (Degrees)=   47.8

f (Hz)=  4500
Eout/Ein=     1.0
Phase Angle (Degrees)=   42.9

f (Hz)=  5000
Eout/Ein=     1.0
Phase Angle (Degrees)=   38.9
```

Figure 11–6. Survey program for calculating frequency response of RC bass-treble losser-type control.

```
1 LPRINT"RC BASS-TREBLE LOSSER-TYPE TONE CONTROL":PRINT"RC BASS-TREBLE LOSSER TY
PE TONE CONTROL"
2 LPRINT"* Eout/Ein, Phase *":PRINT"* Eout/Ein, Phase *":F=0:N=0
3 LPRINT "(Survey Program)"
4 PRINT "(Survey Program)"
5 LPRINT"":PRINT""
6 INPUT"R1 (Ohms)=";RO
7 LPRINT"R1 (Ohms)=";RO
8 INPUT"R2 (Ohms)=";RT
9 LPRINT"R2 (Ohms)=";RT
10 INPUT"R3 (Ohms)=";RH
11 LPRINT"R3 (Ohms)=";RH
12 INPUT"R4 (Ohms)=";RF
13 LPRINT"R4 (Ohms)=";RF
14 INPUT"C1 (Mfd)=";CO
15 LPRINT"C1 (Mfd)=";CO
16 INPUT"C2 (Mfd)=";CT
17 LPRINT"C2 (Mfd)=";CT:LPRINT"":PRINT""
18 N=N+500:F=N
19 A=RO+RT:B=RO+RT+RH+RF:XO=1/(6.283*F*CO*10^-6): XT=1/(6.283*F*CT*10^-6)
20 AB=(B*B+XT*XT)^.5:BA=-ATN(XT/B)
21 AC=(A*A+XO*XO)^.5:CA=-ATN(XO/A)
22 AD=(RF*RF+XT*XT)^.5:DA=-ATN(XT/RF)
23 AE=AB*AC:EA=BA+CA:AF=AE/A:FA=EA
24 AG=AF*COS(FA):AH=AF*SIN(FA):AJ=AG-A
25 AK=(AJ*AJ+AH*AH)^.5:KA=-ATN(AH/AJ)
26 AN=AD/AK:NA=DA-KA
27 B$="######":LPRINT"f (Hz)=";USING B$;F:PRINT"f (Hz)=";USING B$;F
28 A$="#####.#":LPRINT"Eout/Ein=";USING A$;AN:PRINT"Eout/Ein=";USING A$;AN
29 LPRINT"Phase Angle (Degrees)=";USING A$;-NA*360/6.283:PRINT"Phase Angle (Degr
ees)=";USING A$;-NA*360/6.283:LPRINT"":PRINT""
31 IF F>20000 THEN 33
32 GOTO 18
33 END
```

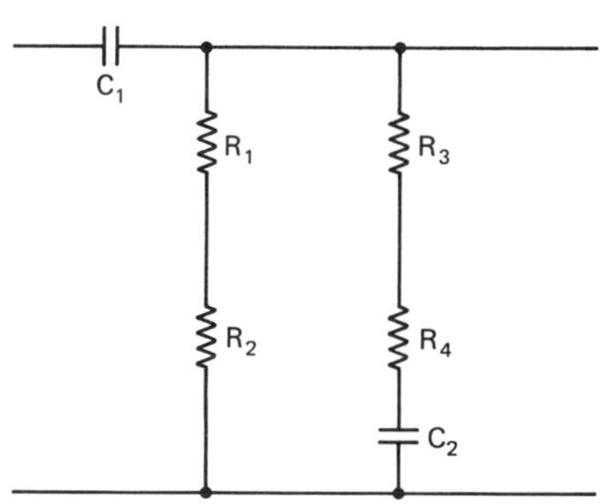

example of Figure 11–8, a bass tone control is placed between device 1 and device 2; a treble tone control follows device 2. Maximum bass boost is obtained in the Figure 11–8 network when the control is set to the top of R2. The frequency characteristic is determined to some extent by the values of source and load resistance. These are accounted for in the design equations.

```
RC BASS-TREBLE LOSSER-TYPE TONE CONTROL
* Eout/Ein, Phase *
(Survey Program)

R1 (Ohms)= 3000000
R2 (Ohms)= 33000
R3 (Ohms)= 99990
R4 (Ohms)= 10
C1 (Mfd)= .02
C2 (Mfd)= .001

f (Hz)=   500
Eout/Ein=     0.9
Phase Angle (Degrees)=  163.6

f (Hz)=  1000
Eout/Ein=     0.8
Phase Angle (Degrees)=  149.3
Phase-Frequency Value= 2604.9

f (Hz)=  1500
Eout/Ein=     0.7
Phase Angle (Degrees)=  138.2

f (Hz)=  2000
Eout/Ein=     0.6
Phase Angle (Degrees)=  130.0

f (Hz)=  2500
Eout/Ein=     0.5
Phase Angle (Degrees)=  123.8

f (Hz)=  3000
Eout/Ein=     0.5
Phase Angle (Degrees)=  119.2

f (Hz)=  3500
Eout/Ein=     0.4
Phase Angle (Degrees)=  115.6

f (Hz)=  4000
Eout/Ein=     0.4
Phase Angle (Degrees)=  112.7

f (Hz)=  4500
Eout/Ein=     0.3
Phase Angle (Degrees)=  110.4

f (Hz)=  5000
Eout/Ein=     0.3
Phase Angle (Degrees)=  108.5
```

Figure 11–7. Modification of preceding program to analyze low bass response of tone control network.

NOTE: Low bass response of tone-control network is analyzed by programming for an initial frequency of 50 Hz and incrementing in 50-Hz steps.

This filter network is analyzed to best advantage by means of a survey program, such as provided in Figure 11–9. For the exemplified component values and with the control set for maximum bass boost, it is evident in the RUN that the maximum output voltage occurs at 50 Hz and progressively decreases to a plateau value at 450 Hz. Note that the value of C1 may be independently specified to modify the boost characteristic, without affecting the cut characteristic. (The cut characteristic is modified by changing the value of C2.)

Next, the basic diagram for calculating the frequency response of the bass tone control at its maximum cut setting is shown in Figure 11–10. Observe that C1 has been short-circuited and that the

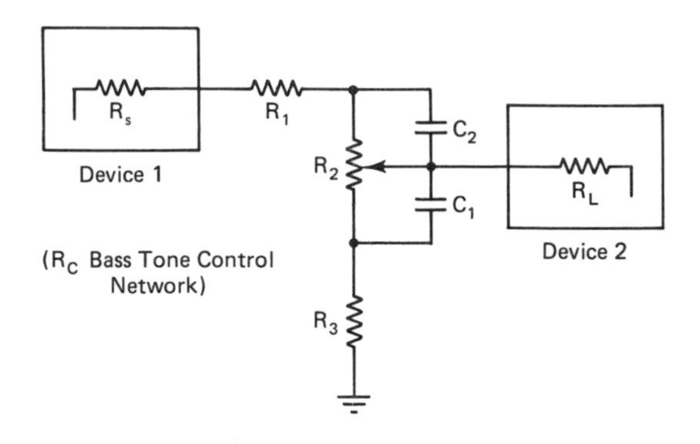

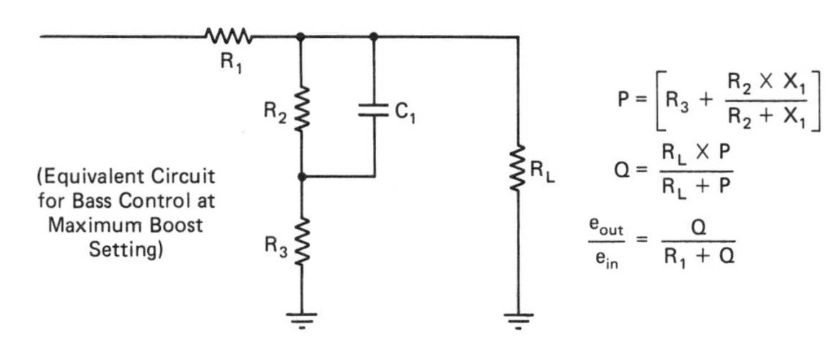

Figure 11–8. Basic diagram for analysis of the frequency response of an amplified RC bass tone control.

NOTE: If a voltage source is utilized, then $R_s = 0$. However, if a voltage source with significant internal resistance is employed, then R1 becomes the sum of its own value plus the value of R_s, insofar as the following program is concerned. The value of RL is the input resistance of device 2.

```
1 LPRINT"RC BASS-SECTION AMPLIFIED TONE CONTROL":PRINT"RC BASS-SECTION AMPLIFIED
 TONE CONTROL"
2 LPRINT "(Survey Program)": PRINT"(Survey Program)"
3 LPRINT"Eout/Ein and Phase Angle":PRINT"Eout/Ein and Phase Angle"
4 PRINT "":LPRINT""
5 INPUT "R1 (Ohms)=";RO
6 LPRINT"R1 (Ohms)=";RO
7 INPUT"R2 (Ohms)=";RT
8 LPRINT"R2 (Ohms)=";RT
9 INPUT "R3 (Ohms)=";RH
10 LPRINT"R3 (Ohms)=";RH
11 INPUT"RL (Ohms)=";RL
12 LPRINT"RL (Ohms)=";RL
13 INPUT"C1 (Mfd)=";CO
14 LPRINT"C1 (Mfd)=";CO
15 LPRINT"":PRINT"":F=0:G=0
16 G=G+1:F=G*50
17 XO=1/(6.283*F*CO*10^-6):AB=RT*XO:BA=-6.283/4:AC=(RT*RT+XO*XO)^.5
18 CA=-ATN(XO/RT):AD=AB/AC:DA=BA-CA:AE=AD*COS(DA):AF=AD*SIN(DA):AG=RH+AE
```

```
19 AH=(AG*AG+AF*AF)^.5:HA=-ATN(AF/AG):AJ=AH*COS(HA):AK=AH*SIN(HA):AL=RL+AJ
20 AM=(AL*AL+AK*AK)^.5:MA=-ATN(AK/AL):AO=RL*AH:OA=HA
21 AP=AO/AM:PA=OA-MA:AQ=(RO*RO+AP*AP)^.5:QA=-ATN(AP/RO):AN=AP/AQ:NA=PA-QA
22 PF=-F*NA:IF F>500 THEN 28
23 A$="######":LPRINT"f (Hz)=";USING A$;F:PRINT"f (Hz)=";USING A$;F
24 B$="####.##":LPRINT"Eout/Ein=";USING B$;AN:PRINT"Eout/Ein=";USING B$;AN
25 LPRINT"Phase Angle (Degrees)=";USING B$;-NA*360/6.283:PRINT"Phase Angle (Degr
ees)=";USING B$;-NA*360/6.283:LPRINT "":PRINT ""
27 GOTO 16
28 END
```

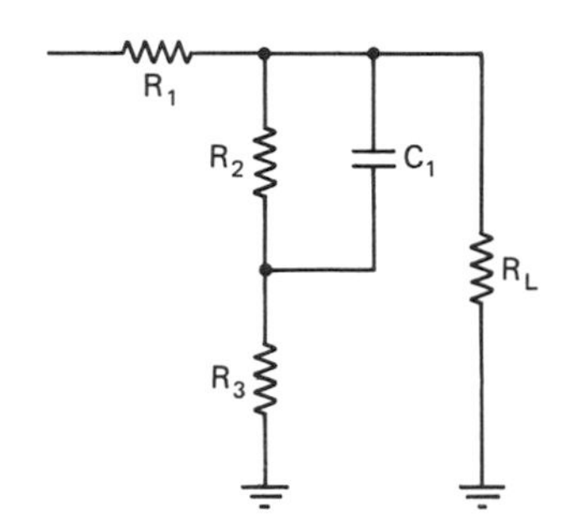


```
RC BASS-SECTION AMPLIFIED TONE CONTROL
(Survey Program)
Eout/Ein and Phase Angle

R1 (Ohms)= 75000
R2 (Ohms)= 200000
R3 (Ohms)= 35000
RL (Ohms)= 100000
C1 (Mfd)= .022

f (Hz)=    50
Eout/Ein=   0.64
Phase Angle (Degrees)=-107.32

f (Hz)=   100
Eout/Ein=   0.56
Phase Angle (Degrees)=-103.81

f (Hz)=   150
Eout/Ein=   0.49
Phase Angle (Degrees)= -91.53

f (Hz)=   200
Eout/Ein=   0.44
Phase Angle (Degrees)= -80.58

f (Hz)=   250
Eout/Ein=   0.41
Phase Angle (Degrees)= -71.86

f (Hz)=   300
Eout/Ein=   0.39
Phase Angle (Degrees)= -64.98

f (Hz)=   350
Eout/Ein=   0.38
Phase Angle (Degrees)= -59.52

f (Hz)=   400
Eout/Ein=   0.37
Phase Angle (Degrees)= -55.11

f (Hz)=   450
Eout/Ein=   0.36
Phase Angle (Degrees)= -51.51

f (Hz)=   500
Eout/Ein=   0.35
Phase Angle (Degrees)= -48.53
```

Figure 11–9. Survey program for calculating the frequency response of an amplified RC bass tone control at its maximum boost setting.

filter characteristic is now determined by the value of C2. A survey program for evaluation of the cut characteristic is provided in Figure 11–11. For the specified component values, it is seen in the RUN that the output voltage is only 9 percent of maximum at 50 Hz, but rises to 23 percent of maximum at 450 Hz. At this point, the frequency characteristic is entering a plateau region.

As previously noted, the bass tone-control stage in this example is followed by a treble tone-control stage with the arrangement shown in Figure 11–12. This filter network is similar to the bass network, with the exception that the positions of the control capacitors and resistors have been interchanged. As before, the provision of two capacitors in the configuration permits the designer to modify the boost and cut characteristics independently.

EQUALIZER FOR HIGH-FREQUENCY CUT

Most equalizers are designed to compensate for an inappropriate amplitude-frequency characteristic. For example, a simple high-frequency cut equalizer for a record player is shown in Figure 11–13. It provides a reasonably uniform amplitude-frequency characteristic for the record-player system over the audio-frequency range. A program for calculation of the frequency response for this equalizer configuration with the specified component values is provided in Figure 11–14. As seen in the RUN, the output at 10 Hz is 58 percent of maximum. The maximum output voltage occurs at zero frequency, or dc. At 100 Hz, the output has declined to 14 percent of maximum, and at 1 kHz the output is 2 percent of maximum.

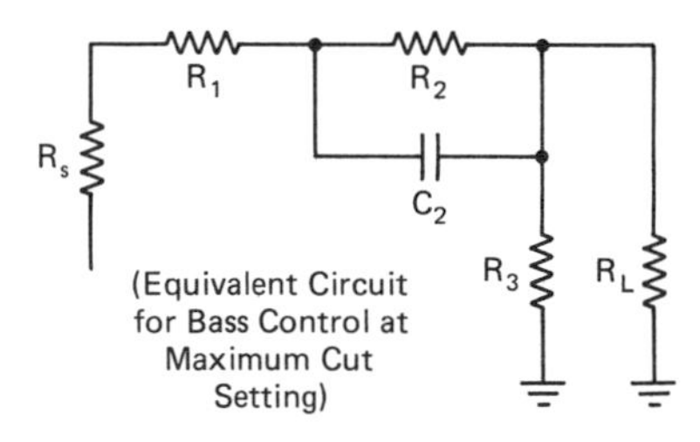

Figure 11–10. Basic diagram for calculating the frequency response of an amplified RC bass tone control (bass cut setting).

NOTE: If a voltage source is used, then $R_s = 0$. However, if a voltage source with significant internal resistance is employed, then R1 becomes the sum of its own value plus the value of R_s, insofar as the following program is concerned. The value of RL is the input resistance of the device following the bass control.

```
1 LPRINT"RC BASS-SECTION AMPLIFIED TONE CONTROL":PRINT"RC BASS-SECTION AMPLIFIED
 TONE CONTROL"
2 LPRINT "(Survey Program)": PRINT"(Survey Program)"
3 LPRINT"Eout/Ein and Phase Angle":PRINT"Eout/Ein and Phase Angle"
4 PRINT "(Response at Maximum Cut Setting)":LPRINT"Response at Maximum Cut Setti
ng)":LPRINT"":PRINT""
5 INPUT "R1 (Ohms)=";RO
6 LPRINT"R1 (Ohms)=";RO
7 INPUT"R2 (Ohms)=";RT
8 LPRINT"R2 (Ohms)=";RT
9 INPUT "R3 (Ohms)=";RH
10 LPRINT"R3 (Ohms)=";RH
11 INPUT"RL (Ohms)=";RL
12 LPRINT"RL (Ohms)=";RL
13 INPUT"C2 (Mfd)=";CT
14 LPRINT"C2 (Mfd)=";CT
15 LPRINT"":PRINT"":F=0:G=0
16 G=G+1:F=G*50
17 XT=1/(6.283*F*CT*10^-6):AB=RT*XT:BA=-6.283/4:AC=(RT*RT+XT*XT)^.5
18 CA=-ATN(XT/RT):AD=AB/AC:DA=BA-CA:AE=AD*COS(DA):AF=AD*SIN(DA):AG=RO+AE
19 AH=RH*RL:AJ=RH+RL:AK=AH/AJ:AL=AG+AK:AM=(AL*AL+AF*AF)^.5:MA=-ATN(AF/AL)
20 AN=AK/AM:NA=-MA
21 PF=-F*NA
22 IF F>500 THEN 28
23 A$="######":LPRINT"f (Hz)=";USING A$;F:PRINT"f (Hz)=";USING A$;F
24 B$="####.##":LPRINT"Eout/Ein=";USING B$;AN:PRINT"Eout/Ein=";USING B$;AN
25 LPRINT"Phase Angle (Degrees)=";USING B$;-NA*360/6.283:PRINT"Phase Angle (Degr
ees)=";USING B$;-NA*360/6.283:LPRINT"":PRINT""
27 GOTO 16
28 END
```

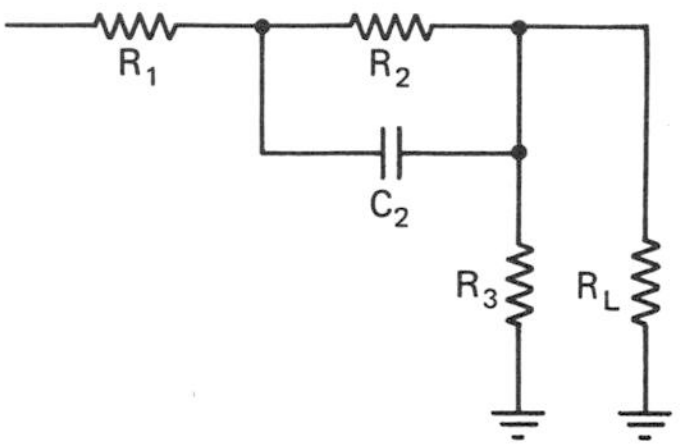

```
RC BASS-SECTION AMPLIFIED TONE CONTROL
(Survey Program)
Eout/Ein and Phase Angle
Response at Maximum Cut Setting)

R1 (Ohms)= 75000
R2 (Ohms)= 200000
R3 (Ohms)= 35000
RL (Ohms)= 100000
C2 (Mfd)= .01

f (Hz)=     50
Eout/Ein=    0.10
Phase Angle (Degrees)=   20.24

f (Hz)=    100
Eout/Ein=    0.13
Phase Angle (Degrees)=   28.63

f (Hz)=    150
Eout/Ein=    0.16
Phase Angle (Degrees)=   29.75

f (Hz)=    200
Eout/Ein=    0.18
Phase Angle (Degrees)=   28.18

f (Hz)=    250
Eout/Ein=    0.20
Phase Angle (Degrees)=   25.85

f (Hz)=    300
Eout/Ein=    0.21
Phase Angle (Degrees)=   23.49

f (Hz)=    350
Eout/Ein=    0.22
Phase Angle (Degrees)=   21.33

f (Hz)=    400
Eout/Ein=    0.23
Phase Angle (Degrees)=   19.42

f (Hz)=    450
Eout/Ein=    0.23
Phase Angle (Degrees)=   17.77

f (Hz)=    500
Eout/Ein=    0.23
Phase Angle (Degrees)=   16.34
```

Figure 11–11. Survey program for calculating the frequency response of an amplified RC bass tone control at its maximum cut setting.

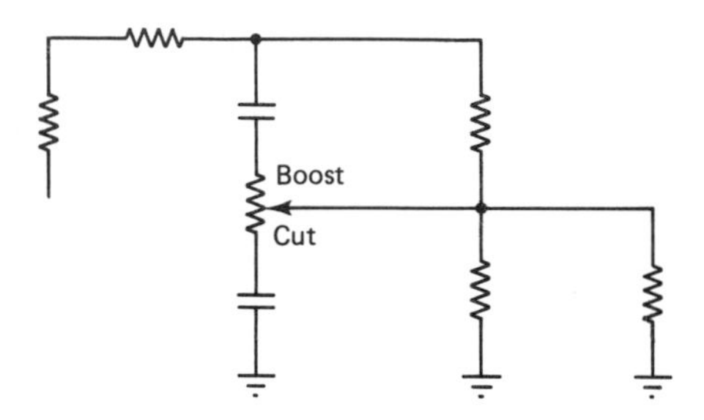

Figure 11–12. Basic diagram for analysis of the frequency response of an amplified RC treble tone control.

NOTE: This RC filter configuration is analogous to the arrangement depicted in Figure 11–8 and can be programmed for analysis in the same general manner.

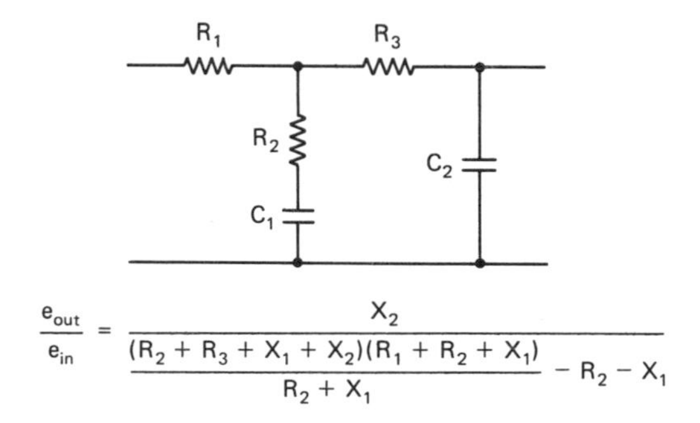

$$\frac{e_{out}}{e_{in}} = \frac{X_2}{\dfrac{(R_2 + R_3 + X_1 + X_2)(R_1 + R_2 + X_1)}{R_2 + X_1} - R_2 - X_1}$$

Figure 11–13. Basic diagram for analysis of frequency response for a high-frequency cut equalizer.

NOTE: This equalizing configuration provides an output/input ratio that starts at 100 percent at zero frequency (dc) and progressively falls to a very small value at 20 kHz. More elaborate filters provide a reduced rate of change in the 1-kHz region. Another elaboration provides rapid attenuation of output below 1 Hz.

LOW-PASS FILTER WITH LOW-FREQUENCY CUTOFF

A bandpass filter comprising a differentiating circuit followed by an integrating circuit was shown in Figure 11–1. Another arrangement that finds useful application consists of an integrating circuit followed by a differentiating circuit. As an illustration, a vertical-sync-integrating circuit in a TV receiver may be followed by a differentiating circuit with a time-constant of approximately 300 μs, which limits the low-frequency response of the network to 50 Hz. In turn, 60-Hz sync pulses are passed, but noise pulses with frequencies below 60 Hz are rejected.

```
1 LPRINT"RC HIGH-FREQUENCY CUT EQUALIZER NETWORK":PRINT"RC HIGH-FREQUENCY CUT EQ
UALIZER NETWORK"
2 LPRINT"Eout/Ein and Phase Angle":PRINT"Eout/Ein and Phase Angle"
3 LPRINT"":PRINT""
4 INPUT "C1 (Mfd)=";CO
5 LPRINT"C1 (Mfd)=";CO
6 INPUT"R1 (Ohms)=";RO
7 LPRINT"R1 (Ohms)=";RO
8 INPUT "R2 (Ohms)=";RT
9 LPRINT"R2 (Ohms)";RT
10 INPUT"C2 (Mfd)=";CT
11 LPRINT"C2 (Mfd)=";CT
12 INPUT "R3 (Ohms)";RH
13 LPRINT"R3 (Ohms)";RH
14 INPUT"f (Hz)=";F
15 LPRINT"f (Hz)=";F
16 XO=1/(6.283*F*CO*10^-6):XT=1/(6.283*F*CT*10^-6):AB=RT+RH:AC=XO+XT:CA=-6.283/4

17 AD=RO+RT:AE=(AB*AB+AC*AC)^.5:EA=-ATN(AC/AB):AF=(AD*AD+XO*XO)^.5:FA=-ATN(XO/AD
)
18 AG=(RT*RT+XO*XO)^.5:GA=-ATN(XO/RT):AH=AE*AF:HA=EA+FA
19 AJ=AH/AG:JA=HA-GA:AK=AJ*COS(JA):AL=AJ*SIN(JA)
20 AM=AJ-RT:AO=AL-XO:AP=(AM*AM+AO*AO)^.5:PA=-ATN(AO/AM):AN=XT/AP:NA=-PA-6.283/4
21 PF=-F*NA:LPRINT"":PRINT""
22 B$="####.##":LPRINT"Eout/Ein=";USING B$;AN:PRINT"Eout/Ein=";USING B$;AN
23 LPRINT"Phase Angle (Degrees)=";USING B$;-NA*360/6.283:PRINT"Phase Angle (Degr
ees)=";USING B$;-NA*360/6.283
```

```
RC HIGH-FREQUENCY CUT EQUALIZER NETWORK
Eout/Ein and Phase Angle
                         ---
C1 (Mfd)= .03
R1 (Ohms)= 330000
R2 (Ohms) 7500
C2 (Mfd)= .002
R3 (Ohms) 27000
f (Hz)= 1000

Eout/Ein=    0.02
Phase Angle (Degrees)= 119.02
```

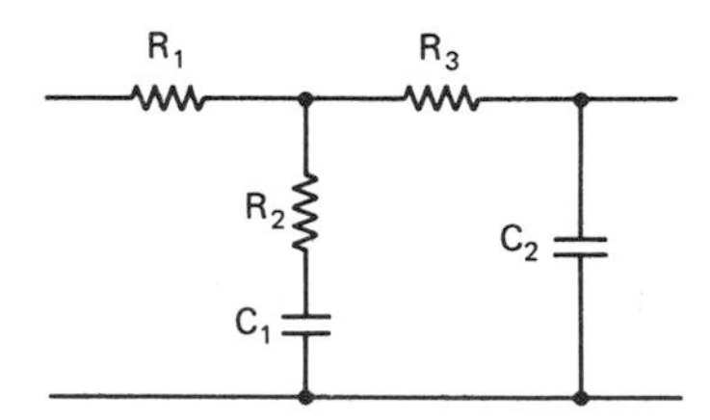

Figure 11–14. Program for calculation of the frequency response for a high-frequency cut equalizer.

BANDPASS FILTER WITH RESISTIVE LOAD

A basic RC bandpass filter configuration was shown in Figure 11–1. If this network is followed by a MOSFET transistor, it can provide the frequency characteristic exemplified in the associated program.

However, if the network is followed by a bipolar transistor, for example, its frequency characteristic will be modified by the load represented by the input resistance of the transistor. In turn, it is helpful for the designer to have an appropriate computer program available, such as shown in Figures 11–15 and 11–16.

The formula given in Figure 11–15 is based on substitution of an equivalent series circuit for the parallel combination of C2 and RL. This substitution is permissible for a single frequency of operation, such as processed in the following program. The advantage of employing this equivalent series circuit is that the original configuration comprises three meshes, whereas the modified network comprises only two meshes. Since it is easier to manipulate two simultaneous equations instead of three, derivation of the programming formula is thereby simplified.

An alternative procedure for avoidance of three simultaneous equations in Figure 11–15 is to regard the impedance of the parallel combination of C2 and R2 as the load, instead of regarding RL alone as the load. In turn, a third mesh equation is not required, and the numerator of the programming formula is expressed in polar form. Observe that to break down the programming formula line by line

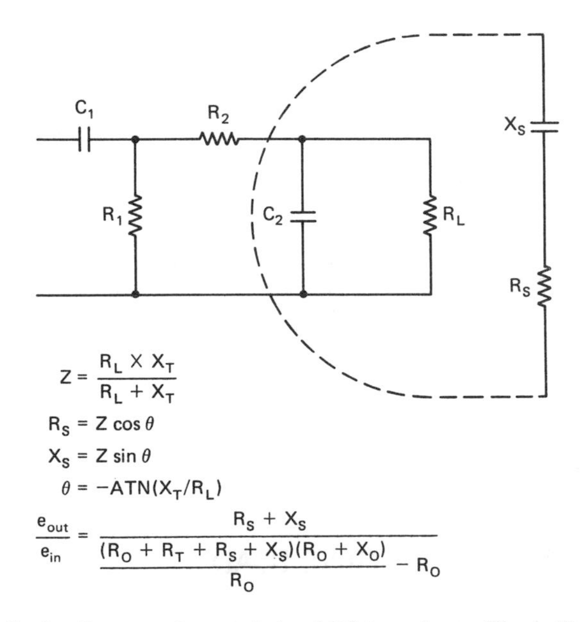

Figure 11–15. Basic diagram for analysis of RC-bandpass filter with resistive load.

NOTE: RS and XS denote the equivalent series resistance and series reactance that correspond (at a single frequency) to shunt resistance RL and shunt reactance X2.

```
1 LPRINT "RC BANDPASS FILTER WITH RESISTIVE LOAD"
2 PRINT "RC BANDPASS FILTER WITH RESISTIVE LOAD"
3 LPRINT "Computes Eout/Ein and Phase":PRINT "Computes Eout/Ein and Phase"
4 LPRINT"":PRINT"":INPUT "C1 (Mfd)=";CO
5 LPRINT "C1 (Mfd)=";CO
6 INPUT "R1 (Ohms)=";RO
7 LPRINT "R1 (Ohms)=";RO
8 INPUT "C2 (Mfd)=";CT
9 LPRINT "C2 (Mfd)=";CT
10 INPUT "R2 (Ohms)=";RT
11 LPRINT "R2 (Ohms)=";RT
12 INPUT "RL (Ohms)=";RL
13 LPRINT "RL (Ohms)=";RL
14 INPUT "f (Hz)=";F
15 LPRINT "f (Hz)=";F:LPRINT"":PRINT"":XO=1/(6.2832*F*CO*10^-6)
16 XT=1/(6.2832*F*CT*10^-6):WX=RL*XT:XW=-6.2832/4:VW=(RL^2+XT^2)^.5
17 WV=-ATN(XT/RL):UV=WX/VW:VU=XW-WV:RS=UV*COS(VU):XS=-UV*SIN(VU)
18 AB=RO+RT+RS:AC=(RS^2+XS^2)^.5:CA=-ATN(XS/RS):AD=(AB^2+XS^2)^.5
19 DA=-ATN(XS/AB):AE=(RO^2+XO^2)^.5:EA=-ATN(XO/RO):AF=AD*AE:FA=DA+EA:AG=AF/RO
20 GA=FA:AH=AG*COS(GA):AJ=AG*SIN(GA):AK=AH-RO:LPRINT"":PRINT""
21 AL=(AK^2+AJ^2)^.5:LA=-ATN(AJ/AK):AN=AC/AL:NA=CA-LA:B$="####.##"
22 LPRINT "Eout/Ein=";USING B$;AN:PRINT "Eout/Ein=";USING B$;AN
23 LPRINT "Phase Angle (Degrees)=";USING B$;-NA*360/6.2832
24 PRINT "Phase Angle (Degrees)=";USING B$;-NA*360/6.2832
25 END
```

```
RC BANDPASS FILTER WITH RESISTIVE LOAD
Computes Eout/Ein and Phase

C1 (Mfd)= .01
R1 (Ohms)= 20000
C2 (Mfd)= .0004
R2 (Ohms)= 150000
RL (Ohms)= 5000000
f (Hz)= 1000


Eout/Ein=    0.69
Phase Angle (Degrees)=    8.93
```

```
RC BANDPASS FILTER WITH RESISTIVE LOAD
Computes Eout/Ein and Phase

C1 (Mfd)= .01
R1 (Ohms)= 20000
C2 (Mfd)= .0004
R2 (Ohms)= 150000
RL (Ohms)= 500000
f (Hz)= 1000


Eout/Ein=    0.56
Phase Angle (Degrees)= 125.78
```

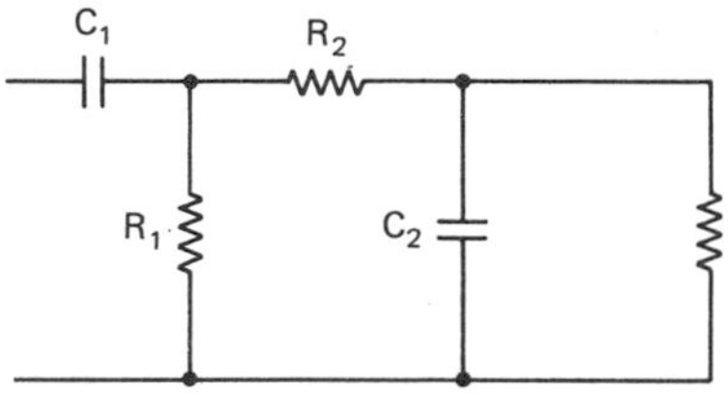

C1 R2 R1 C2 RL

Figure 11–16. Program for calculating the output/input voltage and phase for an RC-bandpass filter with resistive load.

for processing the denominator, the equivalent series resistance and series reactance that correspond to the load impedance at a single frequency must be computed.

Thus, either of the foregoing approaches requires that the equivalent series resistance and series reactance corresponding to the load impedance be calculated. The only distinction is whether this calculation is expressed or whether it is implied in the programming formula. Observe also that a modified diagram based on Thevenin's theorem may be employed. Stated otherwise, the unloaded output voltage of the network and its output resistance can be formulated in accordance with Thevenin's theorem to compute the current through the load resistor.

With reference to Figure 11–16, it is seen that as the load resistance is reduced in value from 5 megohms to 50 kilohms, the output/input voltage decreases from 0.69 to 0.18. This progression is in respect to an operating frequency of 1 kHz and to the specified component values. Note that the effect of network loading is most apparent at low frequencies inasmuch as the reactance of C2 becomes comparatively small at high frequencies and tends to mask changes in load-resistance value.

Note in passing that still another approach to formulation of the network in Figure 11–15 entails assignment of Kirchhoff branch currents with a summation equation for the voltage drops around the periphery of the network. One of the branch currents can be eliminated at the outset by substituting an equivalent series circuit for the parallel combination of C2 and RL. The particular approach that is adopted depends upon the designer's personal preference.

An RC-filter arrangement of considerable importance in various applications is shown in Figure 11–17. This is an RC-frequency-compensation network that works into a capacitive load. It is used as an equalizer between cascaded stages to obtain an overall uniform frequency response, for example. This frequency compensation network is employed in frequency-compensated attenuator and multiplier networks, as in oscilloscope and electronic-voltmeter input systems. When the time constant of the compensation network is made the same as the time constant of the load, a constant voltage division is obtained at any frequency. Note also that when the time constants are equal, both frequency and phase compensation are obtained. In other words, a complex waveform input voltage will be attenuated without any waveshape distortion. (See also Figure 11–18.)

Instructive examples of frequency-compensation network operation with RC sections that have equal time constants and that have unequal time constants are shown in Figure 11–19. Observe that

```
1 LPRINT"RC FREQUENCY COMPENSATION NETWORK WITH CAPACITIVE LOAD":PRINT"RC FREQUE
NCY COMPENSATION NETWORK WITH CAPACITIVE LOAD"
2 LPRINT "(Eout/Ein and Phase Angle)":PRINT"(Eout/Ein and Phase Angle)"
3 LPRINT"":PRINT""
4 INPUT "C1 (Mfd)=";CO
5 LPRINT"C1 (Mfd)=";CO
6 INPUT"R1 (Ohms)=";RO
7 LPRINT"R1 (Ohms)=";RO
8 INPUT "R2 (Ohms)=";RT
9 LPRINT"R2 (Ohms)";RT
10 INPUT"C2 (Mfd)=";CT
11 LPRINT"C2 (Mfd)=";CT
12 INPUT"f (Hz)=";F
13 LPRINT"f (Hz)=";F
16 XO=1/(6.283*F*CO*10^-6):XT=1/(6.283*F*CT*10^-6):AB=RO*XO:BA=-6.283/4
17 AC=(RO*RO+XO*XO)^.5:CA=-ATN(XO/RO):AD=AB/AC:DA=BA-CA
18 AE=RT*XT:EA=-6.283/4:AF=(RT*RT+XT*XT)^.5:FA=-ATN(XT/RT)
19 AG=AE/AF:GA=EA-FA:AH=AD*COS(DA):AJ=AD*SIN(DA)
20 AK=AG*COS(GA):AL=AG*SIN(GA):AM=AH+AK:AN=AJ+AL
21 AO=(AM*AM+AN*AN)^.5:OA=-ATN(AN/AM)
22 AP=AD/AO:PA=DA+OA:PF=-F*PA:LPRINT"":PRINT""
23 B$="####.##":LPRINT"Eout/Ein=";USING B$;AP:PRINT"Eout/Ein=";USING B$;AP
24 LPRINT"Phase Angle (Degrees)=";USING B$; PA*360/6.283:PRINT"Phase Angle (Degr
ees)=";USING B$; PA*360/6.283
```

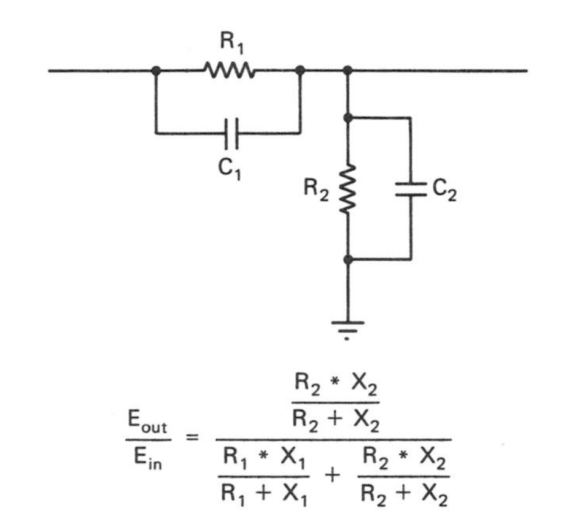

$$\frac{E_{out}}{E_{in}} = \frac{\dfrac{R_2 * X_2}{R_2 + X_2}}{\dfrac{R_1 * X_1}{R_1 + X_1} + \dfrac{R_2 * X_2}{R_2 + X_2}}$$

```
RC FREQUENCY COMPENSATION NETWORK WITH CAPACITIVE LOAD
(Eout/Ein and Phase Angle)

C1 (Mfd)= .16
R1 (Ohms)= 1000
R2 (Ohms) 2000
C2 (Mfd)= .01
f (Hz)= 1000

Eout/Ein=    0.27
Phase Angle (Degrees)= -28.29
```

Figure 11–17. Program for calculating the response of an RC-frequency compensation network with capacitive load.

```
1 LPRINT"RC FREQUENCY COMPENSATION NETWORK WITH CAPACITIVE LOAD AND FINITE SOURC
E IMPEDANCE":PRINT"RC FREQUENCY COMPENSATION NETWORK WITH CAPACITIVE LOAD AND FI
NITE SOURCE IMPEDANCE":LPRINT"":PRINT""
2 LPRINT "(Eout/Ein and Phase Angle)":PRINT"(Eout/Ein and Phase Angle)"
3 LPRINT"":PRINT""
4 INPUT"Rs (Ohms)=";RS
5 LPRINT"Rs (Ohms)=";RS
6 INPUT "Cs (Mfd)=";CS
7 LPRINT"Cs (Mfd)=";CS
8 INPUT "R1 (Ohms)=";RO
9 LPRINT "R1 (Ohms)=";RO
10 INPUT "C1 (Mfd)=";CO
11 LPRINT "C1 (Mfd)";CO
12 INPUT "R2 (Ohms)=";RT
13 LPRINT "R2 (Ohms)=";RT
14 INPUT "C2 (Mfd)=";CT
15 LPRINT "C2 (Mfd)=";CT
16 INPUT "f (Hz)=";F
17 LPRINT "f (HZ)=";F
18 XO=1/(6.283*F*CO*10^-6):XT=1/(6.283*F*CT*10^-6):XS=1/(6.283*F*CS*10^-6)
19 AB=RT*XT:BA=-6.283/4:AC=(RT*RT+XT*XT)^.5:CA=-ATN(XT/RT)
20 AD=RO*XO:DA=-6.283/4:AE=(RO*RO+XO*XO)^.5:EA=-ATN(XO/RO)
21 AF=RS*XS:FA=-6.283/4:AG=(RS*RS+XS*XS)^.5:GA=-ATN(XS/RS)
22 AH=AB/AC:HA=BA-CA:AJ=AD/AE:JA=DA-EA
23 AK=AF/AG:KA=FA-GA:AL=AJ*COS(JA):AM=AJ*SIN(JA)
24 AN=AH*COS(HA):AO=AH*SIN(HA):AP=AK*COS(KA):AQ=AK*SIN(KA)
25 AR=AL+AN+AP:AS=AM+AO+AQ:AT=(AR*AR+AS*AS)^.5:TA=-ATN(AS/AR)
26 AU=AH/AT:UA=HA+TA:PF=-F*UA:LPRINT"":PRINT""
27 A$="####.##":LPRINT"Eout/Ein=";USING A$;AU:PRINT"Eout/Ein=";USING A$;AU
28 LPRINT"Phase Angle (Degrees)=";USING A$;-UA*360/6.283:PRINT"Phase Angle (Degr
ees)=";USING A$;-UA*360/6.283
```

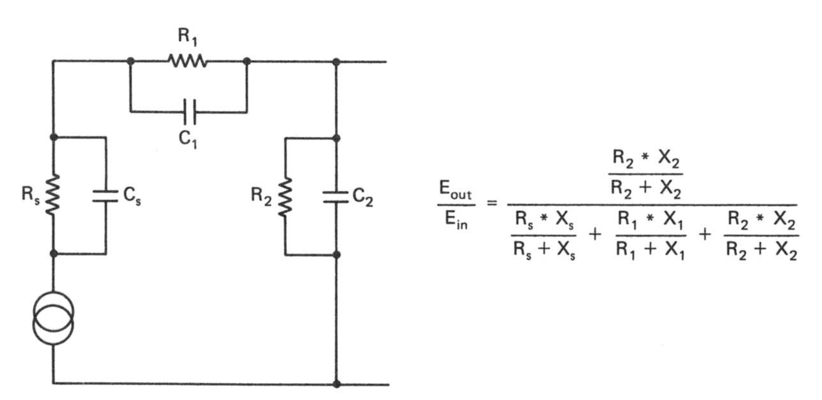

$$\frac{E_{out}}{E_{in}} = \frac{\dfrac{R_2 * X_2}{R_2 + X_2}}{\dfrac{R_s * X_s}{R_s + X_s} + \dfrac{R_1 * X_1}{R_1 + X_1} + \dfrac{R_2 * X_2}{R_2 + X_2}}$$

```
RC FREQUENCY COMPENSATION NETWORK WITH CAPACITIVE LOAD AND FINITE SOURCE IMPEDAN
CE

(Eout/Ein and Phase Angle)

Rs (Ohms)= 500
Cs (Mfd)= .001
R1 (Ohms)= 1000
C1 (Mfd) .1
R2 (Ohms)= 2500
C2 (Mfd)= .01
f (HZ)= 1000

Eout/Ein=    0.66
Phase Angle (Degrees)=  -3.94
```

Figure 11–18. Program for calculating the response of an RC-frequency compensation network with capacitive load and finite source impedance.

when the RC sections have equal time constants, the E_{out}/E_{in} ratio is the same at all frequencies and that the phase-frequency product is the same (zero) at all frequencies. In other words, the network has no filter action, and voltage division is independent of frequency. Also, there is zero-phase shift at any frequency of operation. Thus, voltage division is analogous to that of a resistive voltage divider. However, the analogy is limited to the extent that the input

```
1 LPRINT"RC FREQUENCY COMPENSATION NETWORK WITH CAPACITIVE LOAD AND FINITE SOURC
E IMPEDANCE":PRINT"RC FREQUENCY COMPENSATION NETWORK WITH CAPACITIVE LOAD AND FI
NITE SOURCE IMPEDANCE":LPRINT"":PRINT""
2 LPRINT "(Eout/Ein and Phase Angle)":PRINT"(Eout/Ein and Phase Angle)":LPRINT""
:PRINT""
3 PRINT"(Survey Program)":LPRINT"(Survey Program)"
4 LPRINT"":PRINT""
5 INPUT "Rs(Ohms)=";RS
6 LPRINT"Rs (Ohms)=";RS
7 INPUT "Cs (Mfd)=";CS
8 LPRINT"Cs (Mfd)=";CS
9 INPUT "R1 (Ohms)=";RO
10 LPRINT"R1 (Ohms)=";RO
11 INPUT"C1 (Mfd)=";CO
12 LPRINT"C1 (Mfd)=";CO
13 INPUT"R2 (Ohms)=";RT
14 LPRINT"R2 (Ohms)=";RT
15 INPUT"C2 (Mfd)=";CT
16 LPRINT"C2 (Mfd)=";CT
17 LPRINT""
18 PRINT""
19 G=0:F=0
20 G=G+1:F=G*500
21 XO=1/(6.283*F*CO*10^-6):XT=1/(6.283*F*CT*10^-6):XS=1/(6.283*F*CS*10^-6)
22 AB=RT*XT:BA=-6.283/4:AC=(RT*RT+XT*XT)^.5:CA=-ATN(XT/RT)
23 AD=RO*XO:DA=-6.283/4:AE=(RO*RO+XO*XO)^.5:EA=-ATN(XO/RO)
24 AF=RS*XS:FA=-6.283/4:AG=(RS*RS+XS*XS)^.5:GA=-ATN(XS/RS)
25 AH=AB/AC:HA=BA-CA:AJ=AD/AE:JA=DA-EA
26 AK=AF/AG:KA=FA-GA:AL=AJ*COS(JA):AM=AJ*SIN(JA)
27 AN=AH*COS(HA):AO=AH*SIN(HA):AP=AK*COS(KA):AQ=AK*SIN(KA)
28 AR=AL+AN+AP:AS=AM+AO+AQ:AT=(AR*AR+AS*AS)^.5:TA=-ATN(AS/AR)
29 AU=AH/AT:UA=HA+TA:LPRINT"":PRINT""
30 LPRINT"f (Hz)=";F:PRINT"f (Hz)=";F:IF F >10000 THEN 34
31 A$="####.##":LPRINT"Eout/Ein=";USING A$;AU:PRINT"Eout/Ein=";USING A$;AU
32 LPRINT"Phase Angle (Degrees)=";USING A$;-UA*360/6.283:PRINT"Phase Angle (Degr
ees)=";USING A$;-UA*360/6.283
33 GOTO 20
34 END
```

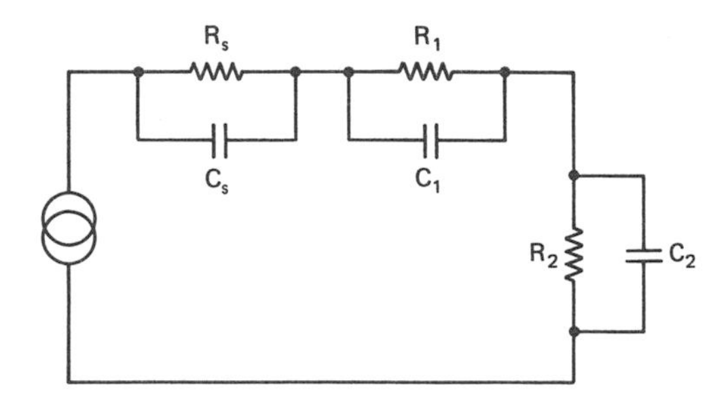

Figure 11–19. (Continues on next page.)

```
RC FREQUENCY COMPENSATION NETWORK WITH CAPACITIVE LOAD AND FINITE SOURCE IMPEDAN
CE

(Eout/Ein and Phase Angle)

(Survey Program)

Rs (Ohms)= 500
Cs (Mfd)= .001
R1 (Ohms)= 1000
C1 (Mfd)= .1
R2 (Ohms)= 2500
C2 (Mfd)= .01


f (Hz)= 500
Eout/Ein=   0.64
Phase Angle (Degrees)=  -2.56

f (Hz)= 1000
Eout/Ein=   0.66
Phase Angle (Degrees)=  -3.94

f (Hz)= 1500
Eout/Ein=   0.68
Phase Angle (Degrees)=  -4.06

f (Hz)= 2000
Eout/Ein=   0.70
Phase Angle (Degrees)=  -3.42

f (Hz)= 2500
Eout/Ein=   0.72
Phase Angle (Degrees)=  -2.43
```

```
RC FREQUENCY COMPENSATION NETWORK WITH CAPACITIVE LOAD AND FINITE SOURCE IMPEDAN
CE

(Eout/Ein and Phase Angle)

(Survey Program)

Rs (Ohms)= 500
Cs (Mfd)= .001
R1 (Ohms)= 1000
C1 (Mfd)= .0005
R2 (Ohms)= 2000
C2 (Mfd)= .00025


f (Hz)= 500
Eout/Ein=   0.57
Phase Angle (Degrees)=   0.00

f (Hz)= 1000
Eout/Ein=   0.57
Phase Angle (Degrees)=   0.00

f (Hz)= 1500
Eout/Ein=   0.57
Phase Angle (Degrees)=   0.00

f (Hz)= 2000
Eout/Ein=   0.57
Phase Angle (Degrees)=   0.00

f (Hz)= 2500
Eout/Ein=   0.57
Phase Angle (Degrees)=   0.00

f (Hz)= 3000
Eout/Ein=   0.57
Phase Angle (Degrees)=  -0.00
```

Figure 11–19. Examples of frequency-compensation network operation with equal and with unequal time constants.

impedance of the frequency-compensation network is not constant and decreases with increasing frequency. This type of network finds extensive application in instrument attenuators, for example.

Optimization of filter networks, particularly with respect to tolerances, is generally facilitated by employing chained geometrical models. Two models should be derived: one for the low-frequency limit of the filter and the other for the high-frequency limit. In other words, a component will frequently require a much tighter tolerance for acceptable performance at the high-frequency end of the band than at the low-frequency end.

12

PRINCIPLES OF TRANSIENT CIRCUIT DESIGN

TRANSIENT RESPONSE VERSUS STEADY-STATE RESPONSE

Steady-state response denotes the output/input relation of a network with respect to a single-frequency (sine-wave) source. Note that a transient response occurs at the time that the sine-wave voltage is applied to the network. However, the network typically proceeds through a short settling period and its output voltage then becomes constant. Whenever a switch is turned on or off in a circuit, a nonsinusoidal waveform is usually produced. This nonsinusoidal waveform is either an exponential or it is a modified exponential waveform. (See Chart 12–1.)

Note that the steady-state response of a network may be described with respect to more than one input frequency. For example, the steady-state response of an amplifier may be described with respect to a fundamental, a third harmonic, and a fifth harmonic frequency, which are simultaneously applied to the input of the amplifier. However, this is merely a generalization of the basic principle that states that a steady-state response is the output/input relation of the network with respect to a sustained single-frequency input.

CHART 12–1

Basic Principles of Transient Versus Steady-State Response

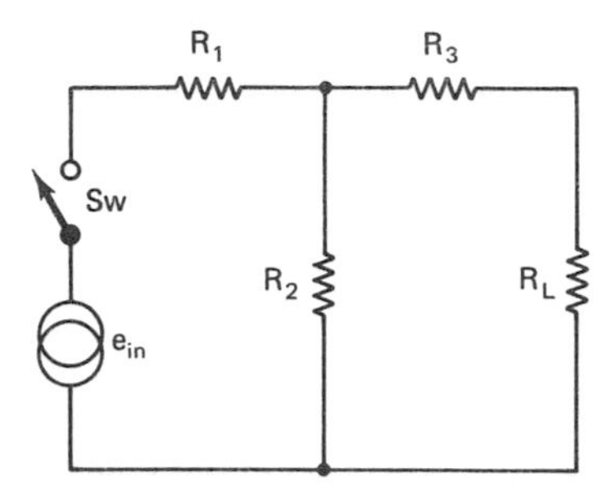

Transient and steady-state responses are the same. When the generator switch is closed, the waveform at the output of the resistive network is the same as the waveform at the input of the network.

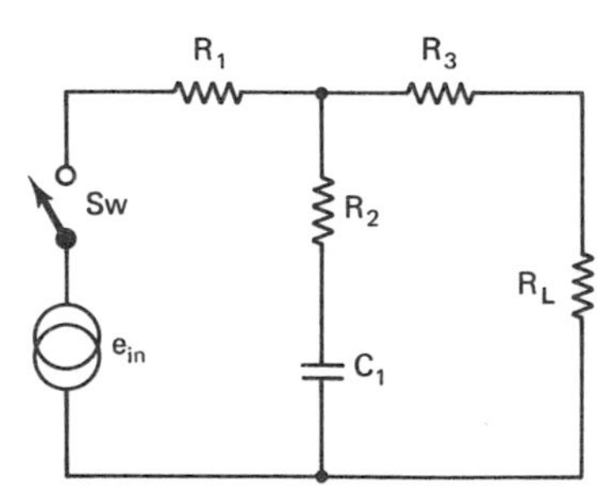

As a general rule, the transient and steady-state responses of this network are not the same. When the generator switch is closed, the waveform at the output of the reactive network usually comprises a decaying exponential waveform with the generator waveform is superimposed. After a brief settling period, the waveform at the output becomes the same as the waveform at the input of the network.

There is a special condition under which the transient and the steady-state responses of the reactive network are the same. If the switch is closed at the instant that the generator phase corresponds to the passage of the capacitor voltage through zero in the steady state, no transient exponential waveform will be produced, and the waveform at the output will be the same as the waveform at the input of the reactive network.

CHART 12–1 (Continued)

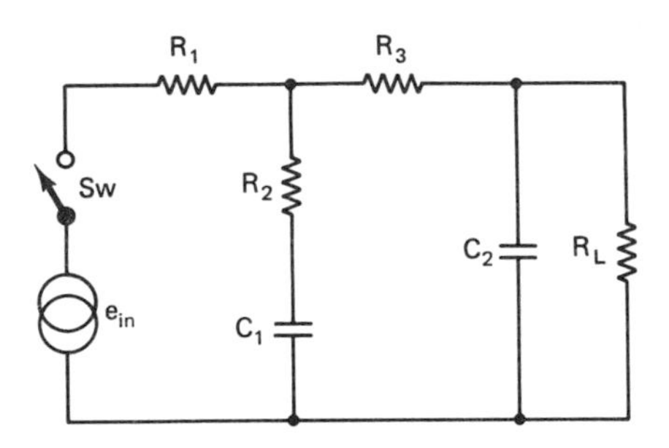

Transient and steady-state responses for this reactive network are not the same, and there is no special condition under which they could be the same. In other words, the voltage drops across C1 and C2 pass through zero at different times. Accordingly, if the generator switch is closed at the instant that the generator phase corresponds to the passage of C1 voltage through zero in the steady state, then the generator phase will not correspond to the passage of C2 voltage through zero in the steady state.

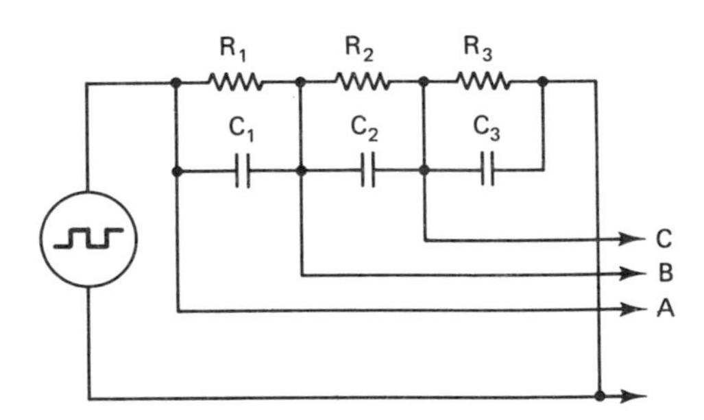

If R1C1 = R2C2 = R3C3 in this reactive network, undistorted square-wave outputs will be obtained at A, B, and C, at three different levels.

At the instant that the generator switch is closed and a voltage (dc or otherwise) is applied to an amplifier, for example, the amplifier briefly responds to an impulse or step input. In theory, this impulse or step comprises all frequencies (an infinite array). Since the amplifier has a limited frequency response, it distorts the impulse or step waveform that is applied to its input terminals. This distortion is characterized by a finite rise time of the output voltage. Rise time is defined as the elapsed time from 10 percent to 90 percent of maximum amplitude on the leading edge of the waveform. (See Figure 12–1.)

As would be anticipated, there is a quantitative relation between the rise time of the output waveform and the bandwidth of an

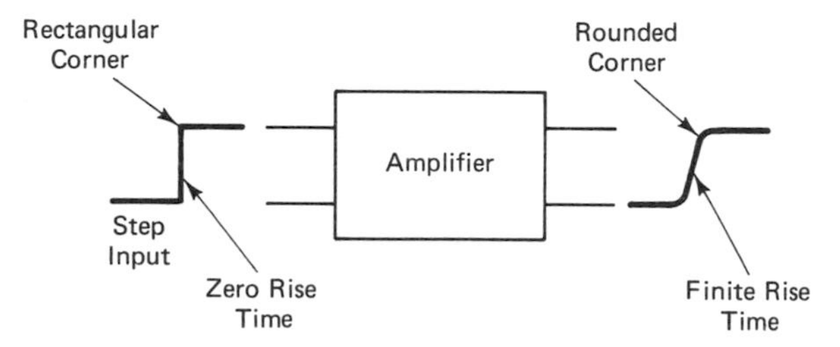

Output rise time is finite and waveform corners are rounded due to lack of infinite bandwidth in the amplifier

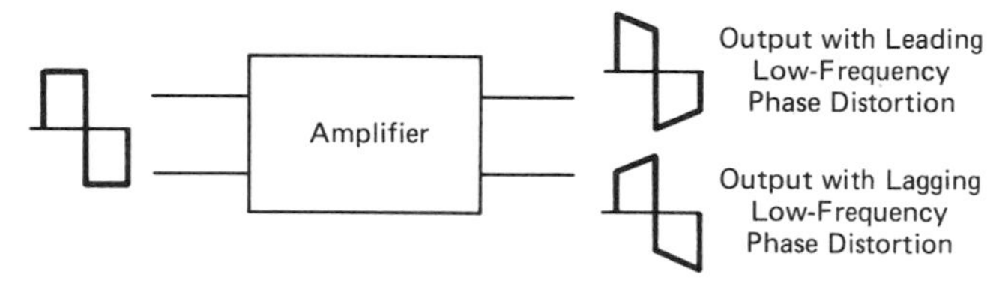

The low-frequency components of the square wave have a comparatively high amplitude. In turn, leading low-frequency phase shift in the amplifier develops down-hill tilt in a reproduced square wave. Conversely, lagging low-frequency phase shift in the amplifier develops up-hill tilt in a reproduced square wave.

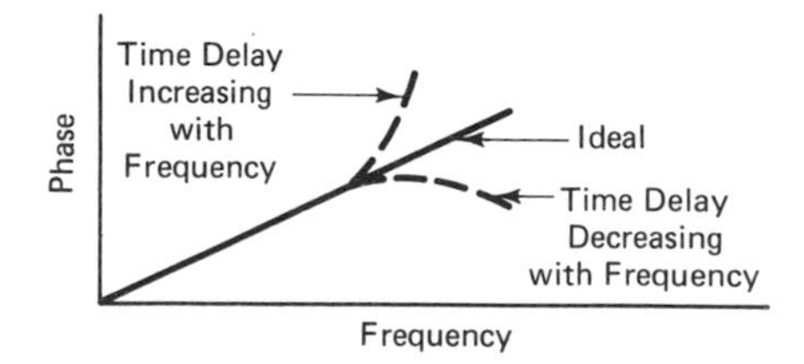

Envelope delay is defined as the time required for a signal with a given frequency to pass from input to output. It is proportional to the slope of the phase-shift curve as a function of frequency. Envelope delay distortion occurs if the delay is not the same at all frequencies in the pass band.

(A)

Figure 12–1(A). Basic amplifier characteristics with respect to transient input.

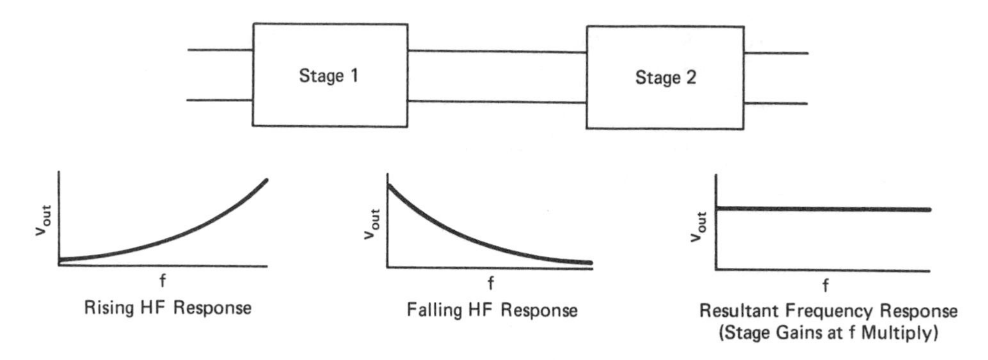

Figure 12–1(B). Transient distortion results from frequency compensation of one stage by a following stage.

Note: Although the resultant frequency response for this arrangement is uniform, it develops distorted transient response. The rising and falling high-frequency characteristics of the stages correspond to significantly nonlinear-phase characteristics that do not combine into a linear-phase characteristic. If optimum transient response is required, both stages should have uniform frequency response.

As explained in Chapter 8, a rising high-frequency response commonly results from a partially bypassed source (or emitter) resistor. A falling high-frequency response frequently results from an excessively high value of load resistance, or from a capacitive load.

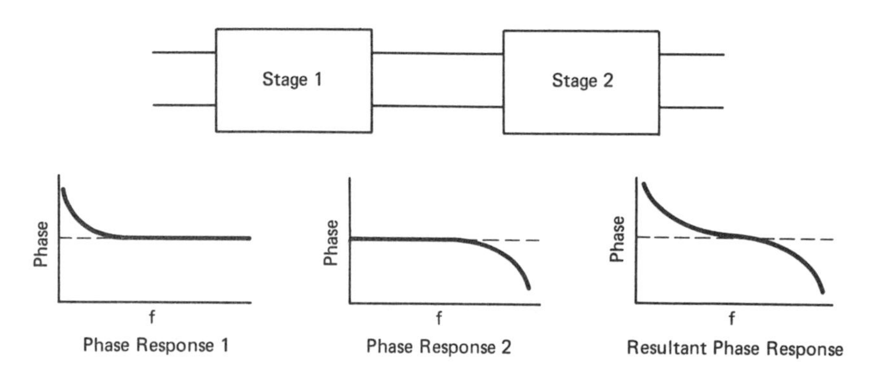

Figure 12–1(C). Basic combination of stage phase shifts versus frequency.

Note: If a stage has a rising high-frequency response, its greatest phase shift occurs at low frequencies. However, when a stage has falling high-frequency response, its greatest phase shift occurs at high frequencies. The phase shift leads in the first situation and lags in the second situation. In turn, the system-phase characteristic becomes more nonlinear, although the system-frequency characteristic becomes more uniform.

A system has a linear-phase characteristic if its product of frequency times phase is a constant. If a phase characteristic is curved, its frequency-phase product is not constant.

amplifier or network. This empirical formula states that the rise time and bandwidth are related as follows:

Bandwidth (Hz) = 0.35/T, approximately
where T is the rise time in seconds

Thus, if an audio amplifier has a bandwidth of 20 kHz, its rise time will be approximately 0.0000175 second, or 17.5 microseconds. Observe that the foregoing relationship is approximate inasmuch as the frequency response of the amplifier or network is assumed to be uniform and that its high-frequency rolloff is not unduly abrupt. (Bandwidth is customarily defined as the frequency span between the half-power points on the frequency-response curve.)

This observation leads to a fundamental principle of amplifier-circuit design for optimum transient response (square wave or pulse response). This principle states that if the rise time of the input waveform is shorter than the rise time of the amplifier, minimum distortion of the output waveform will be realized if the amplifier-frequency response has a Gaussian high-frequency cutoff characteristic. If the high-frequency rolloff is more rapid than Gaussian, the output waveform will exhibit overshoot and ringing distortion. Or, if the high-frequency rolloff is slower than Gaussian, the rise time of the output waveform will be unnecessarily increased.

An amplifier with Gaussian rolloff has a high-frequency attenuation rate of change of approximately 9 dB per octave through its half-power point. As previously noted, the Gaussian rolloff characteristic is approached normally by a series of RC-coupled amplifier stages. Several cascaded and uncompensated stages will provide a reasonable approximation of Gaussian response. Note that if a stage has rising high-frequency response, it will tend to introduce overshoot and ringing into the reproduced waveform. If a stage has a peaked response at any frequency, it will tend to introduce ringing at this frequency into the reproduced waveform.

Consider next how rise times combine. For example, if a pulse generator provides an output waveform with a rise time of 20μs and this waveform is applied to an oscilloscope with a rise time of 20μs, the reproduced waveform has a rise time of approximately 28.3μs.

In other words, rise times combine in accordance with the following empirical formula:

$$T_C = \sqrt{T_1^2 + T_2^2}$$

where T_1 and T_2 are the successive rise times
and T_c is their combined rise time

In other words, rise times combine approximately in accordance with the square root of the sum of the squares formula. This evaluation assumes that the input waveform to the amplifier has a rise time that is less than or equal to that of the amplifier. If the input waveform has a rise time that exceeds the rise time of the amplifier, the rise time of the reproduced waveform will be limited by and equal to the amplifier rise time.

Distortion of nonsinusoidal waveforms in passage through an amplifier is determined to a considerable extent by the amplifier-phase characteristic. (Amplitude nonlinearity and nonuniform and/or limited frequency response are supplemental factors.) With reference to Figure 12–2(A), a typical RC-coupled amplifier-phase characteristic is shown, with an ideal linear-phase characteristic. If an amplifier has a rapid high-frequency rolloff, a rising high-frequency characteristic, or a peak or a dip in its frequency response, its phase characteristic will be less linear than otherwise. Frequency-compensated amplifiers generally have comparatively nonlinear-phase characteristics. Although phase-compensation circuitry can be employed, it can seldom be justified from the viewpoint of production costs.

When a wavefront with a very rapid rise (such as a pulse or square waveform) is applied to an amplifier, the amplifier response is basically an exponential function, such as depicted in Figure 12–2(B). The exponential function expresses the natural law of growth and decay in terms of e, which is the base of the natural system of logarithms. The value of e is approximately 2.71828. In Figure 12–2, curve B illustrates the function $y = e^{-x}$, and curve A illustrates the function $y = (1 - e^{-x})$.

With reference to Figure 12–3, the output from a single-section integrating circuit in response to a step-input voltage is formulated $y = (1 - e^{-x})$. When two integrating sections are cascaded, the output waveform is an exponential function, although it is a

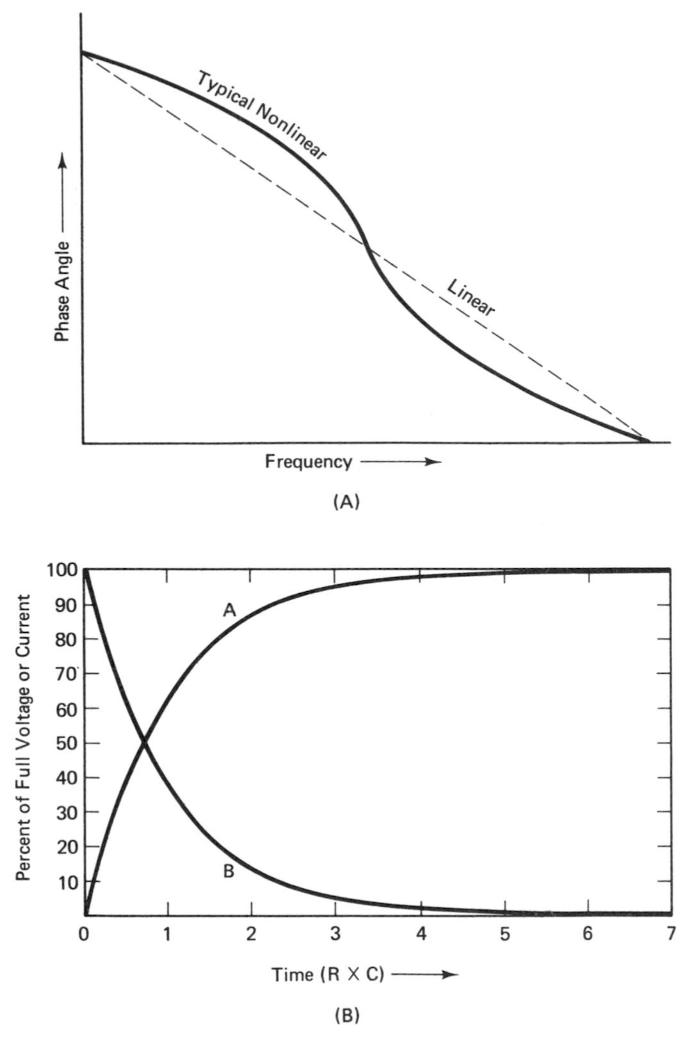

Figure 12–2. Transient-circuit parameters. (A) Linear- and nonlinear-phase characteristics; (B) universal RC time-constant chart.

modified function due to the loading of the first section by the second section. Similarly, when three integrating sections are cascaded, the output waveform becomes a further modified exponential function. The rise time of the output waveform is progressively slowed as more integrating sections are cascaded. Note that the output waveform approaches a Gaussian form as more sections are employed.

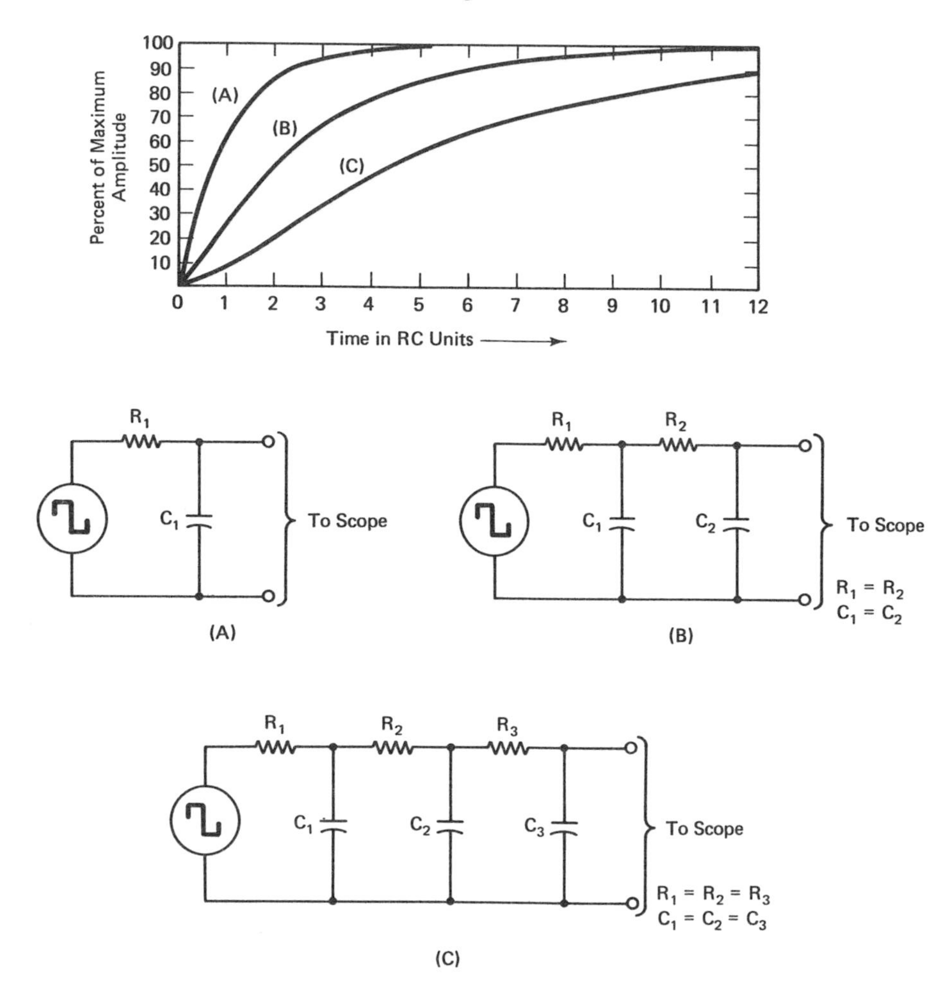

Figure 12–3. Universal time-constant chart for 1-section and symmetrical 2-section and 3-section integrating circuits. (Courtesy of Goodyear Publishing Company)

ASTABLE MULTIVIBRATORS

Free-running (astable) multivibrators are the most basic class of nonsinusoidal oscillators. This class is also designated as relaxation oscillators, which generate rectangular waveforms such as square waves or pulses. A basic astable multivibrator configuration is shown in Figure 12–4. It is a two-stage direct-coupled amplifier with positive feedback from the collector of one transistor to the base of the other transistor.

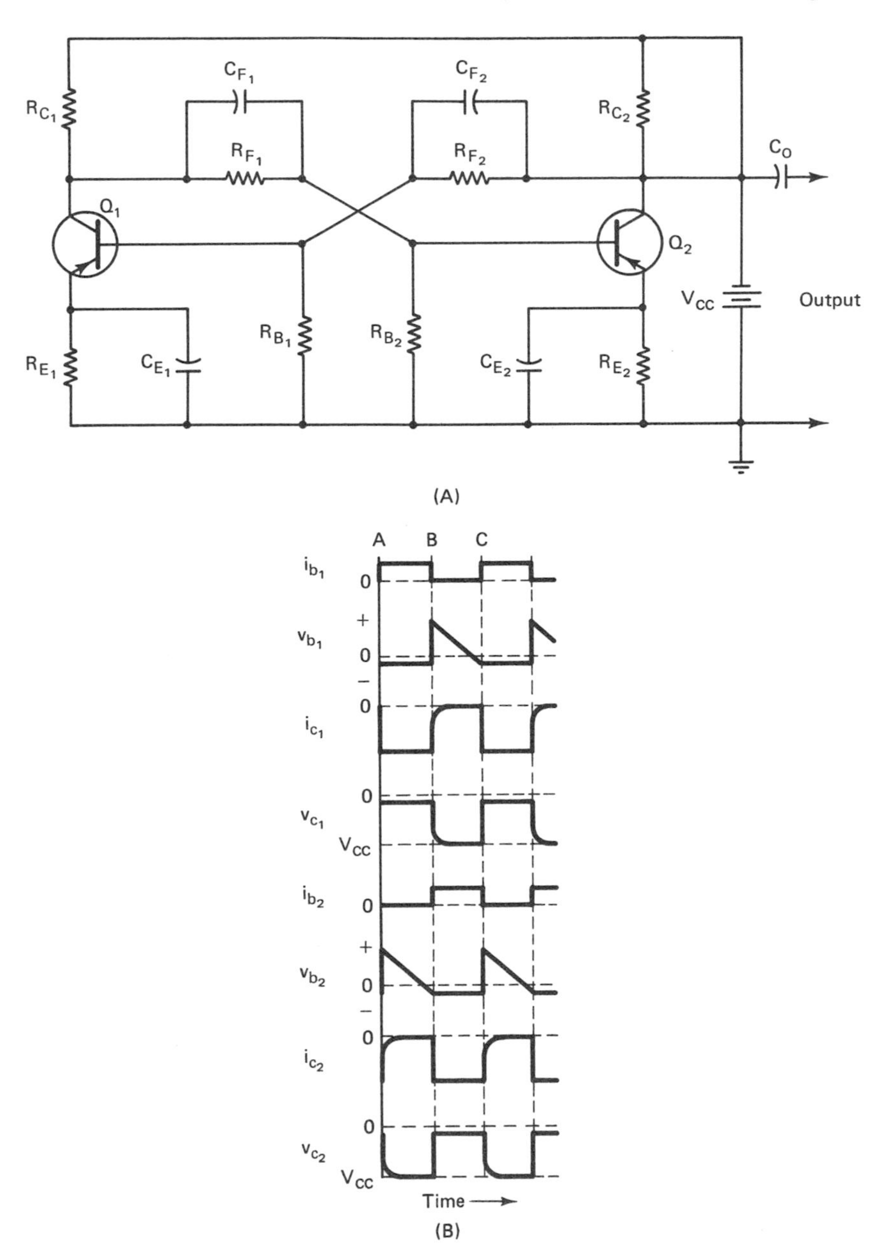

Figure 12–4. Basic astable multivibrator configuration. (A) Circuit; (B) operating waveforms. (Reproduced by special permission of Reston Publishing Company and Campbell Loudoun from *Handbook for Electronic Circuit Design*.)

NOTE: If R_{B1} and R_{B2} are returned to a variable source of negative voltage, the oscillator frequency can be varied over a wide range. The oscillator then operates as a voltage-controlled oscillator (VCO).

From a design viewpoint, the worst-case condition occurs when one RC coupler has positive-tolerance limits and the other coupler has negative-tolerance limits. As an illustration, suppose that C_{F1} and R_{F1} are each 20 percent low. In turn, there will be a difference of more than two to one in the periods of successive half-cycle outputs. Note the C_{E1} and C_{E2} bypass the emitter swamping resistors and have no effect on the output waveform or frequency of oscillation (unless they have an extremely low tolerance limit).

Since the output waveform is not entirely flat topped at the collectors, the designer may elect to pass the output waveform through a clipper circuit to square up the waveshape. Note that multivibrators can be grouped into high-current types and low-current types. A high-current multivibrator utilizes emitter swamping resistors that have approximately half the value of the collector-load resistors. In turn, the transistor collector currents are comparatively high. This design has an advantage in that it is reasonably tolerant of supply-voltage variation.

With reference to Figure 12–4, the value of R_{F1} is typically 10 times the value of R_{C1}. The amplitude of the output waveform in this arrangement is somewhat greater than half the value of V_{CC}. To calculate the maximum power dissipation of each transistor, it is good practice to follow conservative design procedure and to assume that the full supply voltage is applied to the transistor, with current limited only by the series resistance in the circuit. For example, if R_{C1} is 1000 ohms and R_{E1} is 500 ohms and V_{CC} is 10 V, a maximum collector-current flow of approximately 7 mA would be assumed. If it is further assumed that the collector-to-emitter potential has a maximum value of 10 V, the transistor maximum power dissipation would be specified as 70 mW. Conservative design procedure allows reasonable margin for supply-voltage variation and for increased operating temperature derating.

A transistor in a multivibrator operates in the switching mode and worst-case design is based on the premise that overdrive will demand three times the base current that is supplied by the dc bias network. Overdrive is required to obtain fast rise time, although in the limit, rise time is determined by the beta cutoff frequency of the transistor. Rise time is further limited by the interval required for the transistor to come out of saturation (charge storage time). Note that this limitation is taken into account in the rated switching time of the transistor.

Inasmuch as a high-current type of multivibrator is under consideration, values of collector load resistors are comparatively small and do not enter significantly into rise-time parameters. A typical switching-type transistor has the following switching-time ratings:

Delay time = 0.2μs
Rise time = 1μs
Storage time = 1μs
Fall time = 1.2μs

The foregoing ratings and bogie values and apply to the transistor only; they do not take circuit characteristics into account. Thus, when the transistor is cut off and is then suddenly driven into saturation, it will require 0.2μs before the change in base voltage is reflected as a change in collector voltage. Then, it will require 1μs for the leading edge of the output waveform to rise from 10 percent to 90 percent of maximum amplitude. Next, with the transistor in saturation, when the base is suddenly driven into cutoff, storage of charge carriers in the base region will hold the output amplitude at almost its maximum level for 1μs. After this storage time has passed, the output waveform starts to fall, and it will require 1.2μs for the trailing edge of the waveform to fall from 90 percent to 10 percent of its total excursion.

Calculation of bias-circuit value is made on the basis of conventional amplifier operation. The designer assumes that there is no feedback in the circuit and that both transistors are to be forward-biased so that the collector potential will be equal to 0.6 of the supply voltage and the emitter potential will be equal to approximately 0.2 of the supply voltage.

LOW-CURRENT MULTIVIBRATOR

A low-current multivibrator configuration employs collector resistors that have approximately 10 times the value of the emitter swamping resistors. Accordingly, the collector currents have comparatively low values. Note that this design is significantly responsive to supply-voltage variation; an increase in supply voltage causes an increase in operating frequency. Note that both high-current and low-current types of multivibrators operate as pulse generators if the values of C_{F1} and C_{F2} in Figure 12–4 are suitably proportioned. Control of pulse width and duty cycle can be provided by returniing R_{B1} and R_{B2} to variable negative voltage sources.

Observe also that multivibrators can be cascaded and locked in harmonic-frequency relation. In other words, if a multivibrator generates a fundamental frequency and is followed by a multivibrator that generates the second-harmonic frequency, a small amount of capacitive coupling from the first multivibrator to the second will result in operation of the second multivibrator as a locked oscillator. This basic arrangement has numerous applications in consumer-electronic products.

FUNCTION GENERATION

Function generators provide both sinusoidal and nonsinusoidal waveform outputs. A typical function generator has sine-wave, square-wave, and triangular-wave outputs. More elaborate function generators also provide pulse waveform outputs. Amplitude and frequency modulation facilities may also be provided. The basic generating sections in a function generator are a sine-wave oscillator and a multivibrator.

Fundamental waveshaping arrangements for function generation are shown in Figure 12–5. From the viewpoint of ideal waveforms, a square wave has zero rise time and its derivative is an impulse with zero width and infinite amplitude. Insofar as op-amp waveshapers are concerned, the differentiated square wave has finite amplitude and finite width. Nevertheless, an op-amp-differentiating circuit is a closer approximation to the mathematical differentiating function than is an RC-differentiating circuit.

When the square waveform is integrated by an op-amp-integrating circuit, a triangular or ramp output waveform is obtained. The ramp is comparatively linear and is a closer approximation to the mathematical-integrating function than is an RC-integrating circuit. Observe that if an impulse waveform is integrated, a square waveform is obtained. If a triangular waveform is differentiated, a square waveform is obtained. If a triangular waveform is integrated, a parabolic waveform is obtained.

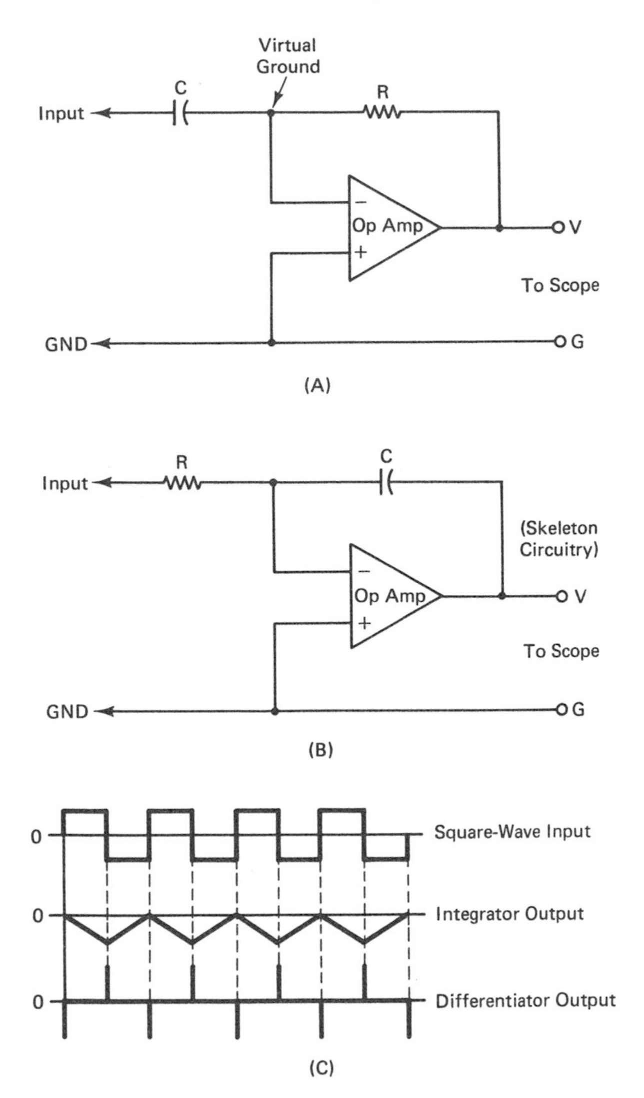

Figure 12–5. Waveshaping arrangements for function generation. (A) Op-amp differentiating circuit; (B) op-amp integrating circuit; (C) integrator and differentiator output waveforms. (Reproduced by special permission of Reston Publishing Company and Douglas Bapton from *Modern Oscilloscope Handbook*.)

Instead of utilizing individual op amps and discrete components in function-generator design, specialized integrated circuits may be preferred. If you want to know more about function generator ICs, refer to *Encyclopedia of Integrated Circuits* by Walter H. Buchsbaum, Sc.D.

13

ELEMENTS OF LINEAR RC-VOLTAGE BOOTSTRAP CIRCUITRY

OVERVIEW

Linear RC-voltage bootstrap circuitry provides an output voltage that exceeds the input voltage at the resonant frequency of the network. The simplest RC-integrating circuit arrangement that has a bootstrap function is shown in Figure 13–1. Although there is a power insertion loss imposed by the network, a voltage insertion gain is provided. The two RC sections may be symmetrical or unsymmetrical. Maximum voltage gain is provided by a two-section network when the second section has the same time constant as the first section, but with a much higher impedance (such as 10 to 100 times). Additional sections may be employed for higher voltage gain.

OPERATION

The locus of operation for a series-RC circuit is shown in Figure 13–2. This locus is a semicircle with the input voltage as its diameter. It is instructive to first examine the operation of a two-

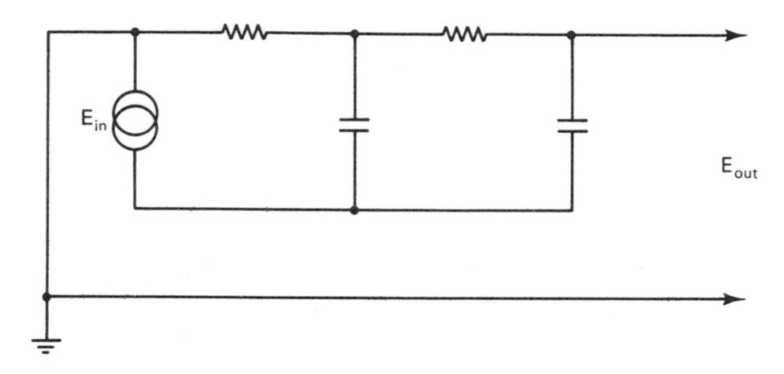

Figure 13–1. Simplest RC-integrating circuit with an output voltage greater than its input voltage at its resonant frequency.

NOTE: E_{out} is very small at low frequencies; as the frequency is increased, E_{out} rises to a peak value that exceeds E_{in}; as the frequency is further increased, E_{out} decreases to the value of E_{in} and then remains at this level at still higher frequencies.

This type of circuitry may be described as a voltage bootstrapping arrangement or as a pseudoactive RC filter.

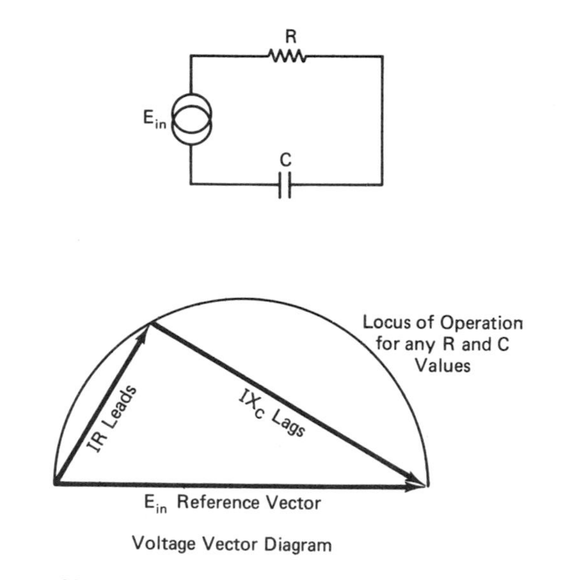

Figure 13–2. Locus of operation for a series-RC circuit.

section RC-integrating circuit at a frequency that makes the absolute values of resistance and reactance equal (phase angle of 45°). Consider a symmetrical two-section RC-integrating circuit in which the second RC section has a much higher impedance than the first RC section.

When this condition is employed, the loading action of the section on the first section can be neglected. In turn, the responses of the two sections may be considered independently. With reference to Figure 13–3, if the first section has a phase angle of 45° and the second section has negligible loading, E_{C1} has a phase angle of 45° with respect to the source. In turn, if the second section has the same time constant as the first section, its phase angle will also be 45°. The locus of operation for the second section is also a semicircle with a diameter equal to E_{C1}.

As shown in Figure 13–4, the output voltage from a symmetrical two-section RC-integrating circuit arranged for voltage bootstrap operation has a maximum value of 1.078 volts with respect to an input of 1 volt. Because the first RC section is substantially loaded by the second RC section, the voltage gain of the network is comparatively small. In other words, the loading action of the second section significantly reduces maximum available voltage across the capacitor in the first section.

To approach the maximum available voltage across the capacitor in the first section, the impedance of the second section may be made 100 times greater than the impedance of the first section. If the two sections have the same time constant, the network vector diagram may be drawn as shown in Figure 13–5. In this example, the output voltage will be 1.118 times the input voltage at an operating frequency that provides a 45° phase angle (absolute values of resistance and reactance equal).

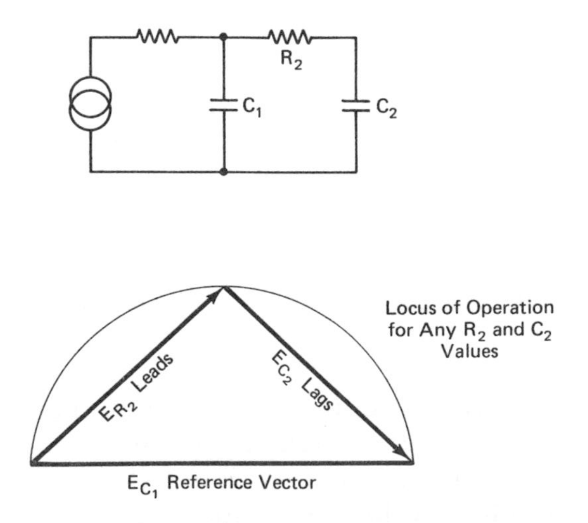

Figure 13–3. Locus of operation for the second RC section in an integrating network.

```
1 LPRINT"* SYMMETRICAL TWO-SECTION RC NETWORK WITH VOLTAGE GAIN *"
2 PRINT"* SYMMETRICAL TWO-SECTION RC NETWORK WITH VOLTAGE GAIN *"
3 LPRINT"(Integrating Sections)":PRINT"(Integrating Sections)"
4 LPRINT"* Eout/Ein and Phase vs. F *":PRINT"* Eout/Ein and Phase vs. F *"
5 LPRINT"(Survey Program)":PRINT"(Survey Program)":LPRINT"":PRINT""
6 INPUT"R (Ohms)=";R
7 LPRINT"R (Ohms)=";R
8 INPUT"C (Mfd)=";C
9 LPRINT"C (Mfd)=";C
10 INPUT"Starting f (Hz)=";H
11 F=0:N=0:G=H:LPRINT"":PRINT""
12 F=G+N:A$="#######":LPRINT"f (Hz)=";USING A$;F:PRINT"f (Hz)=";USING A$;F
13 X=1/(6.283*F*C*10^-6):AB=(R*R+4*X*X)^.5:BA=-ATN(2*X/R):AC=(R*R+X*X)^.5
14 CA=-ATN(X/R):BC=AB*AC:CB=BA+CA:AD=BC/X:DA=CB+6.283/4:BD=AD*COS(DA)
15 DB=AD*SIN(DA):CD=DB+X:DE=(BD*BD+CD*CD)^.5:ED=-ATN(CD/BD):AE=X/DE
16 EA=-6.283/4-ED:N=N+100:PU=AE*COS(EA):UP=AE*SIN(EA):SH=1+PU
17 HS=(SH*SH+UP*UP)^.5:AA=-ATN(UP/SH):C$="####.#"
18 B$="##.###":LPRINT"Eout/Ein=";USING B$;HS:PRINT"Eout/Ein=";USING B$;HS
19 LPRINT"Phase Angle (Degrees)=";USING C$;-AA*360/6.283
20 PRINT"Phase Angle (Degrees)=";USING C$;-AA*360/6.283:LPRINT"":PRINT""
21 IF N>20000 THEN 23
22 GOTO 12
23 END
```

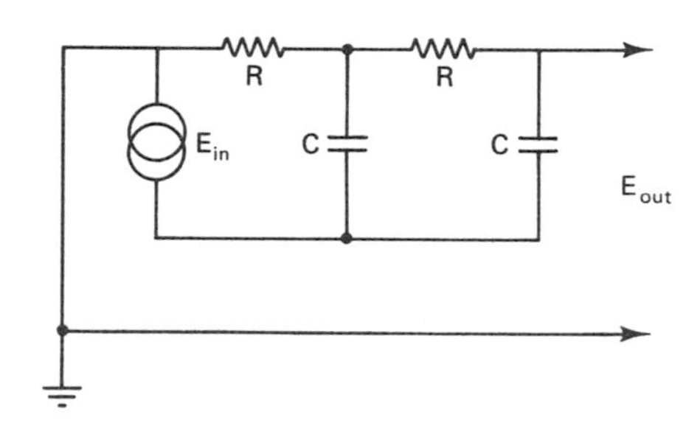


```
* SYMMETRICAL TWO-SECTION RC NETWORK WITH VOLTAGE GAIN *
(Integrating Sections)
* Eout/Ein and Phase vs. F *
(Survey Program)

R (Ohms)= 1000
C (Mfd)= .15

f (Hz)=     100
Eout/Ein= 0.274
Phase Angle (Degrees)= -75.9

f (Hz)=     200
Eout/Ein= 0.507
Phase Angle (Degrees)= -63.2

f (Hz)=     300
Eout/Ein= 0.681
Phase Angle (Degrees)= -52.7

f (Hz)=     400
Eout/Ein= 0.803
Phase Angle (Degrees)= -44.3

f (Hz)=     500
Eout/Ein= 0.887
Phase Angle (Degrees)= -37.8

f (Hz)=     600
Eout/Ein= 0.945
Phase Angle (Degrees)= -32.5

f (Hz)=     700
Eout/Ein= 0.985
Phase Angle (Degrees)= -28.3

f (Hz)=     800
Eout/Ein= 1.013
Phase Angle (Degrees)= -24.9

f (Hz)=     900
Eout/Ein= 1.033
Phase Angle (Degrees)= -22.1

f (Hz)=    1000
Eout/Ein= 1.047
Phase Angle (Degrees)= -19.7

f (Hz)=    1100
Eout/Ein= 1.058
Phase Angle (Degrees)= -17.7
```

Figure 13–4. Survey program for computation of the output voltage and phase angle for a symmetrical two-section RC-integrating circuit arranged for voltage bootstrap function.

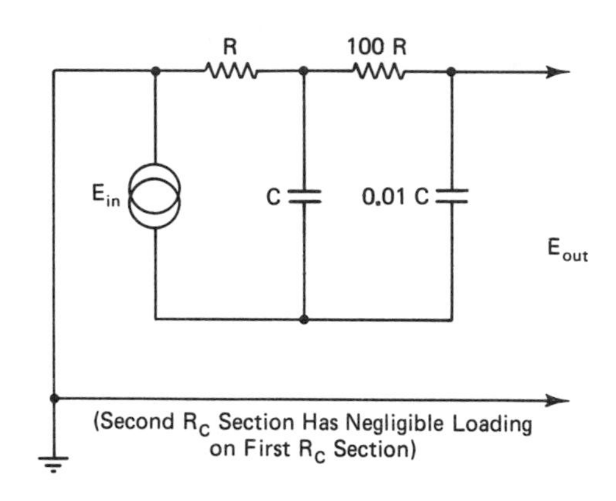

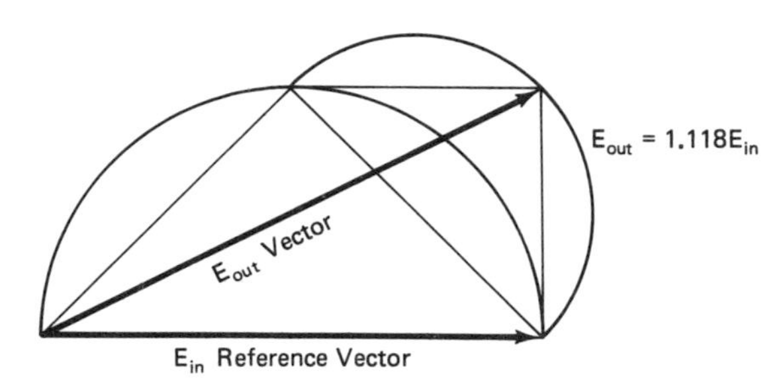

Voltage Vector Diagram for Complete Network
(Operating Frequency for 45° Phase Shifts)

Figure 13–5. Example of an RC-voltage bootstrap configuration in which the second section has little loading action on the first section.

The output voltage from an unsymmetrical two-section RC-integrating circuit arranged for voltage bootstrap operation has a value of 1.118 volts with respect to an input of 1 volt when the operating frequency provides 45° phase angles, as was depicted in Figure 13–5. However, the RUN also shows that this network provides an output voltage of 1.153 volts with respect to an input voltage of 1 volt when the operating frequency is 2250 Hz. This maximum bootstrap action is in consequence of the vector relations shown in Figure 13–6. Stated otherwise, the E_{out} vector in this configuration has a maximum value when its phase angle is 16° with respect to E_{in}.

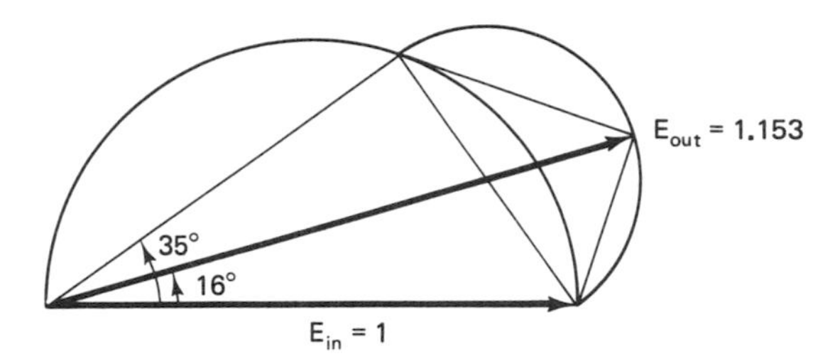

Figure 13–6. Maximum voltage bootstrap action is obtained when the E_{out} vector has a phase angle of 16° with respect to E_{in}.

NOTE: When the second RC section has the same time constant as the first RC section and has sufficiently high impedance that its loading action is negligible, the maximum voltage gain is 15.3 percent at a frequency that makes the reactance equal to 0.707 of the resistance. It follows that the resonant frequency of the configuration is:

$$f_o = \frac{1}{8.88RC}$$

If the absolute value of sectional reactance is less than the absolute value of sectional resistance, the phase angle of the output section voltage is then greater than 90° with respect to the input voltage. In turn, the real part of the output section voltage is then negative with respect to the input voltage, insofar as complex algebraic calculation is concerned. If the output section voltage and the input voltage are terms in an equation, their algebraic signs must be observed accordingly.

OPERATION OF A THIRD RC SECTION

If a suitable third RC-integrating section is added to the configuration depicted in Figure 13–5, the maximum available bootstrap output voltage becomes 1.240 volts with respect to an input voltage of 1 volt, as seen in Figure 13–7. Observe that the maximum available gain (MAG) is obtained when the R/X ratio of the third RC section differs from the R/X ratio of the preceding sections, as shown in the diagram. (See Figure 13–7.) This is just another way of saying that the optimum phase angle for the third section is not the same as the optimum phase angle for the first and second sections.

Although a fourth RC integrating section may be added to obtain a still larger MAG, it is evident from the vector diagram that its utility is marginal. In other words, the output voltage from the

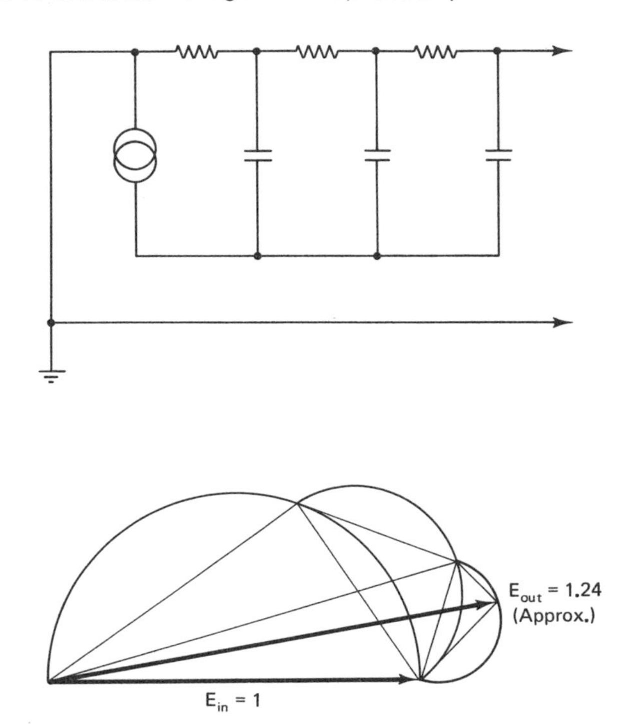

Figure 13–7. Maximized voltage bootstrap operation for an optimized three-section RC-integrating network.

NOTE: When the first and second RC sections have the same R/X ratio and the third RC section has a different R/X ratio with optimum value, the output voltage can be approximately 24 percent greater than the input voltage at the peak frequency. The third RC section has negligible loading action if its impedance is 100 times the impedance of the second RC section.

fourth section per se is sufficiently small that its contribution to the bootstrap voltage output is marginal. However, this increase is of theoretical interest, inasmuch as it shows how the MAG approaches a limiting value as the number of RC sections is increased.

VOLTAGE BOOTSTRAP OPERATION OF RC-DIFFERENTIATING CIRCUIT

The simplest RC-differentiating circuit arrangement that provides voltage bootstrap action is shown in Figure 13–8. We will find that operation of this network is comparable in all basic aspects to its

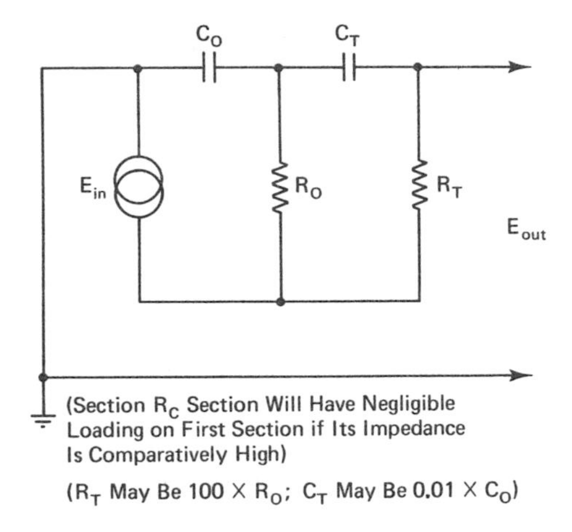

E_{in} Reference Vector

E_{out} Vector

$E_{out} = 1.118E_{in}$

Voltage Vector Diagram Complete Differentiating Network
(Operating Frequency for 45° Phase Shifts)

Figure 13–8. Simplest RC-differentiating circuit arrangement with voltage bootstrap function.

RC integrator counterpart, with one notable exception. In other words, the MAG of the differentiating bootstrap network occurs at a frequency lower than that at which the absolute values of resistance and reactance are equal. By way of comparison, the MAG of the integrating bootstrap network occurs at a frequency higher than that at which the absolute values of resistance and reactance are equal.

Observe also that at very low operating frequencies the output from the differentiating bootstrap network approaches the input voltage. On the other hand, at very high operating frequencies the output from the integrating bootstrap network approaches the input voltage. At very high operating frequencies the output from the differentiating bootstrap network approaches zero, whereas at very low operating frequencies the output from the integrating bootstrap network approaches zero.

A survey program for computation of the output voltage and phase angle for a symmetrical two-section RC-differentiating circuit arranged for voltage bootstrap action is provided in Figure 13–9. The maximum output voltage occurs at approximately 950 Hz in this example. Since the first RC section is substantially loaded by the second section, the voltage gain is only 8 percent over the input voltage.

At zero frequency (dc), the capacitors in the differentiating sections are open circuits, and at infinite frequency they are short

```
1 LPRINT"* * SYMMETRICAL TWO-SECTION RC NETWORK WITH VOLTAGE GAIN * *"
2 PRINT"* * SYMMETRICAL TWO-SECTION RC NETWORK WITH VOLTAGE GAIN * *"
3 LPRINT"Differentiating Sections":PRINT"Differentiating Sections"
4 LPRINT"Eout/Ein versus Frequency"
5 PRINT"Eout/Ein versus Frequency"
6 LPRINT"(Survey Program)":PRINT "(Survey Program)":LPRINT"":PRINT""
7 INPUT"R (Ohms)=";R
8 LPRINT"R (Ohms)=";R
9 INPUT"C (Mfd)=";C
10 LPRINT"C (Mfd)=";C
11 INPUT"Starting f (Hz)=";H
12 LPRINT"Starting f (Hz)=";H
13 F=0:N=0:G=H:LPRINT"":PRINT""
14 F=G+N:A$="#######":LPRINT"f (Hz)=";USING A$;F
15 PRINT"f (Hz)=";USING A$;F
16 X=1/(6.283*F*C*10^-6):B$="####.##"
17 AB=(4*R*R+X*X)^.5:BA=-ATN(X/(2*R))
18 AC=(R*R+X*X)^.5:CA=-ATN(X/R)
19 AD=AB*AC:DA=BA+CA:AE=AD/R:EA=DA
20 AF=AE*COS(EA):AG=AE*SIN(EA):AH=AF-R
21 AJ=(AH*AH+AG*AG)^.5:JA=-ATN(AG/AH)
22 AK=R/AJ:KA=-JA:AL=AK*COS(KA):AM=AK*SIN(KA)
23 IF X>R THEN AN=1+AL
24 IF X<R THEN AN=1-AL
25 AO=(AN*AN+AM*AM)^.5:PRINT"AO=";AO
26 LPRINT"Eout/Ein=";USING B$;AO:PRINT"Eout/Ein=";USING B$;AO
27 IF F>20000 THEN 29
28 LPRINT"":PRINT"":N=N+100:GOTO 14
29 END
```

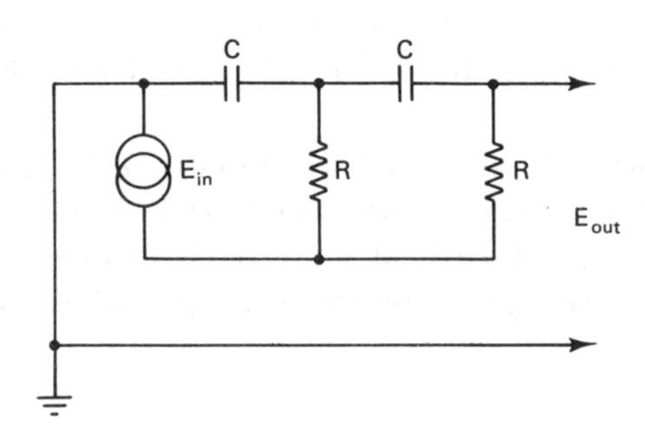

```
* * SYMMETRICAL TWO-SECTION RC NETWORK WITH VOLTAGE GAIN * *
Differentiating Sections
Eout/Ein versus Frequency
(Survey Program)

R (Ohms)= 100
C (Mfd)= 1
Starting f (Hz)= 100

f (Hz)=    100
Eout/Ein=   1.00

f (Hz)=    200
Eout/Ein=   1.01

f (Hz)=    300
Eout/Ein=   1.03

f (Hz)=    400
Eout/Ein=   1.04

f (Hz)=    500
Eout/Ein=   1.05

f (Hz)=    600
Eout/Ein=   1.06

f (Hz)=    700
Eout/Ein=   1.07

f (Hz)=    800
Eout/Ein=   1.08

f (Hz)=    900
Eout/Ein=   1.08

f (Hz)=   1000
Eout/Ein=   1.08

f (Hz)=   1100
Eout/Ein=   1.08

f (Hz)=   1200
Eout/Ein=   1.07

f (Hz)=   1300
Eout/Ein=   1.07

f (Hz)=   1400
Eout/Ein=   1.07

f (Hz)=   1500
Eout/Ein=   1.06

f (Hz)=   1600
Eout/Ein=   1.05

f (Hz)=   1700
Eout/Ein=   1.05

f (Hz)=   1800
Eout/Ein=   1.04

f (Hz)=   1900
Eout/Ein=   1.03

f (Hz)=   2000
Eout/Ein=   1.02

f (Hz)=   2100
Eout/Ein=   1.01

f (Hz)=   2200
Eout/Ein=   1.00

f (Hz)=   2300
Eout/Ein=   1.00

f (Hz)=   2400
Eout/Ein=   0.99

f (Hz)=   2500
Eout/Ein=   0.98

f (Hz)=   2600
Eout/Ein=   0.97
```

```
* * SYMMETRICAL TWO-SECTION RC NETWORK WITH VOLTAGE GAIN * *
Differentiating Sections
Eout/Ein versus Frequency
(Survey Program)

R (Ohms)= 100
C (Mfd)= 1
Starting f (Hz)= 10000

f (Hz)=  10000
Eout/Ein=   0.44

f (Hz)=  10100
Eout/Ein=   0.44

f (Hz)=  10200
Eout/Ein=   0.43

f (Hz)=  10300
Eout/Ein=   0.43
```

```
* * SYMMETRICAL TWO-SECTION RC NETWORK WITH VOLTAGE GAIN * *
Differentiating Sections
Eout/Ein versus Frequency
(Survey Program)

R (Ohms)= 100
C (Mfd)= 1
Starting f (Hz)= 20000

f (Hz)=  20000
Eout/Ein=   0.23

f (Hz)=  20100
Eout/Ein=   0.23
```

Figure 13–9. (Continues on next page.)

```
* * SYMMETRICAL TWO-SECTION RC NETWORK WITH VOLTAGE GAIN * *
Differentiating Sections
Eout/Ein versus Frequency
(Survey Program)

R (Ohms)= 100
C (Mfd)= 1
Starting f (Hz)= 50000

f (Hz)=  50000
Eout/Ein=    0.10


* * SYMMETRICAL TWO-SECTION RC NETWORK WITH VOLTAGE GAIN * *
Differentiating Sections
Eout/Ein versus Frequency
(Survey Program)

R (Ohms)= 100
C (Mfd)= 1
Starting f (Hz)= 500000

f (Hz)= 500000
Eout/Ein=    0.01
```

Figure 13–9. Survey program for computation of the output voltage for a symmetrical two-section RC-differentiating circuit arranged for bootstrap voltage operation.

circuits. At 950 Hz (in this example), although the series capacitors shunt the output circuit, the circuit-phase relations are such that voltage bootstrapping action occurs, and the output voltage exceeds the input voltage.

When the second section in an RC differentiating voltage bootstrap network has significantly higher impedance than the first section, a voltage gain of approximately 15 percent is provided. If the second section has twice the impedance of the first section, a voltage gain of approximately 9 percent is provided. If the second section has half the impedance of the first section, a voltage gain of only 4 percent is provided.

If an unsymmetrical RC-differentiating bootstrap network has a second section with a different time constant from the first section, the program provided in Figure 13–10 may be employed. (This routine uses the output/input phase angle as a processing criterion, instead of relative resistance and reactance values.) The first RUN illustrates a high-impedance second section with a time constant 10 times that of the first section. It provides a voltage gain of approximately 6 percent. The second RUN exemplifies a low-impedance second section with a time constant less than that of the first section. It provides a voltage gain of approximately 4 percent.

```
1 LPRINT"* UNSYMMETRICAL TWO-SECTION RC NETWORK WITH VOLTAGE GAIN *"
2 PRINT"* UNSYMMETRICAL TWO-SECTION RC NETWORK WITH VOLTAGE GAIN"
3 LPRINT"Differentiating Sections; Equal Time Constants":PRINT"Differentiating S
ections; Equal Time Constants"
4 LPRINT"* Eout/Ein vs. Frequency *":PRINT"* Eout/Ein vs. Frequency *"
5 LPRINT"(Survey Program)":PRINT"(Survey Program)":LPRINT"":PRINT""
6 INPUT"R1 (Ohms)=";RO
7 LPRINT"R1 (Ohms=";RO
8 INPUT"R2 (Ohms)=";RT
9 LPRINT"R2 (Ohms)=";RT
10 INPUT"C1 (Mfd)=";CO
11 LPRINT"C1 (Mfd)=";CO
12 INPUT"C2 (Mfd)=";CT
13 LPRINT"C2 (Mfd)=";CT
14 INPUT"Starting f (Hz)=";H
15 LPRINT"Starting f (Hz)=";H
16 F=0:N=0:G=H:LPRINT"":PRINT""
17 F=G+N:A$="#######":LPRINT"f (Hz)=";USING A$;F:PRINT"f (Hz)=";USING A$;F
18 XO=1/(6.283*F*CO*10^-6):XT=1/(6.283*F*CT*10^-6):B$="####.##"
19 AB=RO+RT:AC=(AB*AB+XT*XT)^.5:CA=-ATN(XT/AB):AD=(RO*RO+XO*XO)^.5
20 DA=-ATN(XO/RO):AE=AC*AD:EA=CA+DA:AF=AE*COS(EA):AG=AE*SIN(EA):AH=RO*RO
21 AJ=AF-AH:AK=(AJ*AJ+AG*AG)^.5:KA=-ATN(AG/AJ):AL=RO*RT:AM=AL/AK:MA=-KA
22 AN=AM*COS(MA):AO=AM*SIN(MA)
23 IF XO>RO THEN AP=1+AN
24 IF XO<RO THEN AP=1-AN
25 AQ=(AP*AP+AO*AO)^.5
26 LPRINT"Eout/Ein=";USING B$;AQ:PRINT"Eout/Ein=";USING B$;AQ
27 IF F>20000 THEN 29
28 LPRINT"":PRINT"":N=N+100:GOTO 17
29 END
```

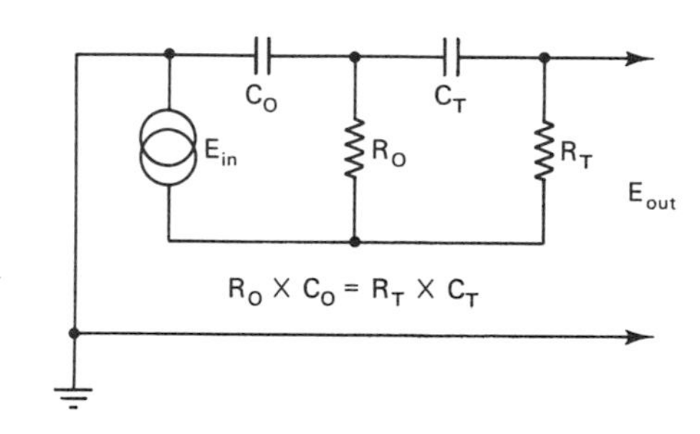


```
* UNSYMMETRICAL TWO-SECTION RC NETWORK WITH VOLTAGE GAIN *
Differentiating Sections; Any Time Constants
* Eout/Ein vs. Frequency *
(Survey Program)

R1 (Ohms= 100
R2 (Ohms)= 100000
C1 (Mfd)= 1
C2 (Mfd)= .01
Starting f (Hz)= 100

f (Hz)=      100        f (Hz)=      400        f (Hz)=      700
Eout/Ein=    1.03       Eout/Ein=    1.05       Eout/Ein=    1.00

f (Hz)=      200        f (Hz)=      500        f (Hz)=      800
Eout/Ein=    1.05       Eout/Ein=    1.04       Eout/Ein=    0.98

f (Hz)=      300        f (Hz)=      600        f (Hz)=      900
Eout/Ein=    1.06       Eout/Ein=    1.02       Eout/Ein=    0.96
```

Figure 13–10. (Continues on next page.)

```
f (Hz)=   1000
Eout/Ein=   0.93

f (Hz)=   1100
Eout/Ein=   0.90

f (Hz)=   1200
Eout/Ein=   0.88

f (Hz)=   1300
Eout/Ein=   0.85

f (Hz)=   1400
Eout/Ein=   0.83

f (Hz)=   1500
Eout/Ein=   0.80

f (Hz)=   1600
Eout/Ein=   0.78

f (Hz)=   1700
Eout/Ein=   0.75

f (Hz)=   1800
Eout/Ein=   0.73

f (Hz)=   1900
Eout/Ein=   0.71

f (Hz)=   2000
Eout/Ein=   0.68

f (Hz)=   2100
Eout/Ein=   0.66

f (Hz)=   2200
Eout/Ein=   0.64

f (Hz)=   2300
Eout/Ein=   0.63

* UNSYMMETRICAL TWO-SECTION RC NETWORK WITH VOLTAGE GAIN *
Differentiating Sections; Any Time Constants
* Eout/Ein vs. Frequency *
(Survey Program)

R1 (Ohms= 300
R2 (Ohms)= 100
C1 (Mfd)= .5
C2 (Mfd)= 1
Starting f (Hz)= 100

f (Hz)=   100
Eout/Ein=   1.01

f (Hz)=   200
Eout/Ein=   1.02

f (Hz)=   300
Eout/Ein=   1.03

f (Hz)=   400
Eout/Ein=   1.03

f (Hz)=   500
Eout/Ein=   1.04

f (Hz)=   600
Eout/Ein=   1.04

f (Hz)=   700
Eout/Ein=   1.04

f (Hz)=   800
Eout/Ein=   1.04

f (Hz)=   900
Eout/Ein=   1.04

f (Hz)=   1000
Eout/Ein=   1.03

f (Hz)=   1100
Eout/Ein=   1.03

f (Hz)=   1200
Eout/Ein=   1.03

f (Hz)=   1300
Eout/Ein=   1.02

f (Hz)=   1400
Eout/Ein=   1.02

f (Hz)=   1500
Eout/Ein=   1.02

f (Hz)=   1600
Eout/Ein=   1.01

f (Hz)=   1700
Eout/Ein=   1.01

f (Hz)=   1800
Eout/Ein=   1.00

f (Hz)=   1900
Eout/Ein=   1.00

f (Hz)=   2000
Eout/Ein=   0.99

f (Hz)=   2100
Eout/Ein=   0.99

f (Hz)=   2200
Eout/Ein=   0.98
```

Figure 13–10. Survey program for computation of the output voltage for an unsymmetrical two-section RC-differentiating circuit with any sectional time constants, arranged for bootstrap voltage operation.

If it is desired to know the phase of the output voltage with respect to the input voltage, the foregoing program may be elaborated, as shown in Figure 13–11. It will be noted that, in this example, the output phase angle changes sign with respect to the input voltage at a frequency between 1200 and 1300 Hz and then approaches −90° as the operating frequency increases. The phase angle approaches zero as the operating frequency approaches zero.

ALTERNATIVE BOOTSTRAP-CASCADED CONFIGURATION

An alternative voltage bootstrap-cascaded RC configuration is shown in Figure 13–12. This arrangement employs four RC-differentiating sections. However, the network is implemented as a dual two-section configuration wherein the output voltage from the first of the dual sections is utilized as the input voltage to the second of the dual sections. If symmetrical RC circuitry is employed, with progressively higher impedances to minimize loading effects, a voltage gain of approximately 32 percent is provided at the resonant frequency.

```
1 LPRINT"* UNSYMMETRICAL TWO-SECTION RC NETWORK WITH VOLTAGE GAIN *"
2 PRINT"* UNSYMMETRICAL TWO-SECTION RC NETWORK WITH VOLTAGE GAIN"
3 LPRINT"Differentiating Sections; Any Time Constants":PRINT"Differentiating Sec
tions; Any Time Constants"
4 LPRINT"* Eout/Ein and Phase vs. F *":PRINT"* Eout/Ein and Phase vs. F *"
5 LPRINT"(Survey Program)":PRINT"(Survey Program)":LPRINT"":PRINT""
6 INPUT"R1 (Ohms)=";RO
7 LPRINT"R1 (Ohms=";RO
8 INPUT"R2 (Ohms)=";RT
9 LPRINT"R2 (Ohms)=";RT
10 INPUT"C1 (Mfd)=";CO
11 LPRINT"C1 (Mfd)=";CO
12 INPUT"C2 (Mfd)=";CT
13 LPRINT"C2 (Mfd)=";CT
14 INPUT"Starting f (Hz)=";H
15 LPRINT"Starting f (Hz)=";H
16 F=0:N=0:G=H:LPRINT"":PRINT""
17 F=G+N:A$="#######":LPRINT"f (Hz)=";USING A$;F:PRINT"f (Hz)=";USING A$;F
18 XO=1/(6.283*F*CO*10^-6):XT=1/(6.283*F*CT*10^-6):B$="####.##"
19 AB=RO+RT:AC=(AB*AB+XT*XT)^.5:CA=-ATN(XT/AB):AD=(RO*RO+XO*XO)^.5
20 DA=-ATN(XO/RO):AE=AC*AD:EA=CA+DA:AF=AE*COS(EA):AG=AE*SIN(EA):AH=RO*RO
21 AJ=AF-AH:AK=(AJ*AJ+AG*AG)^.5:KA=-ATN(AG/AJ):AL=RO*RT:AM=AL/AK:MA=-KA
22 AN=AM*COS(MA):AO=AM*SIN(MA):Z=SGN(MA)
23 IF Z=1 THEN AP=1+AN
24 IF Z=-1 THEN AP=1-AN
25 AQ=(AP*AP+AO*AO)^.5:QA=-ATN(AO/AP)
26 LPRINT"Eout/Ein=";USING B$;AQ:PRINT"Eout/Ein=";USING B$;AQ
27 LPRINT"Phase Angle (Degrees)=";USING B$;-QA*360/6.283
28 PRINT"Phase Angle (Degrees)=";USING B$;QA*360/6.283
29 IF F>20000 THEN 31
30 LPRINT"":PRINT"":N=N+100:GOTO 17
31 END
```

Figure 13–11. (Continues on next page.)

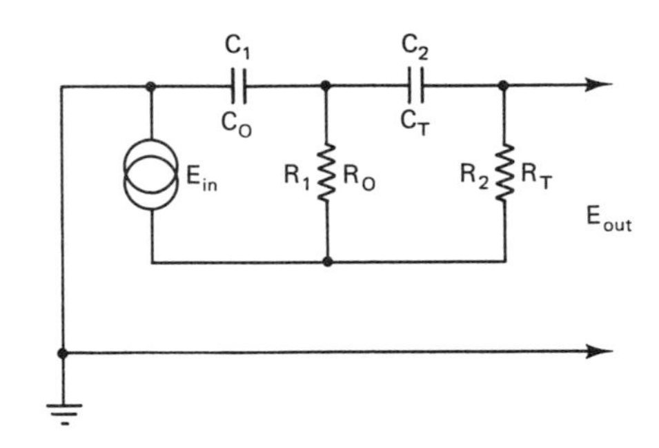


```
* UNSYMMETRICAL TWO-SECTION RC NETWORK WITH VOLTAGE GAIN *
Differentiating Sections; Any Time Constants
* Eout/Ein and Phase vs. F *
(Survey Program)

R1 (Ohms= 100
R2 (Ohms)= 300
C1 (Mfd)= 1
C2 (Mfd)= .5
Starting f (Hz)= 100

f (Hz)=     100
Eout/Ein=    1.01
Phase Angle (Degrees)=   0.06

f (Hz)=     200
Eout/Ein=    1.02
Phase Angle (Degrees)=   0.46

f (Hz)=     300
Eout/Ein=    1.04
Phase Angle (Degrees)=   1.36

f (Hz)=     400
Eout/Ein=    1.06
Phase Angle (Degrees)=   2.78

f (Hz)=     500
Eout/Ein=    1.08
Phase Angle (Degrees)=   4.58

f (Hz)=     600
Eout/Ein=    1.10
Phase Angle (Degrees)=   6.66

f (Hz)=     700
Eout/Ein=    1.10
Phase Angle (Degrees)=   8.88

f (Hz)=     800
Eout/Ein=    1.11
Phase Angle (Degrees)=  11.17

f (Hz)=     900
Eout/Ein=    1.11
Phase Angle (Degrees)=  13.47

f (Hz)=    1000
Eout/Ein=    1.11
Phase Angle (Degrees)=  15.74

f (Hz)=    1100
Eout/Ein=    1.10
Phase Angle (Degrees)=  17.96

f (Hz)=    1200
Eout/Ein=    1.09
Phase Angle (Degrees)=  20.12

f (Hz)=    1300
Eout/Ein=    1.08
Phase Angle (Degrees)= -22.22

f (Hz)=    1400
Eout/Ein=    1.07
Phase Angle (Degrees)= -24.24

f (Hz)=    1500
Eout/Ein=    1.05
Phase Angle (Degrees)= -26.18

f (Hz)=    1600
Eout/Ein=    1.04
Phase Angle (Degrees)= -28.05

f (Hz)=    1700
Eout/Ein=    1.02
Phase Angle (Degrees)= -29.85

f (Hz)=    1800
Eout/Ein=    1.01
Phase Angle (Degrees)= -31.57
```

```
* UNSYMMETRICAL TWO-SECTION RC NETWORK WITH VOLTAGE GAIN *
Differentiating Sections; Any Time Constants
* Eout/Ein and Phase vs. F *
(Survey Program)

R1 (Ohms= 100
R2 (Ohms)= 300
C1 (Mfd)= 1
C2 (Mfd)= .5
Starting f (Hz)= 500000

f (Hz)= 500000
Eout/Ein=   0.01
Phase Angle (Degrees)= -89.71
```

Figure 13–11. Survey program for computation of output voltage and phase for an unsymmetrical two-section RC-differentiating circuit with any time constants, arranged for voltage bootstrap operation.

NOTE: The sign change of the phase angle from a frequency of 1200 Hz to 1300 Hz relates mathematically to parameters in the corresponding conventional differentiating circuit wherein the real part of the output voltage changes sign from one quadrant to the next. When the conventional circuit is rearranged for voltage bootstrap operation, the phase of the input voltage is effectively reversed. In turn, the effective phase of the bootstrap output voltage is zero at very low frequency and approaches 90° at very high frequency. There is no sign change of the phase angle when the program is coded to observe the effect of circuit rearrangement for bootstrap operation.

An analogous alternative voltage bootstrap-cascaded RC configuration is illustrated in Figure 13–13. This network utilizes four RC-integrating sections, implemented as a dual two-section arrangement. Thus, the output voltage from the first of the dual sections is employed as the input voltage to the second of the dual sections. As before, if symmetrical RC circuitry is used, with progressively higher impedances to minimize loading action, a voltage gain of approximately 32 percent is obtained at the resonant frequency.

From the viewpoint of fundamental network theory, there is an important basic distinction between networks configured in simple cascaded arrangements as exemplified in Figure 13–7, versus networks configured in alternative cascaded arrangements as exemplified in Figure 13–13. The essential distinction is that the former arrangements develop an output voltage that approaches a limit as the number of cascaded sections is indefinitely increased. On the other hand, the latter arrangements develop an output voltage that does not approach a limit and that increases indefinitely as the number of cascaded sections is indefinitely increased.

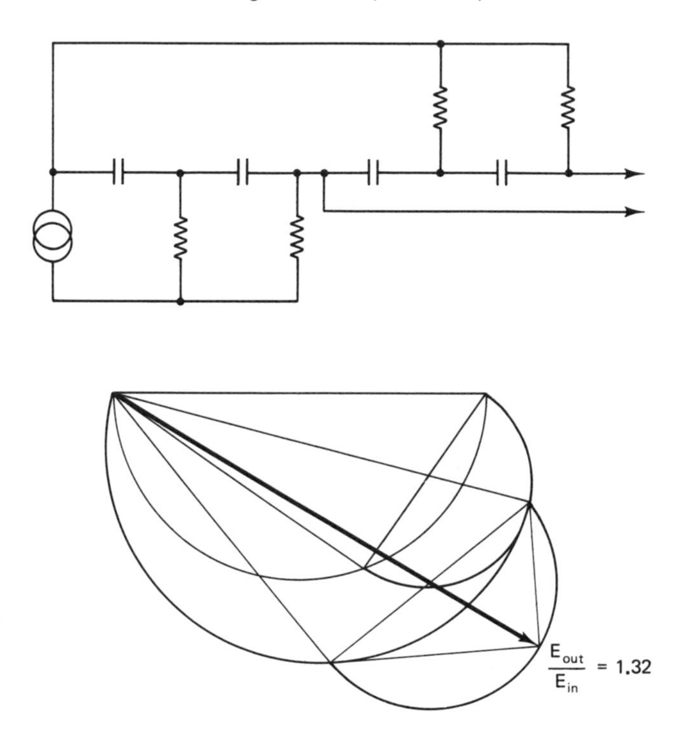

Figure 13–12. Configuration and vector diagram for a dual two-section RC-differentiating network arranged for voltage bootstrap operation.

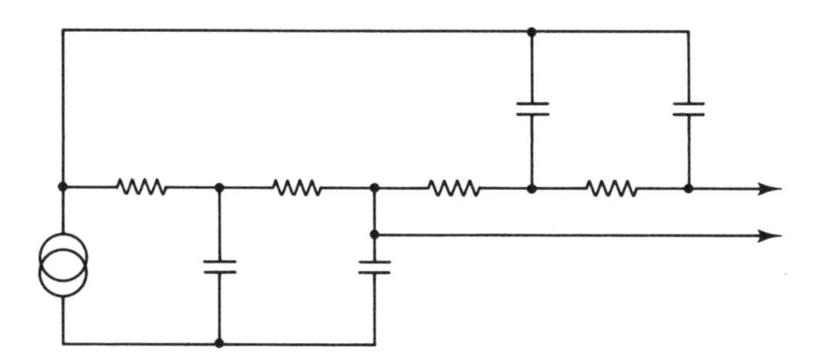

Figure 13–13. Configuration for a dual two-section RC-integrating network arranged for voltage bootstrap operation.

NOTE: As an illustration of the loading imposed by the second dual section upon the first dual section, observe that if the first dual section comprises two 100-ohm resistors and two 1μF capacitors, a load resistance of 1500 ohms across the output of the first dual section will reduce its output voltage by approximately 10 percent at the resonant frequency.

Note also that from the viewpoint of practical application, neither class of linear RC voltage bootstrap circuitry is of general practical utility. In other words, the first class of simple cascaded network is objectionably limited in output voltage capability. The second class of alternative cascaded network requires a large number of resistors and capacitors to provide a substantial boost voltage. Although the required number of resistors and capacitors can be reduced somewhat by employing progressively higher impedance in successive cascaded sections, the network output impedance then becomes very high.

APPENDIX 1: INTERNATIONAL SYSTEM OF UNITS

Basic Units

Quantity	Unit	Symbol	
Length	meter	m	
Mass	kilogram	kg	
Time	second	s	
Temperature	degree Kelvin	° K	
Electric current	ampere	A	
Luminous intensity	candela	cd	

Supplementary Units

Quantity	Unit	Symbol	
Plane angle	radian	rad	
Solid angle	steradian	sr	

Derived Units

Quantity	Unit	Symbol	
Area	square meter	m^2	
Volume	cubic meter	m^3	
Frequency	hertz	Hz	(s^{-1})
Density	kilogram per cubic meter	kg/m^3	
Velocity	meter per second	m/s	
Angular velocity	radian per second	rad/s	
Acceleration	meter per sec. squared	m/s^2	
Angular acceleration	radian per sec. squared	rad/s^2	
Force	newton	N	$(kg \cdot m/s^2)$
Pressure	newton per sq. meter	N/m^2	
Kinematic viscosity	sq. meter per second	m^2/s	
Dynamic viscosity	newton-second per sq. meter	$N \cdot s/m^2$	
Work, energy, quantity of heat	joule	J	$(N \cdot m)$
Power	watt	W	(J/s)

Derived Units

Quantity	Unit	Symbol	
Electric charge	coulomb	C	(A·s)
Voltage, potential difference, electromotive force	volt	V	(W/A)
Electric field strength	volt per meter	V/m	
Electric resistance	ohm	Ω	(V/A)
Electric capacitance	farad	F	(A·s/V)
Magnetic flux	weber	Wb	(V·s)
Inductance	henry	H	(V·s/A)
Magnetic flux density	tesla	T	(Wb/m^2)
Magnetic field strength	ampere per meter	A/m	
Magnetomotive force	ampere	A	
Flux of light	lumen	lm	(cd·sr)
Luminance	candela per sq. meter	cd/m^2	
Illumination	lux	lx	(lm/m^2)

Note that in the international system of units, electric current and mass are defined as basic units, instead of electric charge and force. In turn, force and charge are defined as derived units.

APPENDIX 2: MULTIPLE AND SUBMULTIPLE PREFIXES

MULTIPLE AND SUBMULTIPLE PREFIXES

Multiple/ Submultiple	Prefix	Symbol	Multiple/ Submultiple	Prefix	Symbol
10^{12}	tera	T	10^{-2}	centi	c •
10^{9}	giga	G	10^{-3}	milli	m
10^{6}	mega	M	10^{-6}	micro	μ
10^{3}	kilo	k	10^{-9}	nano	n
10^{2}	hecto	h	10^{-12}	pico	p
10	deka	da	10^{-15}	femto	f
10^{-1}	deci	d	10^{-18}	atto	a

CLASSES OF NUMBERS

Counting Numbers: 1, 2, 3, 4, 5, 6, 7. . . .
Prime Numbers: 1, 2, 3, 5, 7, 11, 13. . . .
Whole Numbers: 0, 1, 2, 3, 4, 5, 6. . . . .
Digits: 0, 1, 2, 3, 4, 5, 6, 7, 8, 9
Negative Numbers: −1, −2, −3, −4, −5. . . .
Rational Numbers: 1/2, −1/3, 2/2, −9/3 . . .
Irrational Numbers: $\sqrt{2}$, $-\sqrt{2}$, $\sqrt{1/2}$, $-\sqrt{1+\sqrt{2}}$, π, . . .
Real Numbers: The rationals and the irrationals.
Ordinal Numbers: 1st, 2nd, 3rd, 4th, 5th. . . . (What order?)
Cardinal Numbers: 0, 1, 2, 3, 4, 5. . . . (How many?)
Transfinite Cardinal Number: ∞
Binary Numbers: 0, 1
Imaginary Numbers: $\sqrt{-1}$, $-\sqrt{-1}$, $2\sqrt{-1}$, $-3\sqrt{-1}$, ...

Complex Numbers: $2 + j3$, $7 - j4$, $\pi + j9$

APPENDIX 3: PROGRAM CONVERSIONS

(WRITTEN BY ROBERT L. KRUSE)

Commodore 64:

The Commodore 64 conversion for the IBM PC program in Figure 1–16(A) may be written as shown in Figure A3–1. Note the following points:

1. Altthough not absolutely essential, it is advisable to decide at the outset whether you wish to work with the video display or to work with the printer. In turn, the coding will be considerably simplified. Separate programs for the video display and for the printer are shown in Figure A3–1.

2. Observe that if you prefer to work with the printer, the first program line should be coded "OPEN1,4." The INPUTS are the same in both programs. However, following the INPUTS in the printer program, a print line "PRINT#1," should be coded. Similarly, each line that contains a variable to be printed must include the code "PRINT#1,." A video-printer program is shown in Figure A3–2.

3. If you are working with the printer, the program should conclude with the coded line "CLOSE1,4."

4. Observe also that to limit the printed-out values to two decimal places (for example), the INT function must be employed in either program. Thus, the line for printing the computed value of X is coded: "X

```
1 REM FIG. 1-16(A)**WITH PRINTER**
5 OPEN1,4
10 INPUT "HFE=";A
20 INPUT "HOE=";B
30 INPUT "HIE=";C
40 INPUT "HRE=";D
50 INPUT "RL=";E
60 INPUT "RG=";F:PRINT ""
62 PRINT#1,"  RUN  "
65 PRINT#1,"  ---  "
70 G=A*E/((C*B-A*D)*E+C)
80 Q=.01*INT(G*100):PRINT#1,"AV=";Q
90 H=A/(B*E+1)
100 T=.01*INT(H*100):PRINT#1,"AI=";T
110 I=A*A*E/((B*E+1)*((C*B-A*D)*E+C))
120 X=.01*INT(I*100):PRINT#1,"GP=";X
130 J=(C+(B*C-A*D)*E)/(1+B*E)
140 Y=.01*INT(J*100):PRINT#1,"RI=";Y
150 K=(C+F)/(B*C-D*A+B*F)
160 Z=.01*INT(K*100):PRINT#1,"RO=";Z
165 CLOSE1,4

READY.

  RUN
  ---
AV= 476.19
AI= 38.46
GP= 18315.01
RI= 1211.53
RO= 85714.28

1 REM FIG. 1-16(A)**W/O PRINTER**
10 INPUT "HFE=";A
20 INPUT "HOE=";B
30 INPUT "HIE=";C
40 INPUT "HRE=";D
50 INPUT "RL=";E
60 INPUT "RG=";F:PRINT ""
62 PRINT "  RUN  "
65 PRINT "  ---  "
70 G=A*E/((C*B-A*D)*E+C)
80 Q=.01*INT(G*100):PRINT "AV=";Q
90 H=A/(B*E+1)
100 T=.01*INT(H*100):PRINT "AI=";T
110 I=A*A*E/((B*E+1)*((C*B-A*D)*E+C))
120 X=.01*INT(I*100):PRINT "GP=";X
130 J=(C+(B*C-A*D)*E)/(1+B*E)
140 Y=.01*INT(J*100):PRINT "RI=";Y
150 K=(C+F)/(B*C-D*A+B*F)
160 Z=.01*INT(K*100):PRINT "RO=";Z

READY.
```

```
**************************
EXAMPLE OF INPUTTED VALUES
**************************
     HFE=50

     HOE=.000020

     HIE=1500

     HRE=.0005

     RL=15000

     RG=1500
```

Figure A3–1. Programs exemplifying the chief differences in coding for the IMB PC and Commodore 64 computers.

```
10 REM FIG. 1-16(A)
30 INPUT "HFE=";A
40 INPUT "HOE=";B
50 INPUT "HIE=";C
60 INPUT "HRE=";D
70 INPUT "RL=";E
80 INPUT "RG=";F
100 G=A*E/((C*B-A*D)*E+C):H=A/(B*E+1)
110 I=A*A*E/((B*E+1)*((C*B-A*D)*E+C))
120 J=(C+(B*C-A*D)*E)/(B*E+1)
130 K=(C+F)/(B*C-D*A+B*F)
135 PRINT "":PRINT "INPUTTED VALUES"
136 PRINT "---------------":PRINT ""
137 OPEN1,4:PRINT#1,"":PRINT#1,"INPUTTED VALUES":CLOSE1,4
138 OPEN1,4:PRINT#1,"---------------":PRINT#1,"":CLOSE1,4
140 PRINT "HFE=";A:OPEN1,4:PRINT#1,"HFE=";A:CLOSE1,4
150 PRINT "HOE=";B:OPEN1,4:PRINT#1,"HOE=";B:CLOSE1,4
160 PRINT "HIE=";C:OPEN1,4:PRINT#1,"HIE=";C:CLOSE1,4
170 PRINT "HRE=";D:OPEN1,4:PRINT#1,"HRE=";D:CLOSE1,4
180 PRINT "RL=";E:OPEN1,4:PRINT#1,"RL=";E:CLOSE1,4
190 PRINT "RG=";F:PRINT "":OPEN1,4:PRINT#1,"RG=";F:PRINT#1,"":CLOSE1,4
195 PRINT "  RUN  ":OPEN1,4:PRINT#1,"  RUN  ":CLOSE1,4
197 PRINT "  ---  ":PRINT "":OPEN1,4:PRINT#1,"  ---  ":PRINT#1,"":CLOSE1,4
198 XX=.01*INT(G*100)
200 PRINT "AV=";XX:OPEN1,4:PRINT#1,"AV=";XX:CLOSE1,4
205 YY=.01*INT(H*100)
210 PRINT "AI=";YY:OPEN1,4:PRINT#1,"AI=";YY:CLOSE1,4
215 ZZ=.01*INT(I*100)
220 PRINT "GP=";ZZ:OPEN1,4:PRINT#1,"GP=";ZZ:CLOSE1,4
225 TT=.01*INT(J*100)
230 PRINT "RI=";TT:OPEN1,4:PRINT#1,"RI=";TT:CLOSE1,4
235 QQ=.01*INT(K*100)
240 PRINT "RO=";QQ:OPEN1,4:PRINT#1,"RO=";QQ:CLOSE1,4

READY.

INPUTTED VALUES
---------------

HFE= 50
HOE= 2E-05
HIE= 1500
HRE= 5E-04
RL= 15000
RG= 1500

  RUN
  ---

AV= 476.19
AI= 38.46
GP= 18315.01
RI= 1211.53
RO= 85714.28
```

Figure A3–2. Example of a combined video-printer program for the Commodore 64.

= .01 * INT(I * 100): PRINT#1,"GP = " ;X." By way of comparison, the A$ function is utilized for this purpose by the IBM PC computer.

5. Another difference in coding between the IBM PC computer and the Commodore 64 is the exponential symbol. Thus, the PC employs a caret, whereas the 64 utilizes an "up arrow" to call the exponentiation function.

Common Conversion "Bugs"

Error messages are frequently vague, and the programmer must carefully "proofread" his or her routine. Since there is a tendency for a programmer to repeat "pet" errors, it is helpful to have an assistant "proofread" the routine. Typical "bugs" are:

1. Numeral 0 typed in instead of capital O.
2. Two-letter variable reversed, such as EO for OE.
3. Semicolon typed in instead of colon, or vice versa.
4. Complete program line overlooked.
5. "Illegal" word processing, which leaves a "bug" hidden in the program memory. If this trouble is suspected, retype the complete line.
6. Substitution of a plus sign (or a blank space) for a required minus sign.
7. Unsuitable units employed for INPUT variables. For example, numerical values can be maintained within an acceptable range by choosing compatible units when coding programs. Thus, the programmer has a choice of farad, microfarad, or picofarad units, for example.
8. Incorrect factors, such as 10^6 for 10^{-6}, or 6.283/360 for 360/6.283.

Note that when a RUN stops at some point and an error message (or no error message) is displayed, the programmer can operate the computer in its calculator mode to successively display the value of any variable that has been processed. In turn, obvious clues may appear, such as an extremely large value or a zero value for a particular variable. Again, the value of the variable may be greater than 1, whereas its correct value must be less than 1 (or vice versa). Such clues help the programmer to "zero in" on subtle coding errors.

Remember that a program may appear to have a coding bug when the fault is actually an input error. If the programmer erroneously inputs 1500 instead of 15000, for example, the routine will be falsely blamed. To avoid this possibility of misdirection, it is good practice to rerun the program to determine whether a different answer may then be computed.

One Line at a Time

When a program runs without an error message but computes incorrect answers, check each line carefully for faults—particularly for transposed variables such as UV for VU, or R0 for RO, and so on. If no faults can be found, it is advisable to retype the program one line at a time. One line at a time permits the programmer to logically follow the action without confusion. If the program suddenly starts to compute different answers (that are still incorrect), the programmer has something to go on, and can reason why the computed answer was changed. Typically, the programmer will eventually see the light—for example, he or she may observe that a negative variable has been programmed as a positive variable.

On the other hand, the programmer may be unable to run down the fault, and wrong answers are being computed though the program has been completely retyped. In such a case, it is advisable for the programmer to ask an assistant to retype the program to obtain a different viewpoint. This is just another way of saying that a programmer is likely to have pet errors that are difficult or impossible for him or her to recognize until they are pointed out.

IBM PC Jr.

All the foregoing IBM PC programs will run without conversion on the IBM PC Jr.

APPLE IIe and II+

When an IBM PC program is converted for the Apple computer, it is impractical to convert the LPRINT USING A$ command. However, the number of decimal places displayed in the answer can be controlled by using the INT function essentially as previously exemplified for Commodore 64 conversions.

Note that although the INT function ordinarily controls the number of decimal places in the answer computed by either the Commodore 64 or the Apple IIe and II+, this is not universally true.

In some cases, the anticipated formatting might not occur, and the practical procedure is to ignore occasional format busts. Observe that the INT function will display 3.70 as 3.7 (omitting display of the zero in the second decimal place), unless several additional processing lines are included. In most situations, the practical procedure is to ignore this minor format bust.

INDEX